DIFFUSE MATTER FROM STAR FORMING REGIONS TO ACTIVE GALAXIES

ASTROPHYSICS AND SPACE SCIENCE PROCEEDINGS

DIFFUSE MATTER FROM STAR FORMING REGIONS TO ACTIVE GALAXIES

A VOLUME HONOURING JOHN DYSON

Edited by

T.W. HARTQUIST

*School of Physics and Astronomy,
University of Leeds, UK*

J.M. PITTARD

*School of Physics and Astronomy,
University of Leeds, UK*

and

S.A.E.G. FALLE

*Department of Applied Mathematics,
University of Leeds, UK*

A C.I.P. Catalogue record for this book is available from the Library of Congress.

ISBN-10 1-4020-5424-6 (HB)
ISBN-13 978-1-4020-5424-2 (HB)
ISBN-10 1-4020-5425-4 (e-book)
ISBN-13 978-1-4020-5425-9 (e-book)

Published by Springer,
P.O. Box 17, 3300 AA Dordrecht, The Netherlands.

www.springer.com

Printed on acid-free paper

All Rights Reserved
© 2007 Springer
No part of this work may be reproduced, stored in a retrieval system, or transmitted
in any form or by any means, electronic, mechanical, photocopying, microfilming, recording
or otherwise, without written permission from the Publisher, with the exception
of any material supplied specifically for the purpose of being entered
and executed on a computer system, for exclusive use by the purchaser of the work.

Contents

John Dyson – A Biographical Sketch ix

Preface xv

List of Contributors xix

Part I: Star Forming Regions 1

Numerical Simulations of Star Formation
S.A.E.G. Falle..................................... 3

Molecular Astrophysics of Star Formation
D. A. Williams 19

Dusty Plasma Effects in Star Forming Regions
T.W. Hartquist, O. Havnes........................ 45

Massive Star Formation
M.G. Hoare, J. Franco............................. 61

Spectropolarimetry and the Study of Circumstellar Disks
R.D. Oudmaijer 83

How to Move Ionized Gas: An Introduction to the Dynamics of H$_{II}$ Regions
W.J. Henney..................................... 103

MHD Ionization Fronts
R.J.R. Williams 129

Herbig-Haro Jets from Young Stars
T.P. Ray 145

Hypersonic Molecular Shocks in Star Forming Regions
P.W.J.L. Brand 163

Part II: The Effects of Evolved Stars on Their Environments 181

Wind-Blown Bubbles around Evolved Stars
S.J. Arthur 183

Do Fast Winds Dominate the Dynamics of Planetary Nebulae?
J. Meaburn 205

Spectral Studies of Supernova Remnants
J.C. Raymond 223

Part III: Multicomponent Flows and Cosmic Rays 243

Mass-Loaded Flows
J.M. Pittard 245

The Effects of Cosmic Rays on Interstellar Dynamics
T.W. Hartquist, A.Y. Wagner, S.A.E.G. Falle 269

The Status of Observations and Speculations Concerning Ultra High-Energy Cosmic Rays
A.A. Watson 283

Part IV: Starburst Galaxies and Active Galactic Nuclei 305

The Messier 82 Starburst Galaxy
A. Pedlar, K.A. Wills 307

Active Galactic Nuclei
S.L. Lumsden 325

John E. Dyson

John Dyson – A Biographical Sketch

T. W. Hartquist, J. M. Pittard, and S. A. E. G. Falle

On 1 October 2006, John Edward Dyson will become a Research Professor in the School of Physics and Astronomy of the University of Leeds where he has been the Professor of Astronomy since 1 April 1996. To celebrate John's transition to his administration-free status, we have produced the Festschrift in which this biographical sketch appears. The kind cooperation of so many of our colleagues in the preparation of the volume reflects the scientific influence that John has had and the value that the contributors place on their friendships with him.

John was born on 7 January 1941 in Meltham Mills, West Yorkshire. He was the second of four children. His father, Jack, was a progress chaser at David Brown's Tractor Works in Huddersfield. His mother, Ivy, worked as a weaver in a woollen mill in Halifax.

Ill health caused Jack to change career and become a cricket and rugby groundsman. Consequently, the family moved to Harrogate in 1948. The proudest moment of John's early life occurred when, at the age of eight, he met Freddie Trueman, who was then an up and coming Yorkshire cricket player and went on to become one of England's greatest fast bowlers of all time.

John sang in the St. Mary's Church Choir and, as a boy soprano, won certificates for solo performances at the Harrogate Music Festival. One of the pieces he sang had lyrics set to the Brahms Opus 39, Number 15 Waltz in A Flat.

After spending one year at Harrogate Grammar School, John moved to Leeds where his father became the groundsman at the Hunslet Rugby League Club. John attended Cockburn High School, then off Dewsbury Road. He did the standard combination of Maths, Physics, and Chemistry to Scholarship level, which was beyond A level. In his final year, he was head boy.

He learned Chemistry from a particularly good teacher, Mr. M. S. Grant. However, he found Physics more interesting. As John was growing up, Sidney

Smith, who was a good friend of Jack and a civil servant, was one of John's closest friends. Sidney had been at school with Edward Appleton, who received the Nobel Prize for Physics.

After hearing a radio lecture given by D. M. McKay of Kings College London, John decided to use the state scholarship that he had won to study at that institution. He lived near Alexandra Palace during his first year and in Bounds Green later. As a good Northerner, John is proud of the fact that, though he did live in London, he has never lived south of the Thames.

A couple of times per term John hitchhiked up The Great Northern Road to Leeds. A return rail ticket would have cost £4, a far too large fraction of the £317 he received per year from his scholarship. His mother had been ill for some time and died during John's second year in London. His father died, unexpectedly, shortly afterwards. John continued to travel regularly to Leeds to see his remaining family and friends. He met his future wife, Rita, in that city on 8 September 1961 in the Majestic Ball Room in City Square.

In 1962, John received a first class B.Sc. Special Honours Degree in Physics. In October of that year, he began postgraduate study in the University of Manchester Department of Astronomy. He had been awarded a prestigious University Scholarship. On 15 December 1962, he and Rita married.

He had planned to work with Zdenek Kopal, the Head of Astronomy. The only guidance that John received was in the form of a question asked by Professor Kopal, "Why do you not devise a new method for measuring the temperature of O stars?" As John has commented, "This was old-style supervision." Of his own volition, John began to work on Balmer decrements by including collisional excitation and de-excitation. After making considerable progress, he found that a paper on similar work had appeared a few years earlier.

Then one day after John had been in Manchester for about a year, Franz Kahn poked his head around the door and asked John about his work. Franz then said that he had a problem based on an idea about the Orion Nebula. Münch (1958) had reported evidence of transonic turbulence in that source. John investigated the possibility that the lines were so broad due to flows driven by the photoevaporation of globules. The resulting paper (Dyson 1968) has been cited frequently in the literature concerning proplyds in Orion. In 1965 John realized that a study of molecular photodissociation fronts was necessary, but at that time the mechanism of H_2 photodissociation was just being identified and had not been studied in much quantitative detail.

John's Ph. D. thesis is entitled "The Age and Dynamics of the Orion Nebula". He passed his oral examination, for which Stuart Pottasch was the external examiner, on 28 February 1966. For several months, John held the post of Assistant Lecturer in Manchester.

Then in June 1966, he moved to the University of Wisconsin where he held a Wisconsin Alumni Research Foundation Fellowship and a Fulbright Fellowship. He learned that the Green Bay Packers were the greatest American football team ever and was impressed by the breadth of interests of the high quality astrophysicists working there. He took the opportunity to learn a great deal about nebular astrophysics. On his own, he identified problems on radio recombination lines as the focus of his work.

John accepted an appointment as a Lecturer in Manchester. Officially, the post began in October 1967, but he was granted a leave to continue in Wisconsin. In December 1967, he returned to England, in the same fashion as he had left, by ship. John and Rita had noticed a considerable improvement in their standard of living upon arriving in the States but found that it immediately dropped back to its ground state when they were in Manchester again.

John's interest in dynamics reasserted itself and grew, and in a few years, he turned from the purely spectroscopic themes he had addressed in Wisconsin. In 1968, he read Pikel'ner's (1968) seminal paper entitled "Interaction of Stellar Wind with Diffuse Nebulae" and was very impressed and influenced by it. At about this time, John Meaburn, also in Manchester, began to discover high velocities in HII regions. John Dyson suspected that wind-blown bubbles were responsible. Dyson & de Vries (1972) introduced similarity solutions for wind-blown bubbles and were the first to compare observational results with those of such models.

In his theoretical efforts on the nebulae of evolved stars (Smith, Pettini, Dyson, & Hartquist 1984) and the origins of AGN line-forming regions (Perry & Dyson 1985) and in his introduction of the working surface model of Herbig-Haro objects (Dyson 1987), John has applied the wind-blown bubble picture more widely than anyone else. In his research on Herbig-Haro objects and other jet-related sources, he has emphasised the similarity between the structures of jet-driven flows and wind-blown bubbles.

John's first published work on AGNs (Dyson, Falle, & Perry 1979) was begun in 1977 and 1978 when he took a sabbatical in the Max Planck Institut für Astrophysik near Munich. Sam Falle had worked in Manchester in Franz's and John's group, and John met Judith Perry in Munich. Tom Hartquist is another important collaborator whom John met in Munich when John, Sam, and Tom were spending time there in the Summer of 1980. Tom met John and Sam through the Heimerl family, who remain mutual friends. In 1978, shortly after returning to Manchester from his sabbatical, John became a Senior Lecturer. Subsequent promotions to Reader and Professor occurred in 1984 and 1993, respectively.

In addition to working on jets in the context of Herbig-Haro objects, John has considered the possible role of jets in the production of narrow-line regions in Seyfert galaxies (Taylor, Dyson, & Axon 1992). He has suggested that the forbidden line regions of the radio Seyferts may be due to the cooling of ambient gas shocked by the expansion of plasmons (Pedlar, Unger, & Dyson 1985).

John has not restricted his work on wind-blown bubbles to highly idealized problems. He was the first to argue that the cooperative effects of stellar winds from clusters of OB stars are responsible for supersonic motions in giant extragalactic HII regions, and his conclusion that the OB stars in such a region are coeval was an early recognition that some galaxies experience starbursts (Dyson 1979). He has driven the effort to model diffuse astrophysical sources as multiphase media in studies of their dynamics (Hartquist, Dyson, Pettini, & Smith 1986). He has emphasized that the multiphase structures of the ejecta of evolved stars must be considered in models of the stellar winds (Dyson, Hartquist, Pettini, & Smith 1989) and has suggested that the existence of such structures may indicate the importance of buoyancy-driven MHD instabilities in the ejection process in some stars (Hartquist & Dyson 1997).

In his consideration of the interaction of diffuse flowing media with clumps (e.g. Dyson, Hartquist, & Biro 1993; Dyson & Hartquist 1994), John presented a theory of cometary-like structure formation and applied it to a range of problems including the diagnosis of the wind of Sgr A*, which is thought to contain a supermassive black hole at the Galactic Centre.

John's early work on similarity solutions demonstrated his facility for analytical and semi-analytical problems and a recognition of the need for the exploration of physical processes. His series of papers on MHD ionization fronts (e.g. Williams & Dyson 2001) amply demonstrate that he has maintained that facility and continues to be motivated by that recognition. John also persists in offering ingenious explanations for observed features in real sources as shown by his work on the connection between jets and FLIERS in planetary nebulae (Redman & Dyson 1999).

John's move to Leeds gave him a good opportunity to renew his collaboration with Sam Falle. It also gave him the chance to build his own group, which Tom Hartquist, Melvin Hoare, Stuart Lumsden, Rene Oudmaijer, and Julian Pittard joined in the period from 1996 through 2000. Sam, Tom, and Julian have been collaborating extensively with John in the use of multidimensional hydrodynamic simulations to articulate a number of John's ideas. Examples of the results from this collaboration are found in the Falle et al. (2002) paper on intermediate-scale structures, including tails, that arise in the interaction of different Mach number winds with high density clumps and in the Pittard, Falle, Dyson, & Hartquist (2005) article on the effects of mass injection from multiple embedded high density clumps on a wind that is hypersonic upstream and the possible relevance of those effects on starbursts.

John Dyson's research has concerned star-forming regions, the late phases of stellar evolution, circumstellar nebulae, starbursts, and active galaxies. In addition to influencing these fields through his original contributions, he has helped to educate a large fraction of practising professional astronomers through the two editions of his highly successful book "The Physics of the Interstellar Medium" coauthored with David Williams.

John has also coauthored "Blowing Bubbles in the Cosmos" with Tom Hartquist and Deborah Ruffle and edited or coedited two other books. One is a volume honouring Franz Kahn. The other is entitled "Active Galactic Nuclei".

With the relocation to Leeds, demanding managerial responsibilities arose. John served as Dean of Research for The Faculty of Mathematics and Physical

Sciences for two years starting in 1998 and as Head of Physics and Astronomy for three years beginning in 2000. In the latter role, he did much to prepare the way for the formation of a new group working in fundamental quantum physics, including quantum information. In these positions, John faced many challenging political situations. He is a master in the resolution of awkward standoffs through the use of his charming wit to make powerful points effectively while inducing good humour in previously tense people.

John has also had major national and international responsibilities. Since 1993, he has been either the sole editor-in-chief of Astrophysics and Space Science or shared the editorial responsibilities with one other scientist. His funding council duties have included service on the SERC Astronomy II Committee (1984–86), the SERC Astronomy and Planetary Science Board (1987–91), and the PPARC Education and Training Committee (1994–97) and as Chair of the PPARC Astronomy Theory Research Assessment Panel (1988–91). He was on the RAS Council (1975–77) and President of IAU Division VI (Interstellar Matter) and Commision 34 (2003–06).

John has chaired the Scientific Organizing Committee and been the Scientific Director of numerous conferences, including two particularly enjoyable, as well as edifying, ones held in Cumberland Lodge, Windsor Great Park in 1996 and 2004. These two international meetings were hosted by University College London, and John got involved at the invitation of David Williams and Allan Willis. Most recently, with Viktor Tóth, he has co-chaired the committee organizing a Hungarian-sited workshop on "The Interaction of Stars with Their Environment".

By the end of 2006, three conferences will have been held in John's honour. Two Leeds meetings, in 2001 and 2006, were designed to give a small number of John's closest friends and collaborators based in the UK and Ireland an opportunity to see him and exchange scientific ideas. Jane Arthur organized a major international conference held in Mexico in 2002 to celebrate John and his contributions. She and Will Henney edited the proceedings which appeared as volume 15 of Revista Mexicana de Astronomi'a y Astrofi'sica (Serie de Conferencias).

In addition to being useful in his professional environment as described above, John's wit delights his friends in a variety of social circumstances. Each of them remembers many times when John has instantaneously and effortlessly taken an already enjoyable occasion to a higher level of fun with the sort of succinct comment or pun that could originate from only him. One of Tom Hartquist's favorite examples of the notorious Dyson humour concerns John's invention of the collective noun for professors, a story Tom will relate to anyone who asks. Of course, during discussions involving John about the factual details in this sketch, the quick remarks came ceaselessly. His friends will not be surprised to read that many were selfdeprecatory and reflected the modesty that John always displays.

John and Rita have opened their home to a good many scientists arriving for short stays or from abroad to stay for a few to many years. Their generosity and hospitality are truly enormous. They even helped Tom pick out his home, lent him money for a downpayment, and worked together to build bookshelves for him as well as to supply many other household essentials, including curtains.

John communicates his enthusiasm for English music, his thorough knowledge of many classic films, and his interests in history (including the history of industrial-era villages) and World War I and American Civil War battlefields with joy in a way that enrichs the lives of his friends. He is moved by reflection upon the events that the touching of old stones evokes.

John and Rita live in Todmorden where they enjoy the regular visits of their four children and five grandchildren.

References

Dyson, J.E. 1968 Ap&SS **1**, 388
Dyson, J.E. 1979 A&A **73**, 132
Dyson, J.E. 1987 Circumstellar Matter - Proceedings of IAU Symposium 122, I. Appenzellar & C. Jordan (eds.), Dordrecht, D. Reidel Publishing Company, 159
Dyson, J.E., Falle, S.A.E.G., Perry, J.J. 1979 Nature **277**, 118
Dyson, J.E., Hartquist, T.W. 1994 MNRAS **269**, 447
Dyson, J.E., Hartquist, T.W., Biro, S. 1993 MNRAS **261**, 430
Dyson, J.E., Hartquist, T.W., Pettini, M., Smith, L.J. 1989 MNRAS **241**, 625
Dyson, J.E., de Varies, J. 1972 A&A **20**, 223
Falle, S.A.E.G., Coker, R.F., Pittard, J.M., Dyson, J.E., & Hartquist, T.W. 2002 MNRAS **329**, 670
Hartquist, T.W., Dyson, J.E., Pettini, M., Smith, L.J. 1986 MNRAS **221**, 715
Hartquist, T.W., Dyson, J.E. 1997 A&A, **319**, 589
Münch, G. 1958 Rev Mod Phys **30**, 1035
Pedlar, A., Unger, S.W., Dyson, J.E. 1985 MNRAS **214**, 463
Perry, J.J., Dyson, J.E. 1985 MNRAS **213**, 665
Pikel'ner, S.B. 1968 Ap Lett **2**, 97
Pittard, J.M., Falle, S.A.E.G., Dyson, J.E., Hartquist, T.W. 2005 MNRAS **361**, 1077
Redman, M.P., Dyson, J.E. 1999 MNRAS **302**, 17
Smith, L.J., Pettini, M., Dyson, J.E., Hartquist, T.W. 1984 MNRAS **211**, 679
Taylor, D., Dyson, J.E., Axon, D.J. 1992 MNRAS **255**, 351
Williams, R.J.R., Dyson, J.E. 2001 MNRAS **325**, 293

Preface

T.W. Hartquist[1], J.M. Pittard[1], and S.A.E.G. Falle[2]

[1] School of Physics and Astronomy, University of Leeds
[2] Department of Applied Mathematics, University of Leeds

The chapters are arranged into four areas:

- Star Forming Regions
- The Effects of Evolved Stars on Their Environments
- Multicomponent Flows
- Starburst Galaxies and Active Galactic Nuclei

John Dyson has contributed significantly to the understanding of problems in each of these areas.

At least some of the physical processes and conceptual pictures important in any one of the areas are central to others. For instance, wind-driven and jet-driven double shock structures are common to star forming regions, environments of evolved stars, regions where mass loading occurs, and extragalactic starbursts and active galactic nuclei. Accretion discs figure prominently in at least three of the groups of chapters, just as spectral diagnostics of radiative shocks do in four. Thus, many of the chapters could have been juxtaposed with others. For instance, René Oudmaijer's work on spectropolarimetry is certainly of relevance to star formation, but he also applies the technique to the investigation of discs around evolved stars and the substructure of outflows from such stars. These outflows affect their environments.

The early phases of star formation, which we take to be the creation of giant molecular clouds and translucent clumps within them, involve magnetohydrodynamics (MHD) and fine structure cooling by neutral and singly ionized atomic species. Thus, the first chapter in the star formation group is that by Sam Falle who has addressed the MHD of star formation.

MHD continues to play a dominant role in subsequent phases of star formation, but then molecular processes also become important. They establish the themal and ionization states of the gas, which, in turn, are key for the MHD. Also, they lead to chemical composition and emission features which serve as diagnostics of the evolution from number densities of 10^2 to 10^8 cm^{-3}. David Williams has contributed a review of molecular astrophysics relevant to this phase of evolution.

David has described ways in which the dust grains are important for the molecular astrophysics of star formation. The dust becomes increasingly important dynamically as the number density increases above roughly 10^4 cm^{-3}. Tom Hartquist and Ove Havnes have described multifluid treatments of flows in star forming regions required to investigate the major impact that grains have on the MHD of stellar birth and shocks driven by the outflows of young stars in star forming regions. In his chapter, Peter Brand contributed a discussion of the influence of dust on shocks that complements that of Tom and Ove.

The formation of low-mass stars (i.e. those with masses of several solar masses and less) differs from that of high-mass stars, as Melvin Hoare and Pepe Franco have made clear in their chapter. The effect that a forming massive star has on its environment is potentially capable of altering the stellar birth process, which is not true of a low mass star's effect.

One means of partially moderating the effect of a massive star on its own birth is through the formation of an accretion disc. Discs surrounding A-type and more massive stars are more difficult to detect than those around lower mass stars. René Oudmaijer has reviewed relevant spectropolarimetric studies.

One way in which a young massive star influences its environment is by the formation of H$_{II}$ regions. Will Henney's chapter on H$_{II}$ regions complements those parts of Melvin's and Pepe's chapter concerned with H$_{II}$ regions. In his contribution, Robin Williams has explained the role of MHD in the ionization fronts preceding H$_{II}$ regions.

Stars emit material outflows as well as radiation. As summarized by Tom Ray, jets from young low-mass stars create optical emission regions called Herbig-Haro objects. The creation of the collimated outflows involves magnetized accretion discs.

Young massive stars also possess outflows, the consequences of which have been observed, in some cases, in emission features of molecular hydrogen. As described by Peter Brand, the H$_2$ moves at speeds exceeding the minimum speed of a hydrodynamic shock capable of dissociating it. The structures and properties of hypersonic H$_2$ in weakly ionized media are relevant to the explanation of hypersonic H$_2$ in regions of low-mass star formation and those generating massive stars.

Evolved stars also possess outflows that affect the matter in their vicinities. Jane Arthur has reviewed the fundamentals of wind-blown bubbles and followed the development of one blown by a star with an initial mass of 40 solar masses at a very early point in its evolution through subsequent phases including the Wolf-Rayet phase. She has addressed a number of interesting observations of bubbles around evolved massive stars.

In contrast, John Meaburn has focussed on planetary nebulae, which surround highly evolved, low-mass stars. He has argued that many features in the outer parts of some nebulae point to nearly simultaneous, impulsive ballistic ejections by each central star rather than driving by continuous fast winds.

Supernova remnants are the products of the interactions of the most evolved massive stars with the ambient media. John Raymond has written of them and the spectral features used to diagnose their shocks and the physical processses ocurring in them.

Many of the models of interactions between outflows of stars, whether they are in pre-main sequence, main sequence, or post-main sequence phases, and their environments are based on the assumption that the outflow and ambient medium can be described as a single fluid. In fact, an outflow or an ambient medium usually consists of multiple components. They may include tenuous thermal X-ray emitting plasma, colder embedded clumps of thermal gas, and cosmic rays. While retaining its own identity, a component's coupling to other components can influence its flow in a way that is not adequately described with a single fluid model. This can be true of multicomponent accretion as well as in problems involving outflows encountering initially static material. The flows described in the chapter by Tom Hartquist and Ove Havnes and by Peter Brand are multicomponent in nature. Julian Pittard has addressed the simplest models of multicomponent flows, those in which one component consists of sufficiently massive objects that the single other component has little effect on them, but the massive objects serve as distributed sources of material entering the other component and affecting its dynamics.

Tom Hartquist, Alex Wagner, and Sam Falle have described how the cosmic ray component of an interstellar medium affects the dynamics of the thermal component and, in turn, responds to the dynamics of the thermal component. Shocks, supernova remnants, galactic winds and accretion in clusters of galaxies are subjects upon which they touch.

Some cosmic rays have energies in excess of 3×10^{18} eV each. Alan Watson has reviewed the status of observations and speculation concerning the origin of these ultra-high energy cosmic rays. The Pierre Auger Observatory may facilitate major advances in the subject in the next few years. These cosmic rays are most likely of extragalactic origin and some radio-loud active galactic nuclei (AGNs) may be the sources.

Other than AGNs, starbursts in the central hundreds of parsecs of some galaxies are the most luminous astronomical sources on that and smaller scales. Alan Pedlar and Karen Wills have summarized observations in the optical, radio, infrared, X-ray, and millimeter ranges of M82, one of the closest starburst galaxies. Some data are of sufficient quality that the expansion of each individual supernova remnant in the starburst region has been the subject of proper motion studies.

Stuart Lumsden's chapter is the final one and concerns AGNs. He has reviewed the nature of the broad and narrow emission line regions and efforts to develop a unified model of AGNs. The accretion rate and the ratio of the luminosity to the Eddington luminosity, which depends on the black hole mass, appear to be key for governing the type of AGN a source is.

We, the editors, are pleased that such a range of authors took time to honour John and with the quality of their chapters. We would like to express our gratitude to them, especially as the creation of a coherent volume that covers the breadth of the subject implied by its title required that we receive all of their contributions.

The University of Leeds, UK
July 2006

T.W. Hartquist
J.M. Pittard
S.A.E.G. Falle

List of Contributors

S. J. Arthur
Centro de Radioastronomía y
Astrofísica, UNAM
Morelia, 58090 Morelia, México
j.arthur@astrosmo.unam.mx.*

P. W. J. L. Brand
The University of Edinburgh
Edinburgh, UK
pwb@roe.ac.uk.*

S. A. E. G. Falle
The University of Leeds
Leeds, LS2 9JT, UK
sam@amsta.leeds.ac.uk.*

J. Franco
Instituto de Astronomía, UNAM
México
pepe@astroscu.unam.mx.*

T. W. Hartquist
The University of Leeds
Leeds, LS2 9JT, UK
twh@ast.leeds.ac.uk.*

O. Havnes
The University of Tromsø
Tromsø, N-9037, Norway
ove.havnes@phys.uit.no.*

W. J. Henney
Centro de Radioastronomía y
Astrofísica, UNAM
Morelia, 58090 Morelia, México
w.henney@astrosmo.unam.mx.*

M. G. Hoare
The University of Leeds
Leeds, LS2 9JT, UK
mgh@ast.leeds.ac.uk.*

S. L. Lumsden
The University of Leeds
Leeds, LS2 9JT, UK
sll@ast.leeds.ac.uk.*

J. Meaburn
Instituto de Astronomia, UNAM
Ensenada, BC 22800, México
jm@ast.man.ac.uk.*

R. D. Oudmaijer
The University of Leeds
Leeds, LS2 9JT, UK
roud@ast.leeds.ac.uk.*

A. Pedlar
Jodrell Bank Observatory
The University of Machester
Jodrell Bank, SK11 9DL, UK
ap@jb.man.ac.uk.*

J. M. Pittard
The University of Leeds
Leeds, LS2 9JT, UK
jmp@ast.leeds.ac.uk.*

T. P. Ray
Dublin Insitute for Advanced Studies
Dublin, Ireland
tr@cp.dias.ie.*

J. C. Raymond
Center for Astrophysics, Harvard
Cambridge, MA 02176, USA
jraymond@cfa.harvard.edu.*

A. Y. Wagner
The University of Leeds
Leeds, LS2 9JT, UK
ayw@ast.leeds.ac.uk.*

A. W. Watson
The University of Leeds
Leeds, LS2 9JT, UK
a.a.watson@leeds.ac.uk.*

D. A. Williams
University College London
London, WC1E 6BT, UK
daw@star.ucl.ac.uk.*

R. J. R. Williams
AWE Aldermaston
Reading, RG7 4PR, UK
k.wills@sheffield.ac.uk.*

K. A. Wills
The University of Sheffield
Sheffield, S3 7RH, UK
k.wills@sheffield.ac.uk.*

Part I

Star Forming Regions

Numerical Simulations of Star Formation

S.A.E.G. Falle

Department of Applied Mathematics, University of Leeds, Leeds LS2 9JT
sam@amsta.leeds.ac.uk

1 Introduction

Like most branches of theoretical astrophysics, progress in the theory of star formation requires numerical simulations, even though there are people like John Dyson who can still get amazing results without them. Astrophysical numerical calculations are seldom easy, but star formation is particularly difficult because of the extreme changes in scale: the mean density of a star is $\simeq 1$ g cm^{-3}, whereas the mean density of the interstellar medium out of which they form is $\simeq 10^{-24}$ g cm^{-3}. The associated change in length scale of $\simeq 10^8$ obviously makes following the complete process difficult. Surprisingly, adaptive mesh techniques (AMR) have been used to do this for the formation of the very first stars (e.g. Abel et al. 2002). However, primordial star formation is very different from galactic star formation and, in some sense, simpler.

Even if we confine ourselves to a part of the process, we clearly need numerical methods that can cope with large changes in scale. Since self-gravity plays a crucial role, we also need to handle this efficiently. Furthermore, the observations tell us that the magnetic field is dynamically important in star forming regions, so that we must be able to include it in our simulations. Even worse, radiative transfer is important in the final stages.

These are pretty stringent requirements and there is, as yet, no numerical code that can meet them all. However, it is possible to make some progress for the early stages, especially if one eschews virtual reality and instead tries to understand the fundamental mechanisms. If we confine ourselves to the early stages of star formation, then we have to consider the effects of self-gravity, magnetic fields and thermal instability, but not radiative transfer.

2 Self-Gravity

It has been known for more than a hundred years that thermal pressure cannot support a cloud against gravity if its mass exceeds a certain limit, given approximately by the Jeans Mass (Jeans 1902). Jeans derived this for a spherical isothermal cloud with a prescribed external pressure, but it is simpler to just consider the one dimensional linear stability of a uniform, isothermal, self-gravitating gas. The dispersion relation is

$$\omega^2 = k^2 - \frac{4\pi G\rho}{a^2},$$

where a is the sound speed, ρ is the density, and ω and k are the angular frequency and wave number of a small perturbation.

This gives instability if the wavelength exceeds the Jeans wavelength,

$$\lambda_J = \left(\frac{\pi a^2}{G\rho}\right)^{1/2} = \left(\frac{\pi p}{G\rho^2}\right)^{1/2},$$

where $p = a^2\rho$ is the pressure. One can then derive the Jeans mass,

$$M_J = \frac{4\pi}{3}\left(\frac{\lambda_J}{2}\right)^3 \rho = \frac{\pi}{6}\left(\frac{\pi a^2}{G\rho}\right)^{3/2}\rho$$

by taking it to be the mass of a sphere of radius $\lambda_J/2$.

It is convenient to use the pressure to eliminate the sound speed to get

$$M_J = \frac{\pi}{6}\left(\frac{\pi p}{G\rho^2}\right)^{3/2}\rho. \tag{1}$$

As treated more fully in section 5, almost all of the diffuse primarily neutral, partially ionized, and nearly fully ionized gas in the interstellar medium exists in several phases, all of which have a comparable pressure of the order of 3×10^{-13} erg cm^{-3}. The densities of the warm and cold phases are roughly 5×10^{-25} g cm^{-3} and 5×10^{-23} g cm^{-3} respectively. The corresponding values of λ_J and M_J are very approximately 2 kpc and 5×10^7 $M_\odot$ and 20 pc and 5×10^3 $M_\odot$ respectively. Here we have the essential problem of star formation: the Jeans mass is much larger than a typical stellar mass even for the cold phase, so there must be some process that fragments a cloud into smaller pieces.

Hoyle (1953) argued that as a cloud whose mass is initially of the order of its Jeans mass contracts, the increase in density reduces the Jeans mass and it will fragment into ever smaller objects. The process stops when the gas becomes optically thick and the gravitational binding energy can no longer be radiated away. Unfortunately this does not work because even though the Jeans mass decreases as the cloud collapses, the longest wavelengths still grow fastest so that there is no real tendency for the cloud to fragment. In fact, as we shall see, the observations do not suggest that the first stages of star formation are due to gravitational instability.

We must therefore consider other fragmentation mechanisms, at least for the initial stages. The first step is obviously to look at the conditions in star forming regions.

3 Conditions in Star Forming Regions

Most current star formation occurs in Giant Molecular Clouds (GMCs), such as the Rosette Molecular Cloud (RMC) which has been studied by William, Blitz & Stark (1995). Its properties are summarised in Table 1. Most of the mass of a GMC is in translucent clumps of gas of densities of roughly 10^{-21} g cm^{-3}. The clumps range in mass from tens to thousands of M$_\odot$, with most of the clumps having masses of the order of 100 M$_\odot$ and less. The RMC contains about 70 such clumps. The linewidths within the clumps are superthermal and since the work of Arons & Max (1975) have often been assumed to give lower bounds to the intraclump Alfvén speeds.

Table 1. Properties of Rosette Molecular Cloud.

"Radius"	$\simeq 35$ pc
Mass	$\simeq 10^5$ M$_\odot$
Mean Density	$\simeq 10^{-22}$ gm cm^{-3}
Clump-to-Clump Velocity Dispersion	$\simeq 10$ km s^{-1}
Jeans Mass	$\simeq 10^7$ M$_\odot$ (based on velocity dispersion)

It is evident that self gravity is only marginally important in GMCs such as the Rosette, which suggests that they were not formed by the Jeans instability. Even if they were, the smaller high latitude molecular clouds (Magnani, Blitz & Mundy 1985) could certainly not be formed in this way.

The properties of RMC translucent clumps are given in Table 2. β is the ratio of the thermal pressure to the magnetic pressure.

Table 2. Properties of Translucent clumps in the Rosette Molecular Cloud.

"Radii"	$\simeq 1 - 3$ pc
Masses	$\simeq 30 - 2\,10^3$ M$_\odot$
Densities	10^{-21} g cm^{-3}
Temperature	$\simeq 10$ K $\Rightarrow$ Sound speed $\simeq 0.2$ km s^{-1}
Alfvén Speed	$\simeq 2$ km s^{-1} $\Rightarrow$ magnetic pressure dominates ($\beta \ll 1$)
Velocity Dispersion	$\simeq 1 - 2$ km s^{-1}
Jeans Mass	$3\,10^3$ M$_\odot$ (based on velocity dispersion)

The fact that these clumps are rather smaller than their Jeans mass suggests that they were not formed by gravitational instability.

The translucent clumps also have substructure: they contain dense cores (e.g. Myers 1987). Those associated with the formation of solar-like stars have the properties summarized in Table 3.

Table 3. Properties of Dense Cores.

Radii	$\simeq 0.1$ pc
Masses	$\simeq 3 - 10$ M$_\odot$
Densities	$\simeq 10^{-19}$ g cm^{-3}
Temperature	$\simeq 10$ K $\Rightarrow$ Sound speed $\simeq 0.2$ km s^{-1}
Velocity Dispersion	$\simeq 0.3$ km s^{-1}
Jeans Mass	$\simeq 10$ M$_\odot$ (based on velocity dispersion)

These cores have masses comparable to their Jeans masses and many and possibly most collapse to form stars. Their masses are also of the same order as that of typical stars. However, since they must have formed out of more diffuse material with a larger Jeans mass, it is not obvious that they were formed by gravitational instability.

4 Ambipolar Diffusion

As we have already seen, the magnetic field is dynamically important in star forming regions. However, in contrast to many astrophysical situations, the conductivity is not high enough for ideal MHD to be valid. This is because in the dense regions the ionisation fraction is small enough for ambipolar diffusion (ion–neutral drift) to be important.

The magnetic Reynolds number associated with ambi-polar diffusion is unity for a length scale

$$L = 0.04 \frac{1}{M_A} \left(\frac{B}{10^{-5} \text{ Gauss}} \right) \left(\frac{10^{-6}}{x_i} \right) \left(\frac{10^3 \text{cm}^{-3}}{n} \right)^{3/2} \text{ pc}$$

where M_A is the Alfvén Mach number ($M_A \simeq 1$), B is the magnetic field strength, x_i is the fractional ionization, and n is the number density. This tells us that the magnetic Reynolds number < 100 in the translucent clumps and dense cores, which means that ambi-polar diffusion is important on scales smaller than a GMC. This is fortunate because it has long been realised that there must be some mechanism that reduces the magnetic flux per unit mass from its value in star forming regions to that in a typical star. Mestel & Spitzer (1956) were the first to suggest that ambi-polar diffusion could accomplish this. This, of course, means that numerical calculations for the formation of translucent clumps and dense cores must be able to handle ambi-polar diffusion efficiently.

5 Thermal Instability

In the ISM the heating rate per unit time per unit volume is of the form

$$H = A\rho - \rho^2 \Phi(T)$$

where A is a constant (Wolfire et al. 1995). The first term represents heating by cosmic rays, X-rays, and ultraviolet radiation and the second term radiative cooling.

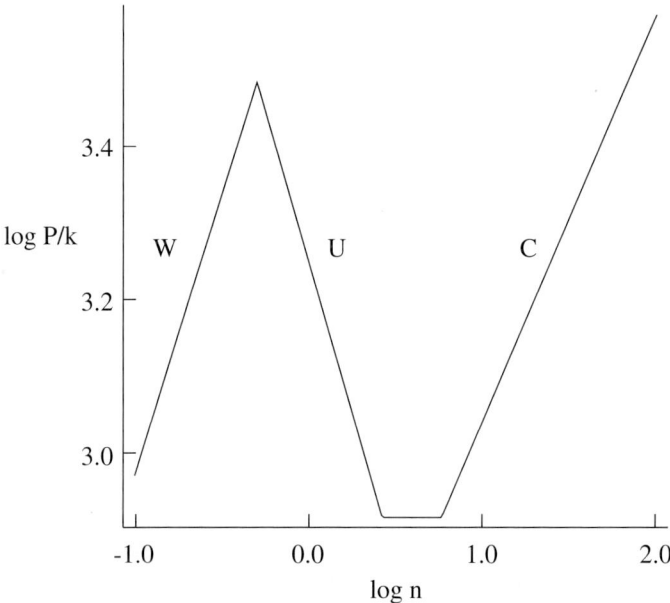

Fig. 1. Pressure as a function of number density for gas in thermal equilibrium. W – warm phase, U – unstable phase, C – cold phase.

The form of the heating function means that the temperature at which the interstellar gas is in thermal equilibrium is a function of density. The determination of this function is quite complicated since it depends upon the strength of the ultraviolet radiation field, but it is possible to construct a simple model that captures the essentials (Sanchez-Salcedo, Vasquez-Semadeni & Gazol 2002). Figure 1 shows the equilibrium pressure as a function of density for this model. It is clear from this that there are two stable phases in which the pressure increases with density and an unstable one in which the pressure decreases with density. For this particular model these phases are: a warm phase for $T > 6102K$); an unstable phase for $313K < T < 6102K$; a cold phase for $T < 313K$.

It has been realised for a long time that thermal instability can account for density inhomogeneities in the ISM and it can also contribute to the formation of regions with low β (see e.g. Lim, Falle & Hartquist 2005). We shall see later that MHD waves propagating in a material in the thermally unstable phase can produce structures very similar to the translucent clumps.

6 Numerical Requirements

We are now in position to specify the requirements for numerical codes for the early stages of star formation. There are three types of problems with somewhat different requirements.

6.1 Type I

On the scale of a GMC or larger, ambipolar diffusion is unimportant so we have ideal MHD. Self-gravity is also not important, but thermal instability is; indeed, we shall argue that it plays a crucial role in the formation of GMCs and translucent clumps. Since the velocities are supersonic, there must be shocks. We, therefore, need a code that can handle MHD shocks, fast cooling and large density contrasts.

6.2 Type II

On the scale of the translucent clumps, ambipolar diffusion is significant, but not so large that we need an explicit scheme to avoid the restriction on the time step that it imposes. Again there are shocks and large density contrasts, but the cooling time is so short that the gas can be assumed to be in thermal equilibrium. Self-gravity may be important, but can probably be neglected as a first approximation.

6.3 Type III

In the dense cores, the magnetic Reynolds number is so small that an explicit scheme would be very inefficient and the only viable option is to use an implicit scheme. Self-gravity now plays a crucial role. As for the translucent clumps, one can assume that the gas is in thermal equilibrium. The density contrasts are now very large, particularly if the evolution is followed to the point where distinct objects with stellar masses are formed.

It is clear that type III problems are the most difficult, so much so that there have, as yet, not been any simulations of these that include all the significant effects. The situation is rather better for type I and II problems, although even here the need to resolve large density contrasts is a stringent requirement.

7 Numerical Methods

7.1 Adaptive Mesh Refinement (AMR)

Shock capturing algorithms for ideal MHD have been available for many years (e.g. Brio & Wu 1988; Stone & Norman 1992; Zachary & Colella 1992; Ryu & Jones 1995 etc). Of these, it is clear that conservative upwind methods are superior to those based on artificial dissipation (Falle 2002). It is not difficult to add an explicit treatment of ambipolar diffusion to these schemes (e.g. Indebetouw & Zweibel 2000; Balsara & Crutcher 2001; Sano & Stone 2002). As we have already pointed out these explicit methods are not efficient unless the magnetic Reynolds number due to ambipolar diffusion is large, but this can readily be overcome, either by an implicit approximation of the dissipative terms (Falle 2003), or by using super time stepping (O'Sullivan & Downes 2006). These methods are therefore eminently suitable for all three types of problem provided they can achieve the required resolution.

As an example of what can be done, consider a medium sized Beowulf cluster with 256 3 GHz processors and 256 Gbytes of Ram. This has sufficient memory for $3.2 \ 10^9$ cells, which corresponds to a uniform grid of 1500^3. Since the CPU time required for one dynamical time would be about 50 hours, such calculations are eminently feasible. Unfortunately, this resolution is inadequate because of the high density contrasts. For example, suppose that 50% of the mass is in sheets with a density contrast of 100. Then these sheets are only $\simeq 15$ cells across, which is not enough for a reliable simulation of any fragmentation process that may be occurring there. The obvious answer is to adapt the grid so that it is concentrated in those regions where high resolution is required.

An Eulerian mesh based code can readily be made adaptive by refining those cells in which the solution varies rapidly. The most usual way of doing this is to use a hierarchy of grids $G^0 \ldots G^N$ such that the mesh spacing on G^n is smaller than that on G^{n-1}. Grids G^0 and G^1 cover the whole domain, but the finer grids need only exist in regions which require high resolution. The grid hierarchy is used to generate an estimate of the truncation error by comparing solutions on grids with different mesh spacings and the grid refines if this error exceeds a given tolerance. It is desirable, but not essential to also refine in time so that finer grids use smaller time steps.

At present there are two different approaches to the refinement, which I shall call structured and unstructured. These are illustrated in Figure 2. In the structured grid, regions requiring high resolution are covered by patches of finer grid whose mesh spacing is smaller than that of its parent grid by a power of 2. In the unstructured grid, refinement is done on a cell by cell basis and always by a factor of 2. Each method has its advantages: the structured grid has a simpler integration algorithm, whereas the fine grids have a smaller filling factor in the unstructured grid.

I assume that for either method, the cost is dominated by the finest grid and the regions requiring high resolution are two dimensional sheets, and

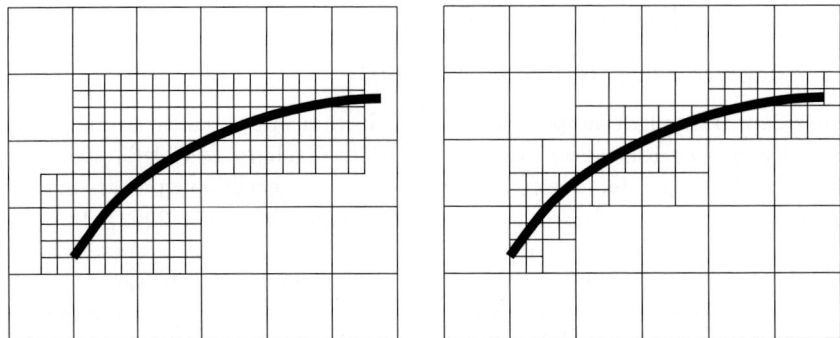

Fig. 2. The two types of AMR grid. The thick line is a region requiring high resolution (e.g. a dense sheet). Left – structured; right – unstructured.

consider a three dimensional calculation. For a uniform grid calculation, the memory costs scales as $(1/\Delta x)^3$ and the CPU time scales as $(1/\Delta x)^4$, where Δx is the mesh spacing. For an AMR calculation, the memory costs and CPU timescales scale as $(1/\Delta x)^2$ and $(1/\Delta x)^3$ respectively, where here Δx is the finest mesh spacing. If the structures that require high resolution are approximately spherical, as in the later stages of star formation, then the gain is even greater.

After a slow start, the use of AMR methods has now become quite widespread in astrophysics (see the review by Norman 2005). In particular, Truelove et al. (1998) have applied a very sophisticated AMR code with self-gravity but without a magnetic field to type III problems and, as I have already mentioned, Abel et al. (2002) have used such a code to study primordial star formation where magnetic fields are unimportant. As we shall see later, these methods are also very effective for type I and II problems.

7.2 Smoothed Particle Hydrodynamics

Before the advent of AMR, the only method that could resolve large density contrasts was smoothed particle hydrodynamics (SPH). This is a Lagrangian method in which the fluid is represented by a set of N particles whose mass and other properties are smoothed with a spherically symmetric kernel (see e.g. Monaghan 1992). For example, the density at a point $\mathbf{r}$ is given by

$$\rho(\mathbf{r}) = \sum_i m_i W(\mathbf{r} - \mathbf{r}_i, h)$$

where $\mathbf{r}_i$ is the position of the ith particle, m_i is its mass and and $W(\mathbf{r}, h)$ is the kernel function. W has the properties

$$W(\mathbf{r}, h) = W(|\mathbf{r}|, h), \qquad \int_{all\ space} W\ dV = 1, \qquad \lim_{h \to 0} W(\mathbf{r}, h) = \delta(\mathbf{r}).$$

The quantity h is called the smoothing length and plays the same role as the mesh spacing in grid based methods.

The equation of motion for a particle is

$$\frac{d^2 \mathbf{r}_i}{dt^2} = -\sum_j m_j \left(\frac{P_i}{\rho_i^2} + \frac{P_j}{\rho_j^2} \right) \left[\nabla W(\mathbf{r} - \mathbf{r}_j, h) \right]_{\mathbf{r} = \mathbf{r}_i}$$
$$+ \text{artificial viscosity, gravity etc.}$$

Here P_i is the pressure associated with the particle. If the gas is isothermal, then $P_i \propto \rho(\mathbf{r}_i)$. Otherwise, it is computed from an energy equation.

The SPH equations for each particle contain sums over all the other particles, which makes the method very expensive unless the kernel is compact i.e. $W(\mathbf{r}, h) = 0$ for $|\mathbf{r}| \geq h_c \propto h$. For a compact kernel, the pressure force on a particle only depends upon a finite number, N_n, of neighbours. When there are large variations in density, h is varied to keep $N_n \simeq \text{const} \simeq 50$. Formally the method converges for $h \to 0$ and $N_n \to \infty$. Lombardi et al. (1999) suggest $N_n \propto N^{0.2}$.

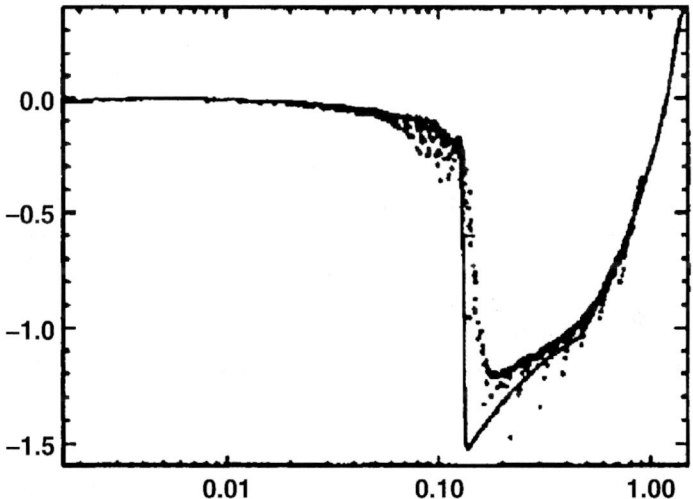

Fig. 3. Velocity as a function of radius for the gravitational collapse of an adiabatic sphere. The points are from the SPH simulation with 28,768 particles and the line from a high resolution, spherically symmetric calculation with a grid code (Steinmetz & Müller 1993).

Accuracy of SPH

There is a considerable literature on the accuracy and reliability of SPH. In particular, Steinmetz & Müller (1993), have carried out a large number of tests on problems of astrophysical relevance. Figure 3 shows one such test, which involves the gravitational collapse of an adiabatic sphere. It can be seen that the accretion shock is very poorly resolved and there is a good deal of noise. However, the resolution is extremely low: there are only about 4 mean smoothing lengths in a radius. A grid code would be even worse at this resolution.

In order to explore the convergence of SPH and compare it with a grid code, we consider a somewhat simpler problem. This is just a hot sphere, with initial radius 4, immersed in a uniform medium with the same density and a pressure 0.02 times that in the sphere. Figure 4 shows that SPH and a grid code give roughly comparable results when the number of mesh points is equal to the number of particles. However, the SPH code, GADGET, is roughly 200 times slower than the grid code. Part of the difference in speed is due to the way in which this version of GADGET is implemented, but much of it is due to the nature of the SPH algorithm. Note that neither method gives very accurate results, even though the resolution is much greater than used in most astrophysical calculations. For example, Bonnell, Bate & Vine (2003) use $5 \, 10^5$ particles to produce 400 stars.

Adaptive SPH

The standard SPH algorithm is not adaptive in the same sense as an AMR code: it simply puts high resolution where the density is large. This is fine for some problems, but it is not much use at shocks, for example. Recently there have been some interesting modifications to SPH in which the particles are redistributed to give a better representation of the solution (e.g. Bøorve, Omang & Trulsen 2001, 2004). Of course, this means that the method is no longer strictly Lagrangian, but the results are very encouraging. In particular, their method works much better for MHD than standard SPH (e.g. Price & Monaghan 2004).

Ambipolar Diffusion

SPH can handle ambipolar diffusion (ion–neutral drift) by including neutral and charged particles in an MHD code (Hosking & Whitworth 2004). Since one can use different time steps for each particle, it seems that there is no need for an implicit scheme.

From the above, it is clear that standard SPH is not really suitable for Type I or Type II problems because it is not satisfactory for MHD and it does automatically refine near shocks. In fact it has mostly been used for simplified Type III problems in which self-gravity is the dominant effect or at least assumed to be.

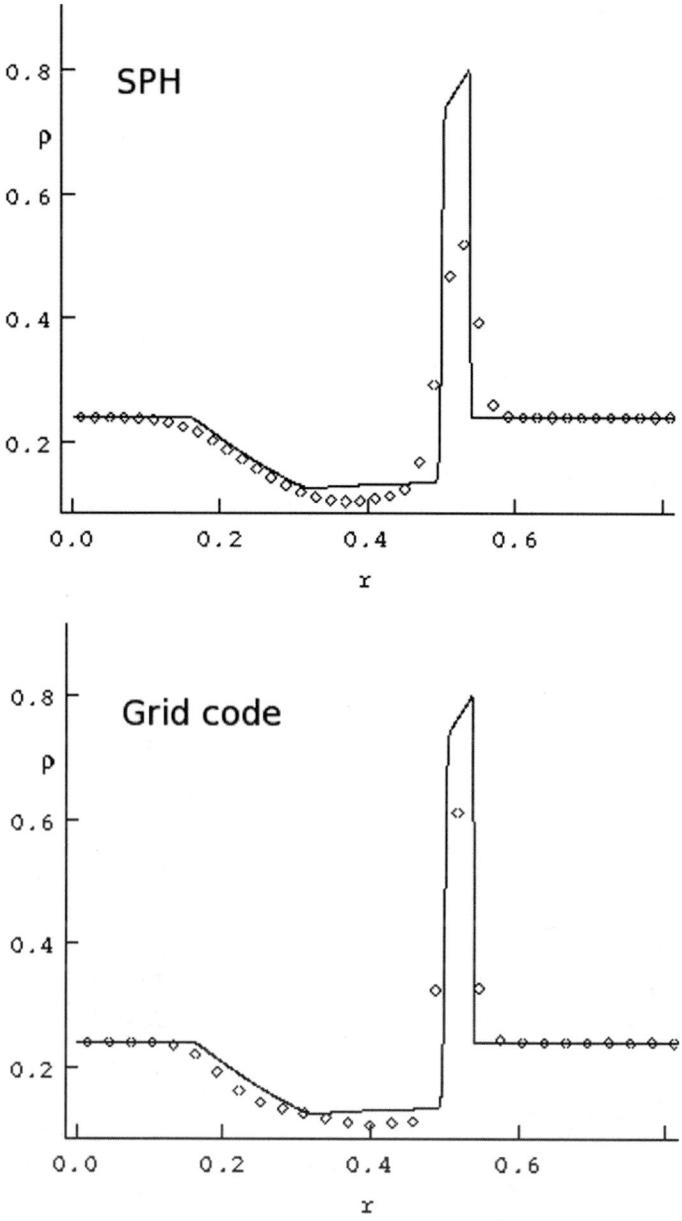

Fig. 4. Density as a function of radius for the expansion of a hot sphere. The SPH calculation used 523305 particles and the three dimensional grid calculation used 523305 cells. The line is from a high resolution, spherically symmetric calculation with a grid code.

8 AMR Calculation of Thermal Instability

In order to illustrate the power of AMR methods, we consider the formation of density inhomogeneities by thermal instability i.e. a Type I problem. Lim, Falle & Hartquist (2005) showed how a fast-mode shock propagating through a cloud in the warm phase with $\beta \simeq 1$ can induce a transition to the unstable phase with $\beta \ll 1$. Their calculation showed the formation of density inhomogeneities due to thermal instability, but they did not follow their further evolution.

Here we consider the effect of a small amplitude fast-mode wave on material in the unstable phase with number density $n = 1$ cm^{-3} and $\beta = 0.05$. The computational domain is 20×20 pc with periodic boundary conditions. The initial disturbance is a sinusoidal fast-mode wave in the x direction with a transverse velocity amplitude of 0.04 times the initial Alfvén speed and a wavelength of 20 pc. This wave is then given a 10 % sinusoidal perturbation in the y direction of its initial x position.

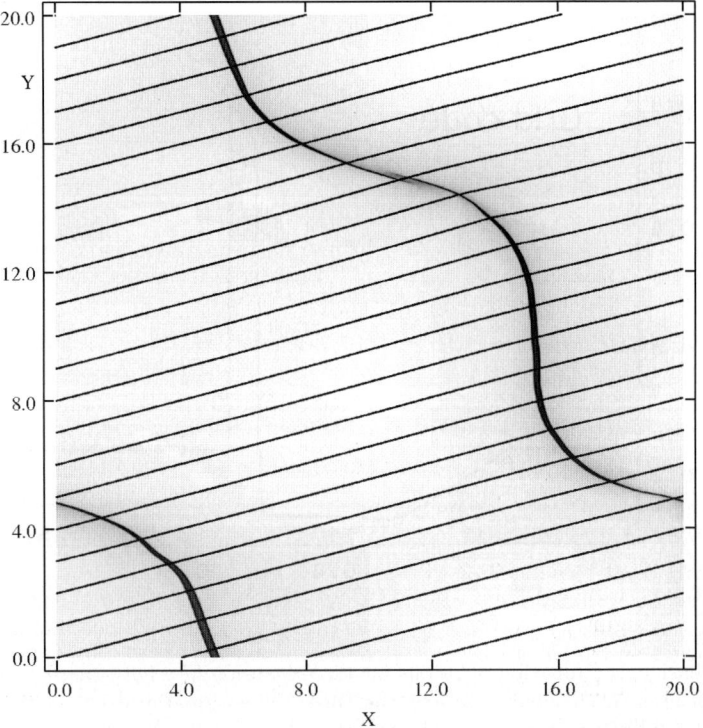

Fig. 5. Log density and magnetic field lines at $t = 6.1\ 10^6$ yrs for the AMR calculation. The density range is $n = 0.4 - 108$ cm^{-3}.

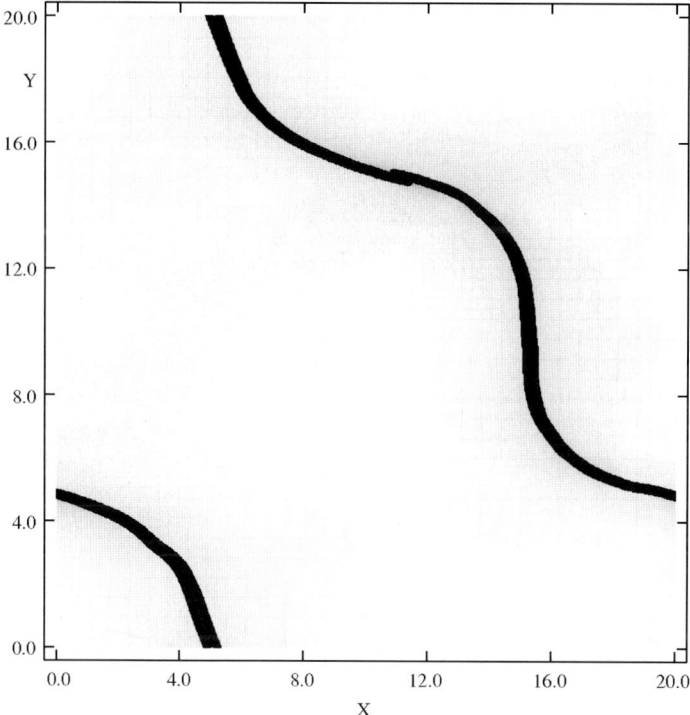

Fig. 6. G^7 grid for the AMR calculation.

Falle & Hartquist (2002) showed that a fast wave propagating at an angle to the magnetic field in low β material with an isothermal equation of state produces dense structures due to the generation of slow-mode waves. The same mechanism operates here, but the effect is much enhanced by the thermal instability.

Figure 5 shows the density and magnetic field lines at $t = 6.1\ 10^6$ yrs. The maximum density is $n = 108$ cm^{-3}, which shows that even very weak disturbances can generate very large density contrasts under these conditions. In fact the properties of these dense regions are very similar to those of the translucent clumps in GMCs: number density $\simeq 200$ cm^{-3}, temperature $\simeq 10$ K, $\beta \simeq 0.1$.

This calculation used an AMR grid with 8 levels, the finest, G^7, being 2560×2560. Figures 6 and 7 show that finest grids are entirely confined to the dense region. From Table 4, which shows the grid filling factors, it can be seen that the filling factor decreases by roughly a factor of 2 of as the grid level increases. This is because the region requiring high resolution is essentially one dimensional. The net result of this is that the calculation is $\simeq 30$ times faster than a uniform grid calculation on G^7 and uses $\simeq 22$ times less memory.

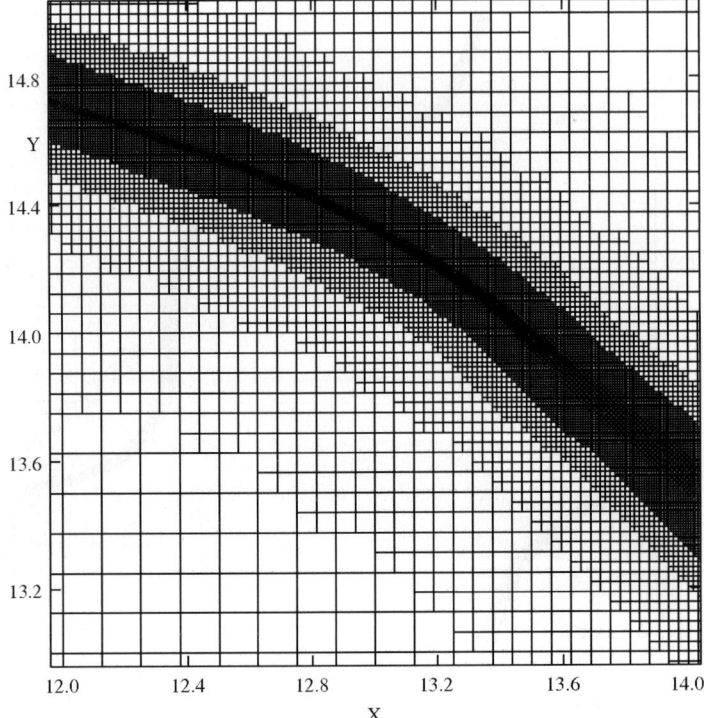

Fig. 7. Blow-up of the region with the maximum density for the AMR calculation showing grids $G^3 - G^7$.

From Figure 8, which shows a blow-up of the region with the largest density, it can be seen that, although the dense regions are sheets, they are not oriented perpendicular to the magnetic field. Also the magnetic field does not show a rapid variation in either magnitude or direction as it crosses the dense sheet. This is consistent with observations that suggest that the magnetic field varies rather weakly with density (Crutcher 1991).

9 Conclusions

The above brief review shows that, although there are numerical techniques that can handle at least some of the problems that arise in star formation, there is still much to be done. Codes for type I and II problems exist: it is just a matter of applying them intelligently to find out exactly how structure arises in the early stages of star formation. The picture is much less rosy for type III problems, particularly the later stages when radiative transfer becomes important. Although efficient methods exist for treating radiative transfer, both in SPH codes (Stamellos & Whitworth 2005) and in AMR

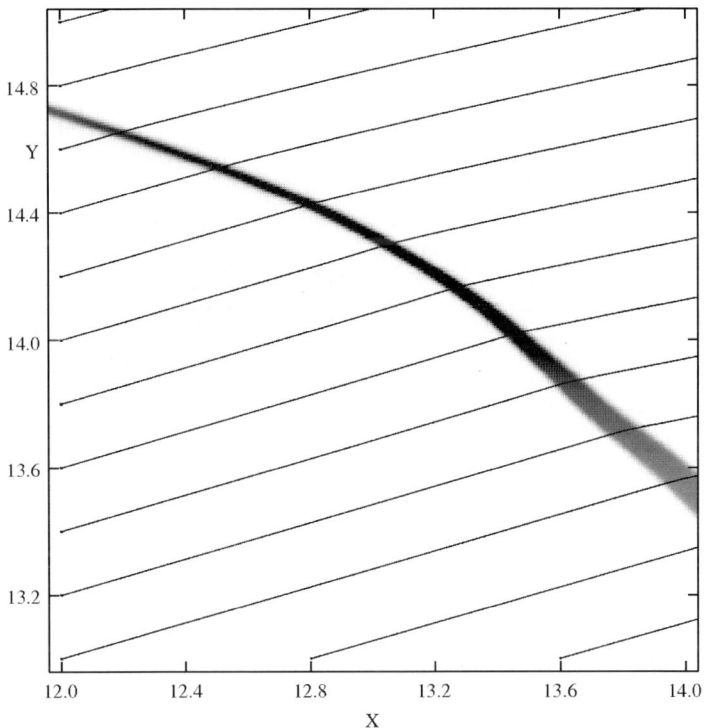

Fig. 8. Blow-up of the region with the maximum density for the AMR calculation. Linear density plot $n = 0.4 - 108$ cm^{-3}.

Table 4. Grid filling factors for the AMR calculation.

Grid	Size	Filling Factor
G^0	20×20	1.0
G^1	40×40	1.0
G^2	80×80	0.518
G^3	160×160	0.265
G^4	320×320	0.138
G^5	640×640	0.0744
G^6	1280×1280	0.0437
G^7	2560×2560	0.0267

codes (Krumholtz, McKee & Klein 2005), the computational cost places a severe limitation on the types of problems that can be tackled.

References

Abel, T., Bryan, G.L., Norman, M.L. 2002 Science **295**, 93
Arons, J., Max, C.E. 1975 ApJ **196**, L77
Balsara, D.S., Crutcher, R.M. 2001 ApJ **557**, 451
Bonnell, I.A., Bate, M.R., Vine, S.G. 2003 MNRAS **343**, 413
Bøorve, S., Omang, M., Trulsen, J. 2001 ApJ **561**, 82
Bøorve, S., Omang, M., Trulsen, J. 2004 ApJS **153**, 447
Brio, M., Wu, C.C. 1988 J. Comput. Phys **75**, 500
Crutcher, R.M. 1991 ApJ **520**, 706
Falle, S.A.E.G. 2002 ApJ **577**, 1123
Falle, S.A.E.G. 2003 MNRAS **344**, 1210
Falle, S.A.E.G., Hartquist, T.W. 2002 MNRAS **329**, 195
Indebetouw, R., Zweibel, E.G. 2000 ApJ **532**, 361
Hosking, J.G., Whitworth, A.P. 2004 MNRAS **347**, 994
Jeans, J.H. 1902 Phil. Trans. A **199**, 1
Krumholtz, M.R., McKee, C.F., KLein, R.I. 2005 ApJ **618**, L33
Lim, A.J., Falle, S.A.E.G., Hartquist, T.W. 2005 ApJ **632**, L91
Lombardi, J.C., Sills, A., Rasio, F.A., Shapiro, S.L. 1999 J. Comput. Phys. **152**, 687
Magnani, L., Blitz, L., Mundy, L. 1985 ApJ **295**, 402
Mestel, L., Spitzer, L. 1956 MNRAS **116**, 503
Monaghan, J.J. 1992 Ann. Rev. Astron. Astrophys **30**, 543
Myers, P.C. In: *IAU Symposium 115 – Star Forming Regions* ed by Peimbert, M., Jugaku (Reidel, D. Publishing Company, Dordrecht 1987) p 307
Norman, M.L. In: *Adaptive Mesh Refinement – Theory and Applications* ed by Plewa, T., Linde, T., Weirs, V.G. (Springer, Berlin Heidelberg New York 2005) p413
O'Sullivan, S. Downes, T.P. 2006 MNRAS, in press
Price, D.J., Monaghan, J.J. 2004 MNRAS **348**, 123
Ryu, D., Jones, T.W. 1995 ApJ **442**, 228
Sanchez-Salcedo, F.J., Vazquez-Semadeni, E., Gazol, A. 2002 ApJ **577**, 768
Sano, T., Stone, J.M. 2002 ApJ **570**, 314
Stamellos, D., Whitworth, A.P. 2005 MNRAS **439**, 153
Steinmetz, M., Muller, E. 1993 A&A **268**, 391
Stone, J.M., Norman, M.L. 1992 ApJS **388**, 791
Truelove, J.K., Klein, R.I., McKee, C.F., Holliman, J.H., Howell, L.H., Greenough, J.A., Woods, D.T. 1998 ApJ, **495**, 821
Williams, J.P., Blitz, L., Stark, A.A. 1995 ApJ **451**, 252
Zachary, A.L., Colella, P. 1992 J. Comput. Phys **99**, 341

Molecular Astrophysics of Star Formation

D. A. Williams

Department of Physics and Astronomy, UCL, Gower Street, London, WC1E 6BT, UK
daw@star.ucl.ac.uk

1 Introduction

John Dyson has been a friend and colleague of mine for pretty well all my working life. We work in contiguous and related fields and yet have only rarely published any research together. (We did, of course, write our remarkably long-lived undergraduate text "Physics of the Interstellar Medium", popular with professionals as well as students since 1980). Nevertheless, John has had a distinct influence on my research during all those years, and not merely through the many alcohol-induced reductions of brain capacity that he encouraged me to try. For John's own research has elegance and purpose. I knew that I would never attain in my research the analytical elegance that he demonstrates in his papers, an elegance of mathematical economy based on firm physical principles. But I also realised that I should at least strive to attain the sense of purpose that his papers show. John's papers always aim clearly to understand a physical situation, to describe its essential features, and to suggest observations that will test that description. Nowhere was this driving motivation more forcefully expressed than in the many informal seminars and colloquia I sat through in Manchester, where John expounded his powerful and compelling theories of galactic dynamics in action.

So, the question: "what's the purpose; what's it all for?" has always been in my mind as I have thought about my own research on molecules in space. During the period of the working lives of John and myself, the study of astronomical molecules and the chemistry that gives rise to them has matured from being a interesting if murky backwater of astronomy and chemistry to becoming an essential and unique tool for the study of the densest and most massive gaseous components of the Universe. This was the area in which I chose to work, and in which John's influence from his own field has been important for me.

The subject of astronomical molecules has various strands, one important one being the identification of the various chemical processes that generate the wealth of detected molecular species, of which more than 120 are currently known. That work involves the interaction of laboratory and theoretical chemists (who provide the fundamental chemical data), as well as modellers (who show how various species react and identify potentially important reactions) and observers (who identify molecules in astronomical sources); I tend to call that whole area of activity "astrochemistry". It requires the study of thousands of reactions, their temperature dependences, their products, and their interaction in complicated networks, and the testing of ideas against observation. It is essentially a question of understanding chemical processes under the extreme conditions of interstellar and circumstellar environments.

Another strand of the study of astronomical molecules concerns the question: "how can I use astronomical molecules to tell me more about astronomy?" Obviously, molecular line observations of several species or of several lines of the same molecule can give reliable estimates of gas density and temperature. But there is a wealth of more detailed information that can be obtained from a deeper study of astronomical molecules and the chemistry that gives rise to them. In such an approach, one need not take on the responsibility of attempting to answer all questions concerning molecules in space. One concentrates on those species that can reveal much of the astronomical truth; these are often the smaller and more abundant species. One tries to ignore a multitude of questions about astrochemistry, as far as one can, to address questions about the macro-physics, i.e. the evolution of astronomical regions. I tend to call this type of research "molecular astrophysics". It obviously overlaps with astrochemistry, to some extent.

Both astrochemistry and molecular astrophysics (defined in this way) are essential and valid areas of research. But the example of John's powerful, elegant, and purposeful research has left me with a preference to try to emulate his work by using molecular astrophysics as a tool to explore astronomical situations and to gain insight into their future developments. The choice is wide, and molecular astrophysics can offer realms of almost unbridled speculation - a delight to those who, like me, find observational data sometimes too constraining.

In fact, almost any region in the Universe in which the density is not too low (say, more than one H atom per cm^3) and the temperature not too great (say, less than a few thousand Kelvin) is likely to contain at least trace molecules. Even sunspots contain water, molecular hydrogen, and other molecules (e.g. Jordan et al. 1977, Sonnabend et al. 2006). But many regions are almost entirely molecular, and from them molecular emissions usually constitute an important energy loss. Such regions include, for example, the most massive accumulations of baryonic matter, giant molecular clouds (containing up to a million solar masses of gas, mostly molecular hydrogen) in our own and other galaxies. It's now clear that molecules play a controlling role in the formation of stars in these great reservoirs of matter. This is true in galaxies in the present epoch, and was true in galaxies at high redshift, and even true also in those primitive galaxies in which the very first stars were formed.

The formation of stars is a topic that lends itself naturally to molecular astrophysics. Star-forming regions are almost entirely molecular; they are dense by the standards of the interstellar medium, and the dust embedded in the gas ensures that very high optical depths arise, so that the material is generally cold and almost entirely neutral. Observationally, these regions can therefore be studied with millimetre-wave emissions from molecules and far-infrared emissions from dust. In this article I would like to give a few illustrative examples of how molecular astrophysics, through a combination of observations and modelling, can help to provide insight into that most basic and fundamental of astronomical processes: the formation of stars.

Finally, a health warning. Do not be distressed if you do not see your favourite papers on star-forming regions referred to here. This subject is one of the most active topics of modern research in astronomy, and I cannot possibly hope to be comprehensive. Therefore, I have simply chosen to refer to some work that I have particularly enjoyed, and which shows - as a tribute to John's insightful approach - the predictive power of the subject.

2 Regions of Formation of Low-Mass Stars

2.1 The Pre-stellar Phase

Observations of Structure

It has long been known that the formation of low mass stars takes place in dense cores of molecular clouds (e.g. Beichman et al. 1986; Benson & Myers 1989). Their properties are summarized in Table 3 of the chapter by Sam Falle. Molecules such as CS, NH_3 and HCO^+ have frequently been used as good tracers of high density molecular gas. The broad picture that emerges from such observations shows an association of dense gas with the infrared sources that are embedded young stars. Evidently, dense cores can collapse under gravity and evolve independently to form low mass stars which then warm their immediate environment. However, closer examination revealed some problems with this general idea. Pastor et al. (1991) and Morata et al. (1997) compared CS (J = 1 - 0) and NH_3 (J,K = 1,1) emissions with similar angular resolutions. They found that there was a separation between the emission peaks of the two molecules, a wider spatial distribution of CS than NH_3, and wider spectral lines of CS than NH_3. These results were unexpected and do not fit easily with the general picture. For example, one would expect on the basis of molecular emission properties that NH_3 should trace lower density gas than CS, and therefore might be more widely distributed. An illustration of the CS and NH_3 maps spatial discrepancy is shown in Fig.1.

An early suggestion to resolve this problem (Taylor et al. 1996, 1998) was that the time-dependence of the chemistry could affect the CS/NH_3 balance. These authors argued that, for uniform density gas, NH_3 should become

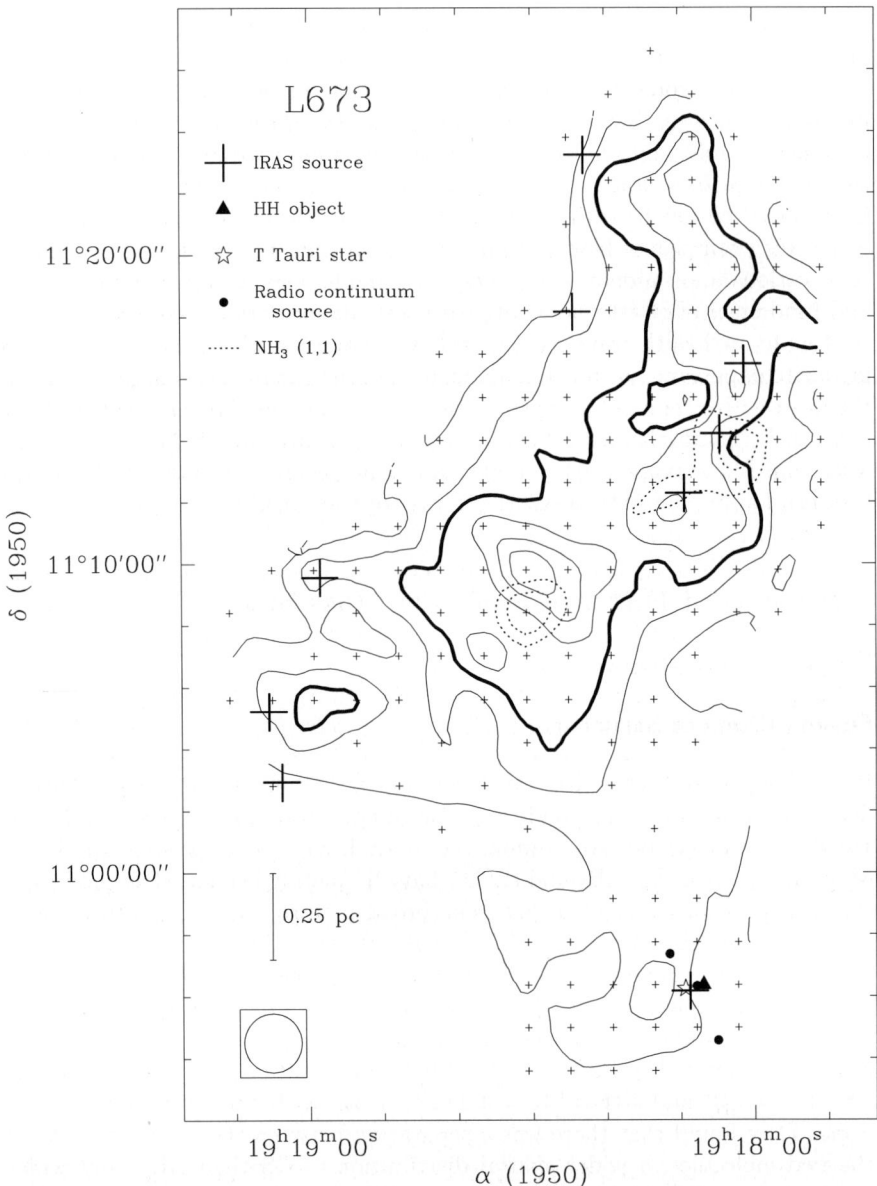

Fig. 1. Contour map of the integrated line intensity of the CS (J = 1 - 0) transition for L673 (Morata et al. 1997). The lowest contour is 0.475 K km s^{-1} and the increment is 0.125 K km s^{-1}. The thick line represents the half-power contour. The half-power contour of the NH$_3$ emission is also shown (Anglada et al. 1997).

abundant later in the chemical evolution than CS. But since chemical rates are strongly density dependent, the chemistry forming NH_3 would be more rapid in denser cores, while CS is made both in lower density gas and in the high density cores. Therefore, the NH_3 traces the denser cores, while the CS traces the general shape of the molecular cloud that contains the cores. These authors suggested that dense cores in a molecular cloud must be transient, otherwise NH_3 and CS emissions would be spatially coincident, and that the cores must be of such a size as to be unresolved with the single-dish radio and millimetre wave telescopes then in use; the implication was that the core sizes should less than about 0.1 pc. Could molecular clouds really consist of ensembles of unresolved and independent transient cores? The timescale of the growth and decay of such cores would be determined by the chemical rates to be on the order of a million years. Evidently, most of the cores would dissipate before high densities could be achieved and substantial NH_3 abundances arise; but an occasional core might, for some reason, achieve a higher than average density, produce the tracer NH_3 molecule, and become gravitationally unstable and form a star.

Observational support for this picture now exists for two interstellar regions. Peng et al.(1998) studied C_2S emissions from the ridge of the well-studied interstellar cloud TMC-1, in which cyanopolyyne emissions were known to peak (this is Core D of Hirahara et al. 1992). The observations of Peng et al. were of high angular resolution, corresponding to 0.02 - 0.04 pc at TMC-1, and identified a total of 45 clumps in C_2S emission within Core D, and computed their physical properties. Thus, TMC-1 Core D is very much more complex than has been understood previously. As Hartquist et al. (2001) have argued, if such sub-structure is not contained by external forces, it will dissipate very rapidly (10 000 y). This would imply a very short timescale for chemical evolution within the sub-structure, and Hartquist et al. suggested that this extreme youth may be the cause of the exceptional chemical richness of TMC-1.

A more detailed study using the BIMA millimetre wave interferometer has been made of part of the dark cloud L673 (Morata et al. 2003). Figs. 2a, b, and c show maps of the integrated emission in the CS (J = 2 - 1), N_2H^+ (J = 1 - 0), and HCO^+ (J = 1 - 0) lines, respectively, for a small region of L673. Several clumps appear in each map, though each transition generally traces different clumps.

A further study combining the BIMA and FCRAO maps (Morata et al. 2005) recovers structure previously undetected or marginally detected in the 2003 paper, and confirms the earlier picture of filamentary structure connecting several intense clumps. The combined maps for HCO^+ and CS are presented for comparison in Figs. 3a and b. These authors found a total of 15 resolved clumps in the small region of L673 observed, with diameters in the range 0.03 - 0.09 pc and estimated masses between 0.02 - 0.2 solar masses, except for the largest clump which has a mass of 1.2 solar masses. The lower masses are below corresponding virial masses, but the larger masses approach

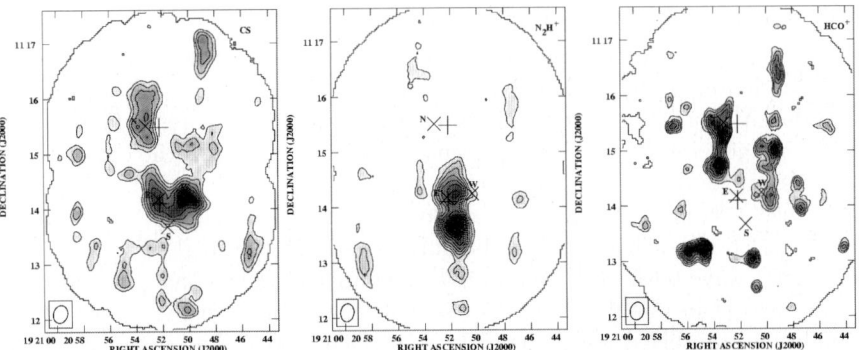

Fig. 2. Integrated emission of (a) CS (J = 2 - 1), V_{LSR} range 5.6 - 9.8 km s^{-1}; (b) N2H$^+$ (J = 1 - 0), V_{LSR} range -1.4 - 14.2 km s^{-1}; and (c) HCO$^+$ (J = 1 - 0), V_{LSR} range as for (a) lines for a starless core in L673 (Morata et al. 2003).

the virial masses. This supports the idea that the clumps are transient, and that only the most massive clumps will be able to collapse to form stars. Therefore, assuming that the characteristics found in L673 are common to molecular clouds, then the factor that determines the mass spectrum of the transient clumps appears also to determine the low mass star formation rate in molecular clouds.

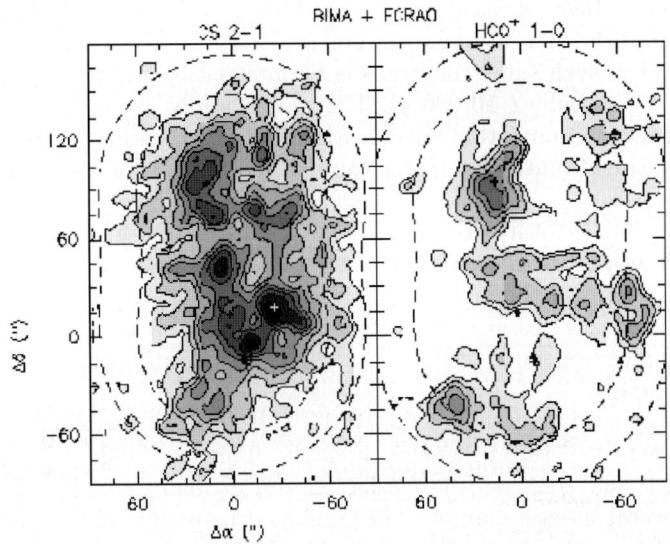

Fig. 3. Combined FCRAO and BIMA maps for (a) CS (2 - 1) and (b) HCO+ (1 - 0) transitions, for V_{LSR} range 6.05 - 8.45 km s$^-$1 (Morata et al. 2005; this figure supplied by O Morata and J M Girart).

Molecular Clouds as Ensembles of Transient Cores

If we make the assumption that the sub-structure of L673, as described by Morata et al. (2003, 2005), applies generally to molecular clouds that are low-mass star-forming regions, then we should be able to model the physical and chemical properties of such clouds in unprecedented detail. Such an approach contrasts sharply with conventional modellings of molecular clouds, which often use static and/or uniform density slabs or spheres. Garrod et al. (2006) have explored the alternative approach of molecular clouds considered as ensembles of transient cores. In their picture, a molecular cloud is considered as a dynamic region in which cores grow from the background gas, attain a maximum density and smallest size, and then decay once again into the background gas. In the Garrod et al. (2006) model, for want of further information the cores are assumed to be all the same and with the same cycle period; they are randomly distributed in space within the cloud, and the phase of each core is random within the cycle period. The chemistry follows the time-dependence of the changing density structure and the consequent changing radiation field within the core. The number of cores per projected unit area of molecular cloud is chosen to be similar to that measured by Morata et al. in L673.

Before developing the full Garrod et al. (2006) model, Garrod et al. (2005) examined the chemical behaviour of single, isolated transient cores. The density structure and time development of a single core was required, so that the core's chemical evolution could be evaluated, prior to considering the ensemble of such cores. Garrod et al.(2005) adopted the ideas of Falle & Hartquist (2002) concerning the passage of one-dimensional magneto-hydrodynamic waves through a cold magnetically dominated plasma (cf. Section 8, paragraph 3 of Sam Falle's chapter). Falle & Hartquist found that modest density perturbations could lead to the growth of significant density inhomogeneities. The density profiles showed that an initially shallow function evolved into a strongly peaked function, while the time evolution of peak density showed a similarly peaked function. Garrod et al. (2005) therefore approximated both the density profile and the time-dependence by Gaussians to avoid the need to include the full MHD equations along with the very extensive chemical rate equations; a complete treatment of that kind is beyond current computational capability. With these space and time functional forms, Garrod et al. (2005) evaluated the time and space dependence of the chemistry within a single core taken through one or more cycles, and considered the sensitivity of that chemistry to the various parameters adopted. They included a large gas-phase network and a rudimentary description of the interaction of gas with surfaces of dust grains, including the accumulation of ice mantles in the denser and darker parts of the transient core and the return of these ices in regions that are less dense and more transparent to the interstellar radiation field.

The main conclusion of this work on a single transient core is that the chemistry of such a dynamic cloud is necessarily "young", i.e. far from steady-state values. This is in agreement with observations, and also resolves a long-standing difficulty concerning the freeze-out of species on to dust. In this dynamic model, there is a natural limit to the growth of ice mantles because of the finite cycle time, whereas in static models ices would accumulate over the entire lifetime of the cloud so that all atoms other than hydrogen and helium would end up in the ice. Any fluid element may find itself in a dense region at one time and in diffuse gas at another. The chemistry in the diffuse regions is enriched by molecules that evaporate from ices as a fluid element becomes of low density. Finally, taking gas through repeated cycles shows that there is a limit cycle in the chemistry, after which the spatial chemistry of the core does not change from cycle to cycle.

Garrod et al. (2006) then considered an ensemble of transient cores to represent a molecular cloud. Each of the cores was assumed to have the properties described in the 2005 paper, and as stated above the number density of cores was chosen to match that measured by Morata et al. (2005). As was also mentioned above, the cores were assumed to be distributed randomly within the space of the model cloud, with random evolutionary phase. Maps were then created of predicted molecular column densities of various species in the model cloud with a crude allowance for optical depth effects. It was assumed that the model cloud was then viewed at either low and high resolution, representing crudely observations using either single-dish or array telescopes. Computed maps for several species are shown in Figs. 4 (low resolution) and 5 (high resolution). The morphological similarities between these maps and the observations of low and high resolution maps of Morata et al. are striking. The theoretical maps broadly reproduce the morphologies, sizes of emitting regions and separation in peak positions found in the observational maps. The agreement depends on the cycle period being about a million years and the interaction of gas and dust in the cycle. Such molecular line maps cannot be generated by simpler models of molecular clouds, and the implication is that molecular clouds are indeed dynamic, with a substructure that forms and dissipates in a cyclic manner.

If the model proposed by Garrod et al. is valid, then it should also explain the behaviour of ice deposition within molecular clouds. Observations (Whittet 1999) show that H_2O ice is not present on dust grains in molecular clouds until the line of sight towards a background star is extinguished by more than a certain critical value; for molecular clouds in the Taurus region this is about 2.5 magnitudes. For lines of sight with greater extinction, then the optical depth in the water ice line at 3 microns increases linearly with optical depth. In the model of Garrod et al. ice is assumed to be deposited in the denser part of the cycle of any core, and removed again as the density decreases. The line of sight may pass through several cores at different points in the cycle. A preliminary examination of the water ice abundance and its dependence on the visual extinction has been made by Garrod, and Fig.6 (R T Garrod, private communication) shows the predicted dependence. This shows that a critical value of visual extinction is obtained, and that a linear dependence on extinction is expected in this model, extending up to around

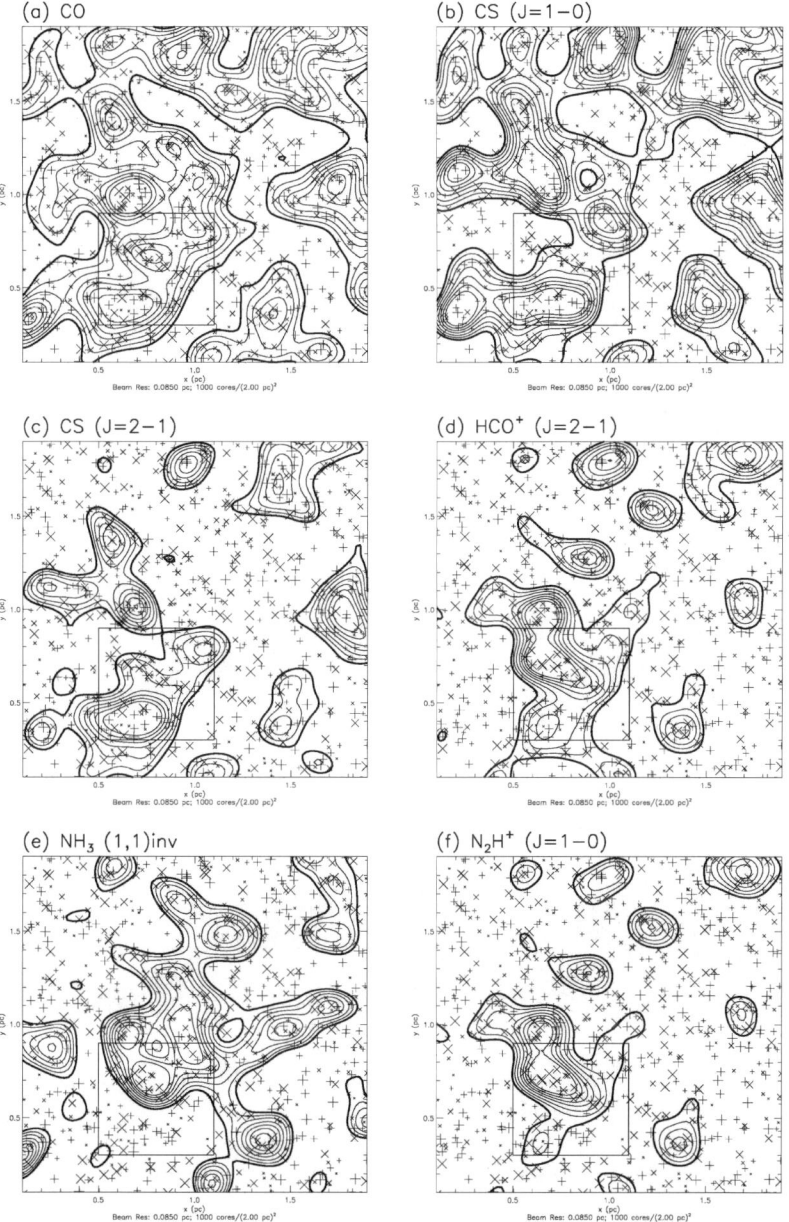

Fig. 4. Low resolution simulated maps for a beam of FWHM = 0.17 pc for several species. Box in lower left indicates region shown in Figure 5 (Garrod et al. 2005).

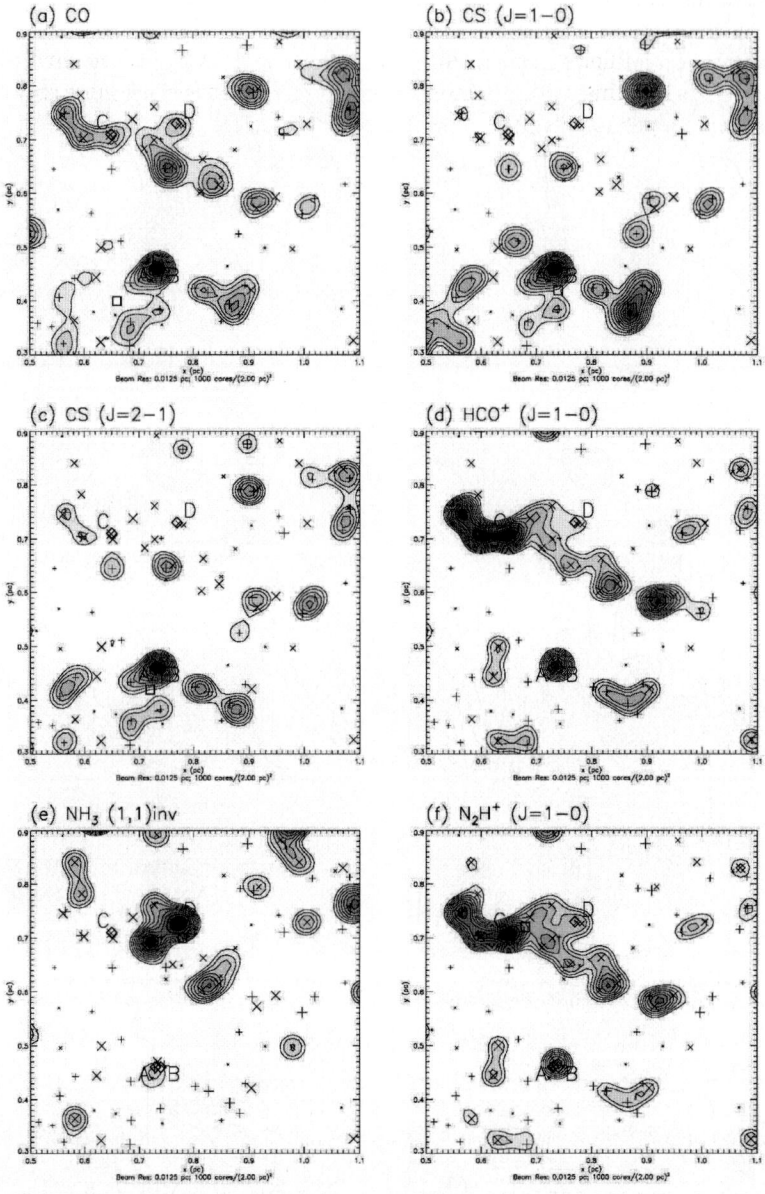

Fig. 5. High resolution maps of same species as in Fig. 4 for the region indicated, beam = 0.025 pc. Square symbol indicates the low resolution peak of the mapped molecule. Diamond symbols refer to the high resolution peaks in (a) CO; (b) CS (1 - 0); (c) CS (2 - 1); (d) HCO$^+$ (1 - 0); (e) NH$_3$ (1, 1)inv; (f) N$_2$H$^+$ (1 - 0) (Garrod et al. 2005).

10 visual magnitudes. The slope is about that obtained for the Taurus clouds. This result tends to suggest that the mechanism of formation of water ice that has been assumed here, i.e. the efficient conversion of O atoms incident on the surface of dust grains to water molecules and the retention of those molecules to form ice, must be approximately correct (cf. Jones & Williams 1984).

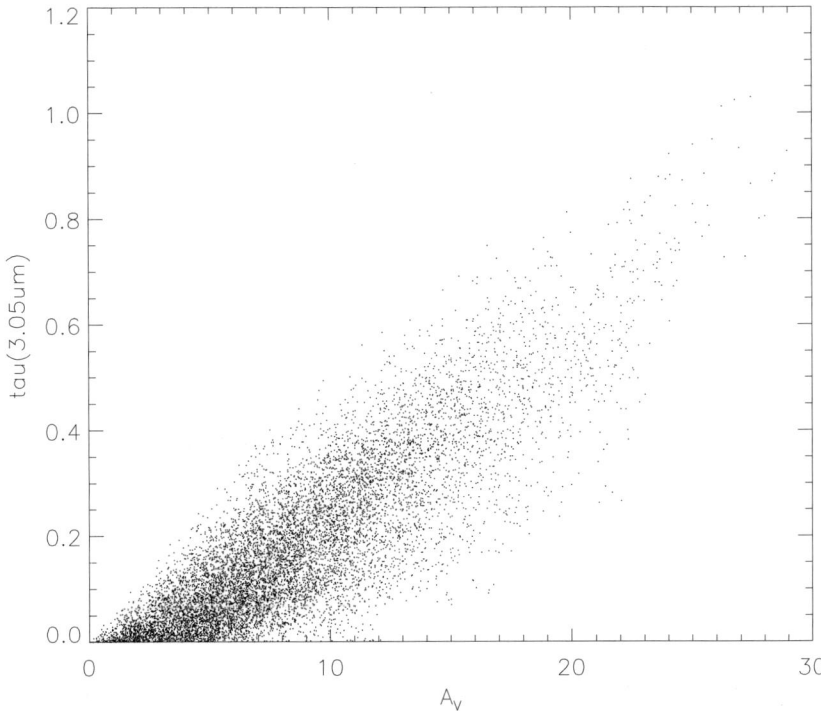

Fig. 6. Correlation of ice absorption with visual extinction, as predicted by the model of Garrod et al. (2005) (R T Garrod, private communication).

Collapse and Freeze-out in a Pre-stellar Core

The formation of a star takes place in a collapsing core; the evolution is from a star-less core into a core possessing an infrared object, and then further evolution takes place as the star becomes detached from its environment and interacts with it (André et al. 1993). Here, we are concerned with the early star-less phase in which collapse, chemistry, and freeze-out of atoms and molecules on to dust are occurring. As has been stressed elsewhere (e.g. Williams 2003; Rawlings 2003), the timescales of dynamics, gas-phase chemistry, and the gas/dust interaction are similar for interstellar material at a density typical of that in dense cores. The collapse phase is relatively short-lived compared the duration of a star, and it is difficult to observe an object in this important transition. The classic signature of infall is that of an optically thick line that is double-peaked with a "blue" wing that is stronger than the "red" (Zhou 1992), together with an optically thin single-peaked line of a higher transition or of a rarer isotopomer. Zhou et al. (1993) were the first to identify infall signatures (of formaldehyde and carbon monosulfide) in the object B335. These are consistent with infall but do not confirm it. The difficulty is that double-peaked lines can arise for a variety of reasons. For example, Redman et al. (2004) showed that the pre-stellar core and infall candidate L1689B has both "blue" and "red" enhanced double-peaked profiles and can be considered as a purely rotating body without infall. Evidently, considerable care is needed in the interpretation of molecular line profiles.

The similarity of the important timescales of dynamics and chemistry means that quite unexpected chemical behaviours may occur. Rawlings et al. (1992) showed that some molecular tracers of infall (such as HCO^+ and N_2H^+) may actually increase during infall of the core, because molecules that control their loss (such as H_2O) are being frozen on to the dust grains. Evidently, the nature of the infall and the details of the chemistry affect the spatial distribution of molecular abundances throughout the core. Since the nature of the infall determines local velocities in the core, the line profiles are affected very sensitively by the dynamics and particularly by the gas/grain interaction (for which we have little reliable information). Rawlings & Yates (2001) have demonstrated this sensitivity by coupling a dynamical/chemical model of an infalling core with a sophisticated radiative transfer calculation. Fig.7 shows the effect on the computed model profiles of the HCO^+ (J = 4 - 3) line for rather modest changes in assumed parameter values in the model. Model 2 reflects a change in sticking probability from 0.3 in model 1 to 0.5; the double-peaked "red" enhanced signature is retained, but the intensity is reduced by a factor of about three. Model 3 reflects a reduction in the cosmic ray ionisation rate by a factor of 2 compared with model 1. In model 4 the infall is interrupted for 0.2 My, whereas in model 1 infall continues uninterrupted. The changes in the line profiles for this transition, and those of other important tracers are affected very significantly by the variations imposed on what are poorly known parameter values. How are we to proceed? Evidently, results for a single line profile cannot give reliable information as to the infall dynamics or of the gas/grain interaction. The implied degeneracy of possible solutions can, however, be removed. We are not restricted to a single line; we have the opportunity of using several transitions of each of a number of molecular species along many lines of sight.

The discussion of Rawlings & Yates makes it clear that the gas/dust interaction may affect our understanding of infall dynamics. Yet there is little reliable information from laboratory studies, partly perhaps because the physical nature of the dust is not well understood (though its chemical structure

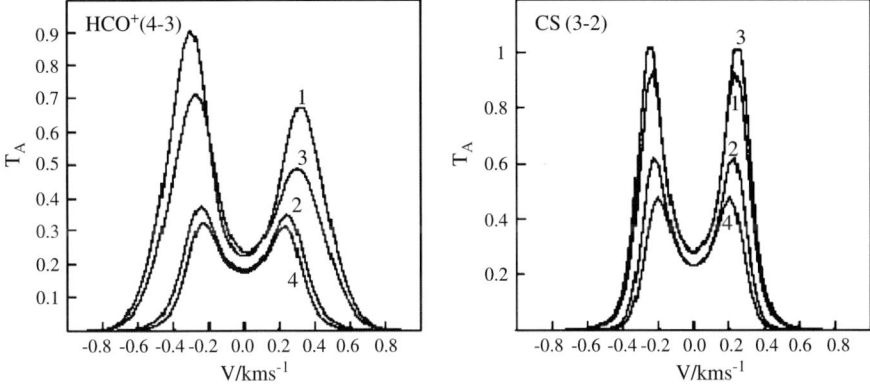

Fig. 7. The HCO$^+$ (4 - 3) and CS (3 - 2) line profiles generated by four slightly differing models (Rawlings and Yates 2001). Relative to model 1, model 2 changes the sticking coefficient from 0.3 to 0.5, model 3 reduces the cosmic ray ionisation rate by a factor of 2; and model 4 introduces a pause in the collapse representing a period of quasi-stability.

can be reasonably well determined; see e.g. Whittet 2002). It is interesting, therefore, that a direct measure of the freeze-out process can be obtained from astronomical observations. Redman et al. (2002) have measured the $^{12}C^{17}O$ (J = 2 - 1) line intensities at various positions in the pre-stellar core L1689B, and have shown that the lines of this minor isotope are optically thin. The relative abundance of $^{12}C^{17}O$ to the major isotope $^{12}C^{16}O$ is known, and therefore the actual CO abundance across the core can be obtained. Alternatively, the total gas density in terms of H_2 molecules cm^{-3} can be computed from SCUBA measurements of dust far-infrared emission across the core, if one assumes that the gas/dust ratio is well determined. The results of this comparison are shown in Fig.8. Redman et al. infer that in the central 5000 AU of L1689B about 90 percent of the CO has frozen on to the dust grains. If desorption is not occurring, then the level of freeze-out can give an indication of core age relative to the free fall times. In L1689B, the level of freeze-out can only be achieved after a duration at least comparable to the free-fall timescale.

Perhaps, the most detailed observational and theoretical study of a collapsing core is that of Evans et al. (2005) of B335. This is a fairly round dark globule that is relatively close. The observational study includes 25 transitions of nine molecules. The data are simulated with a Monte Carlo code which uses various physical models of density and velocity as a function of radius. The dust temperature is calculated self-consistently by a radiative transfer code, and the gas temperature is calculated at each radius, including heating by photoelectrons, cosmic rays, and cooling by molecular lines and gas/dust interactions. In all the cases considered by Evans et al. the collapse dynamics adopted is the "inside-out" model of Shu (1977). In spite of all the uncertainties remaining in the physical and chemical description, as discussed above,

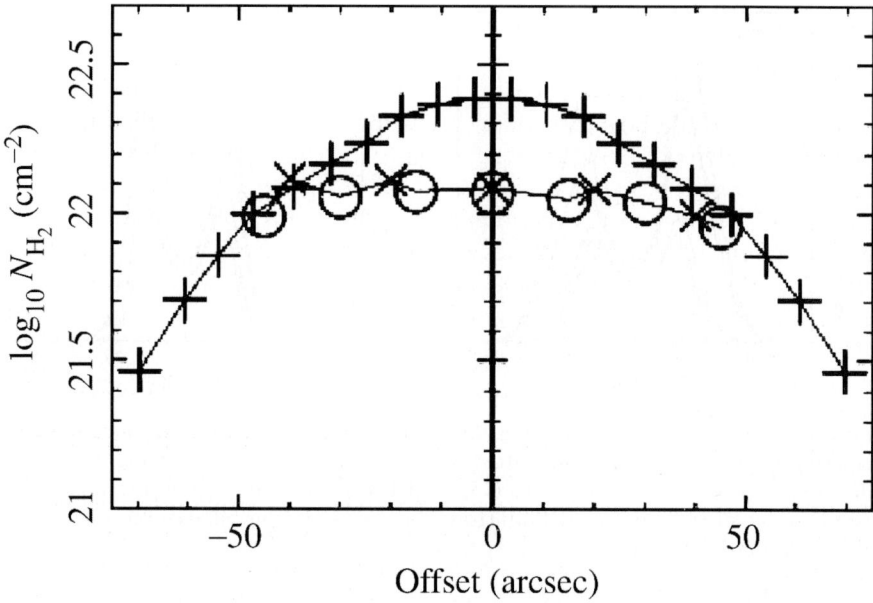

Fig. 8. Column densities of H_2 across L1689B measured using the $C^{17}O$ (1 - 0) transition (crosses, x), $C^{18}O$ data of Jessop and Ward-Thompson (2001) (circles), and SCUBA 850 micron radial data fits of Evans et al. (2001) (pluses, +) (Redman et al. 2002).

these models give good fits to the observed line profiles. These authors point out that the fits are now good enough that one can look in some detail at what might improve the match with observations, both with respect to the chemistry and the dynamics. Thus, the study of the pre-stellar phase of low-mass star formation is now entering a new stage in which the physics is well enough defined that some precision in the matching of model predictions with observations can now be anticipated.

2.2 The Post-stellar Phase

Chemistry Near Embedded Stars

In their earliest stages, newly-formed or forming stars are embedded in the core of dense gas that gave rise to their birth (André et al. 1993). Obviously, the core responds to the presence of the star and to the winds and radiation fields it generates in what was previously a dark and quiescent environment. In Section 3.1 we shall consider the case of the interaction of newly-formed massive stars with their natal core; here we consider the case of relatively low-mass stars. These stars are frequently observed to show, when young, outflows of gas that may be highly collimated (see the Chapters by Peter Brand and Tom Ray).

A strong indication that an interaction was occurring was given in unresolved, single-dish, detections of HCO^+ in star-forming cores. It was found that, while the line profiles of HCO^+ suggested infall (as discussed above) its abundance could be enhanced significantly, possibly by factors of up to a thousand compared to the values normally found in dark clouds. Certainly, something unusual is required to bring this enhancement about. In fact, it is rather difficult to do this by simply adjusting parameters associated with the normal interstellar chemistry. In cold quiescent clouds HCO^+ is mainly formed by the reaction of H_3^+ with CO, where H_3^+ is a direct consequence of cosmic ray ionisation of H_2. But HCO^+ is lost in reactions with electrons, which also derive from cosmic ray ionisation, so a higher ionisation rate affects both formation and destruction of HCO^+; this molecule tends to be rather insensitive to changes in physical parameters.

The well-studied core B335 has an abundance measured by unresolved single-dish measurements (Zhou et al. 1993; Choi et al. 1995) to be at least an order of magnitude larger than that which can be accounted for by cold cloud chemistry. Rawlings, Taylor & Williams (2000) adopted an idea suggested by Taylor & Williams (1996) in a slightly different context and proposed that this enhancement comes about because of chemistry occurring in a turbulent mixing layer which is the interface between the high velocity outflow from the young star and the quiescent or infalling dense gas that is being continually eroded by the outflow. The idea is that CO and H_2O are being released from grain mantles in the cold core and photoprocessed by shock-generated radiation fields. Thus, CO is dissociated and atoms ionised, so that a reaction between C^+ and H_2O can occur, forming HCO^+ directly, in a process that does not normally occur in interstellar clouds since H_2O is a very minor constituent of clouds. However, H_2O can be abundant in localised regions where ice mantles are released. The work of Rawlings et al. (2000) showed that significant enhancements of HCO^+ would plausibly be achieved in this way, and Viti et al. (2002) using a more extensive chemical network showed that similar abundance enhancements should be found for other molecular species, including H_2S, CS, H_2CS, SO, SO_2, and CH_3OH.

With sufficiently high angular resolution the morphology of the interaction can be deduced. For example, Velusamy & Langer (1998) showed that the outflow from the embedded star IRS1 in the cloud B5 is opening a cavity in the infalling core that will widen to a near-spherical outflow on a timescale of about 10000 y. Hogerheijde et al. (1998) used the Owens Valley Millimetre Array to explore the morphology of molecular emissions in a number of young stellar objects. In the case of L1527, they confirmed that the HCO^+ emission originates in the boundary lobes of the outflow, exactly as in the later model of Rawlings et al. (2000). Rawlings et al. (2004) then combined a full three-dimensional radiative transfer code with a spatially dependent chemistry to model the morphology of the outflow in L1527. Their best fit model is shown in Fig.9. They also showed that the line profiles from such an object with

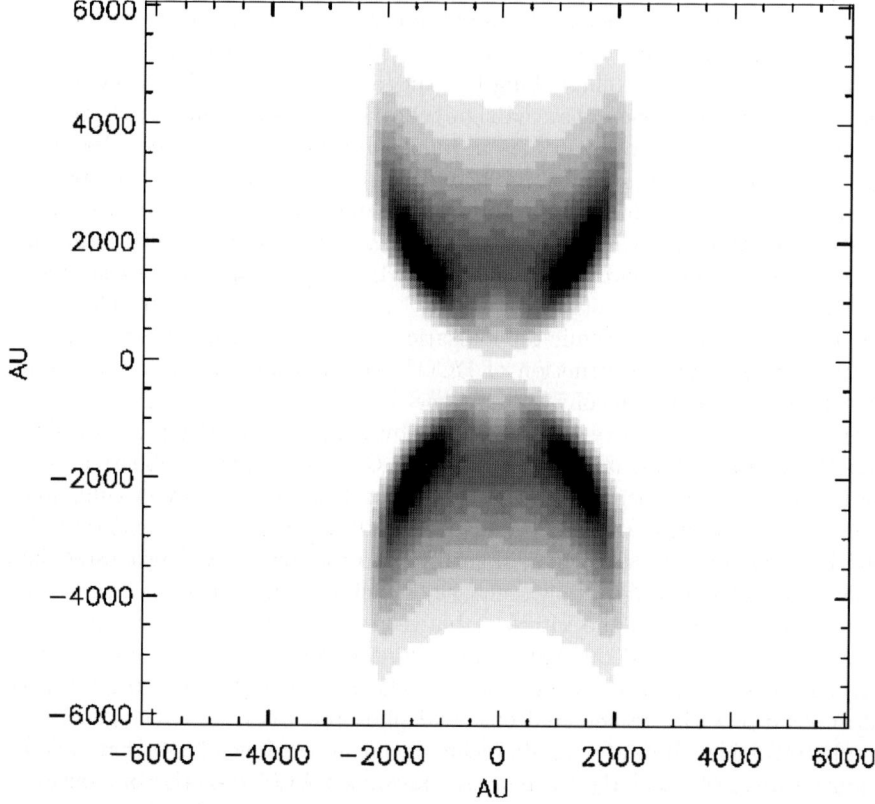

Fig. 9. Linear gray-scale representation of HCO^+ (3 - 2) emission in a bipolar outflow (Rawlings et al. 2004). The peak of the emission is 0.74 K km s^{-1} and the viewing angle is zero degrees.

infall and outflow should show strong variations in the relative strengths of the "red" and "blue" wings.

The interface between different astrophysical media has been a special interest of John Dyson, in diverse astronomical contexts. It is particularly pleasing for me to see that molecular astrophysics can play a role alongside the more formal hydrodynamics studies that he uses in understanding the nature of the turbulent mixing, and in providing molecular tracers of that interaction that are capable of demonstrating its morphology.

Interaction of Stellar Jets from Young Stars with Molecular Clouds

It has long been known that stellar jets are injected at high velocity into the interstellar medium from young stars, and can extend for a parsec or more from the star. Where these jets impinge on interstellar gas they set up a shock, behind which the gas is heated and ionised, and are visible as "fuzzy" radiation sources known as Herbig-Haro objects (see the Chapters by Peter Brand and Tom Ray). The Herbig-Haro shock structure has been one of John Dyson's many interests in research. At first sight, it seems unlikely that such a highly energetic region could have any relevance to molecular astrophysics, or be probed by it. In fact, the role of Herbig-Haro objects (HHOs) in molecular clouds has been an unexpected and fruitful area of study for more than a decade. Here, we give a brief description of some recent work in this area, and show how this jet/cloud interaction can give information about the structure of interstellar clouds.

Many HHOs have associated with them quiescent dense clumps, characterised by low temperatures and narrow molecular line widths and by abundance enhancements, with respect to the ambient molecular cloud, of several species, notably HCO^+ and NH_3 (in the first instance). Although the HHO shocks are close to the clumps (typically less than 0.1 pc) the clumps do not usually show signs of dynamical excitation. Girart et al. (2002), Viti et al. (2003), and references therein, summarise the current observational situation with respect to these neutral molecular clumps. Girart et al. (1994) suggested that the particular chemical signature associated with these clumps might arise because of a radiative interaction between a HHO and a neighbouring clump. Girart et al. (2005) have since shown that a radiative interaction may also be accompanied by dynamical interactions of various kinds. However, we shall limit our discussion here to the radiative interaction.

Girart et al. (1994) suggested that irradiation by the HHO might evaporate icy mantles from dust grains in clumps within the molecular cloud and promote a photochemistry in the chemically enriched gas. Taylor & Williams (1996) investigated this situation and showed that HCO^+ should be enhanced in abundance as a result of reactions between C^+ (from photodissociated CO) and H_2O molecules (from the desorbed ice). This is a situation similar to that discussed in turbulent interfaces, above. Viti & Williams (1999) extended the chemistry and showed that, in addition to HCO^+ and NH_3, enhanced abundances should arise for CH_3OH, H_2CO, SO, SO_2, and CN. Raga & Williams (2000) showed how the morphology of such a clump should be modified as the HHO moved through the molecular cloud.

The predictions of Viti & Williams (1999) were tested by detailed multi-transition observations of the region around the HHO known as HH2 (Girart et al. 2002). A total of 14 species were detected, and the results confirmed that the region had a peculiar chemistry; HCO^+, CH_3OH, and H_2CO were strongly enhanced in abundance, SO, and SO_2 were weakly enhanced, while HCN and CS were underabundant. In qualitative terms, the model results of Viti and Williams were confirmed. A more detailed observational study of the same source (Girart et al. 2005) was able to describe the spatial dependence of molecular abundances as a function of distance from HH2, and the most complex models yet made of this radiative interaction were used to constrain physical and chemical properties of the HH2 region. The conclusions of this work were that the clump must have several density components with a maximum number density of less than one million H_2 molecules cm^{-3}, that the visual extinction through the clump must be less than 3.5 magnitudes, and that the radiation field from the HHO is more intense than the standard interstellar radiation field by a factor less than 1000.

In effect, in these studies the stellar jet is being used as a probe of molecular cloud structure. It is curious that stellar jets that are emitted from newly formed stars in molecular clouds are themselves useful in exploring the nature of neighbouring clouds. Recent observational work by Viti and Hatchell (private communication) shows that nearly all HHOs of a sample of about 20 have associated with them clumps of enhanced HCO^+ emission. If confirmed, this seems to imply that molecular clouds viewed on a scale of about 0.1 pc have a high filling factor of clumps. This preliminary conclusion is so similar to the picture of molecular clouds deduced by Morata et al. (2005) that one is tempted to infer that the clumps irradiated by HHOs are indeed the same as the cores described by Morata et al. This would be an important result, because the high angular resolution studies of molecular clouds are currently limited to TMC-1 and L673, for technical reasons. The fact that a wide sample of molecular clouds, probed by HHOs, gives a similar conclusion suggests that such structures are common in interstellar clouds, and consequently that the spectrum of such structures may indeed be related to the low-mass star-formation rate in molecular clouds.

3 Regions of Formation of High-mass Stars

3.1 Hot cores

The formation of massive stars takes place within dense molecular clouds and the star itself may be obscured by hundreds or thousands of optical depths in the visible region of the spectrum. Tracers of the early phases of massive star formation are necessarily at long wavelengths, and radio and millimetre wave emissions from ultra-compact HII regions (see Section 5 of the Chapter by Melvin Hoare and Pepe Franco) and from hot cores (often associated with UCHIIs) are used to infer the stage of development of massive star formation in a particular region. In this section we discuss hot cores and the information they can provide about massive star formation.

The evolution of hot cores

Hot cores are small (less than 0.1 pc), dense (more than 10 million H nuclei cm^{-3}), relatively warm (200 - 300 K), optically thick and transient objects detected in the close vicinity of young massive stars. They exhibit a characteristic chemistry that is distinct from that of quiescent molecular clouds. Hot cores are rich in small saturated molecules and in large organic species (see, e.g. Walmsley & Schilke 1993), but may differ from each other to a significant extent. It is important, therefore, to understand how hot cores arise, how they may evolve to show a variety of chemistries, to determine their ultimate fate, and to define their use in describing massive star formation.

The characteristic chemistry is believed to arise when the icy mantles that accumulated on dust grains during the collapse of the pre-stellar cloud are heated by the proto-star and evaporated. The molecules that were in the ices, or which have been created by solid-state processing in the ices or during the warming process, are then injected into the gas phase and contribute to the emission from the hot core in rotational lines (see, e.g., Millar 1993; Nomura & Millar 2004). Since the ices were deposited during the pre-stellar collapse, and possibly processed in solid-state chemistry before evaporation, they represent an integrated history of the conditions during the collapse process. It is the role of astrochemistry to interpret that history from the observational record of hot cores.

There is, however, a complication to this picture that must be taken into account. As discussed by Viti & Williams (1999), the warm-up period for the material in a hot core from around 10 K to 200 K is likely to be comparable to the duration of the hot core itself; between ten and a hundred thousand years (Bernasconi & Maeder 1996). Thus, the evaporation cannot be considered to be instantaneous, but gradual. Such a phenomenon is essentially similar to that occuring in laboratory experiments in which slow warming of mixtures is carried out to define the nature of evaporation and thereby to learn something about the nature of the ice. This type of experiment is known as Temperature Programmed Desorption (TPD), though of course the cosmic TPD (in a hot core) is carried out over ten thousand years or more, whereas a laboratory experiment is completed in an hour or two.

TPD experiments with ices of composition relevant to regions of massive star formation have been performed at the University of Nottingham. Fraser et al. (2001) reported a TPD study of pure H_2O ice, Collings et al. (2003) examined the TPD of H_2O/CO ice mixtures, and Collings et al. (2004) made similar studies for a range of other relevant molecules, and by theoretical considerations extended the results to an even wider range of species. Some of the experimental results are shown in Fig. 10. The simplest case is that of pure H_2O. Desorption occurs for pure ice under the conditions of the experiment for a narrow range of temperatures around 160 K. However, CO desorption from a mixed H_2O/CO ice occurs at four main peaks between temperatures of about 30 K and 160 K. In general, astrophysicists have assumed that all CO is desorbed at temperatures near 30 K, but clearly the situation is more complicated than that.

The interpretation of Collings et al. (2003) made of these results is that the evaporation of CO from a mixed ice depends on the morpholgy of the ice. The lowest temperature desorption band corresponds to desorption of pure solid CO on the surface of the ice; the next band is of desorption of CO from the surface of the H_2O ice iself. Then, as the ice continues to warm

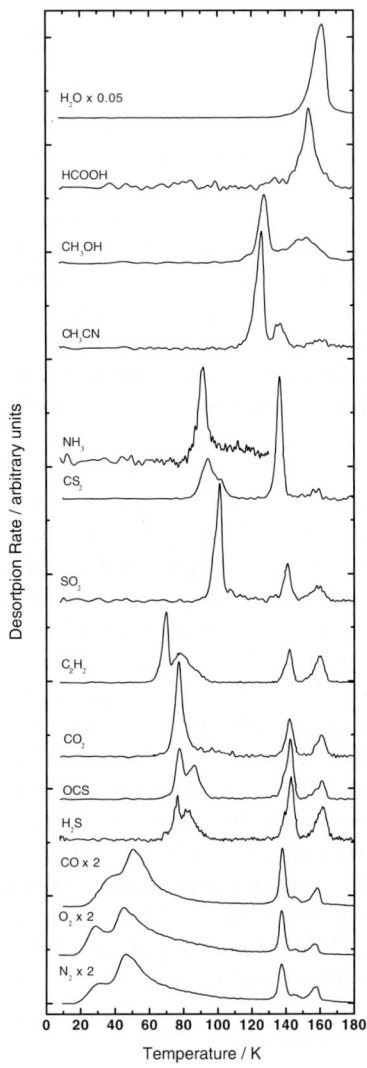

Fig. 10. TPD traces of various species deposited onto a pre-adsorbed H₂O film. Traces are offset for clarity (Collings et al. 2004).

up, a phase change occurs from a highly porous amorphous structure to a less porous amorphous phase; this change traps CO within cavities within the water ice. At a higher temperature, the amorphous ice crystallizes and some of the trapped species are released. Finally, any remaining CO molecules are released when the H₂O ice itself evaporates. It can be seen from Fig. 10 that other species may be CO-like, or H₂O-like, or have an intermediate behaviour with respect to desorption.

Taking account of the much longer timescales available in the cosmic TPD experiment, one can convert the laboratory data for use in hot core models. It is immediately clear that there is an extended warm-up period in which different species contribute to the evaporation, to different extents as the warm-up continues. For example, CO molecules are injected in four "pulses" between 30 K and the temperature at which H_2O evaporates. The implication is that the warm-up period can be tracked by time dependent chemical changes in a hot core, so that the warm-up of the underlying star becomes evident through the chemistry of the star's immediate environment. Viti et al. (2004) have explored this time-dependence for stars of different masses. The most massive stars evolve very rapidly, so that the different chemical phases are poorly distinguished. However, for stars of 5 - 15 solar masses the time-dependence of the warm-up phase should be clearly evident, particularly in sulfur-bearing species.

After the warm-up phase, a hot core enters a relatively stable phase until gas phase reactions begin to modify the chemistry to that appropriate to the density and temperature regime. On about the same timescale, dynamical effects from the winds and expanding UCHII region begin to destroy the hot core. Note, however, that the chemistry of a hot core in which a slow cosmic TPD occurs is not necessarily the same as that in which instantaneous evaporation is assumed to take place.

The model of Viti et al. (2004) can be used to investigate many of the chemical and dynamical parameters. One particular case has been examined: the dynamical timescale (Lintott et al. 2005a). It has already been mentioned above that the chemical and dynamical timescales in star-forming regions are on the same order of magnitude. However, one could imagine that sequential star-formation, known to occur in regions of massive star-formation, might create an external pressure that accelerates the collapse of a star-forming core (cf. Section 1 of the Chapter by Melvin Hoare and Pepe Franco). The consequence of increasing the collapse rate above free-fall is that the gas reaches a higher density at an earlier time. Because of the higher density, the rates of chemical reactions are high and the gas is more chemically-developed before species are affected significantly by freeze-out. Thus, reactions removing N_2H^+, for example, are faster in the accelerated collapse case and reactions forming CS are also faster. Thus, the CS/N_2H^+ ratio may be a signature of the collapse rate. These calculations may offer an explanation of the observations of Pirogov et al. (2003) of high-mass star-forming cores where the CS and N_2H^+ distributions (often used as mass tracers) are quite different.

Hot Cores as Probes of High Redshift Galaxies

The centre of the Milky Way Galaxy is a region of massive star formation. It is known to be a strong source of molecular line emissions characteristic of hot cores. In the range of galaxies that we can observe, the Milky Way is not particularly active; its star formation rate is about one solar mass per year. Starburst galaxies and quasars, by contrast, may have star formation rates thousands or ten thousands of times larger. If the newly-formed stars in these objects are also associated with hot cores, then molecular line emissions from those hot cores may be very intense (though unresolved), and possibly detectable at high red-shifts.

This is the interesting scenario explored by Lintott et al. (2005b). The aim of such hot core observations would be to use existing models of hot cores with the observational data to determine the chemistry in hot cores at high redshifts. From the chemistry one could infer the relative elemental abundances in high redshift galaxies provided by the first generation of stars within those galaxies. There is some disparity between the predictions of existing models for the first generation of stars and the elemental abundances they inject into the interstellar medium from which the second generation of stars (and associated hot cores) will form. Thus, in principle, observations of hot cores could help to determine the nature of the first generation of stars.

This exciting prospect depends on the detectability of high red-shift hot cores in galaxies with high rates of star formation. This is the issue addressed by Lintott et al. (2005b), using the hot core model described in the previous section. In fact, we know little about the nature of hot cores at high redshift. If the metallicity at high red-shift is much less than the galactic value, presumably the dust:gas ratio is also reduced. How does this change the structure of a hot core? One can rely on the fact that hot cores in the Galaxy have visual extinctions of perhaps 1000 associated with them to infer that hot cores at high redshift are moderately well shielded from starlight, but this assumption requires further investigation. Similarly, the nature of the dust at high redshift is completely unknown, and changes in dust size range or in chemical properties will certainly affect the optical properties and may affect the hot core structure. The extent to which molecular icy mantles would be deposited under conditions appropriate to high red-shift galaxies is unknown; if no mantles are formed, then the galactic star formation environment is not replicated at high red-shift.

Nevertheless, making cautious assumptions with respect to dust and hot core structures, Lintott et al. (2005b) estimate that a wide range of hot core molecules should be detectable with existing facilities at high red-shift (and certainly with future facilities such as ALMA) if the number of hot cores exceeds that in the Milky Way by about a factor of 1000. Given the very high star formation rates detected in some high red-shift objects, this seems a modest requirement, and hot core numbers may greatly exceed this number. Lintott et al. (2005b) also evaluate the chemistry arising in these hot cores, on the assumptions of the different models of stellar evolution for the first generation of stars. There is a clear distinction to be expected in the hot core chemistry for the different cases. The prospect of using hot cores as a tool to explore the early Universe is an exciting one.

3.2 Stellar Disks

Disk Chemistry

Stellar disks are locations where a great variety of physical processes may be occurring. For example, in the innermost parts of the disk close to the central star temperatures are high and the dust grains are destroyed. X-rays from the central star may influence chemistry in this region, as well as photoprocessing by stellar UV and optical radiation. In regions outside that central zone the evaporation of icy mantles must be occurring and contributing to the chemistry, as in hot cores. Further out again, the disk may be thick enough to impede the flux of ionising cosmic rays. Temperatures in such regions may be very low, and the role of magnetic fields may be minimised. Densities may be so high that three body reactions play a role, and the conventional databases for astrochemistry need to be extended. Perpendicular to the plane of the disk, density and temperature profiles may have steep gradients so that vertical motions induced by diffusion may play a role in moving material from very cold and dark regions to less cold and more irradiated environments.

Given this complexity it is not surprising that an extensive literature on the subject has developed rapidly, once computer power capable of dealing with hydrodynamics and a complex chemistry became readily available. The aim of these models is to indicate useful observational tracers of the various processes and regions by which tests of the model assumptions may be made. Ultimately, the hope is that a complete understanding of the nature of disks and the implications for planet formation can be obtained. However, we are far from that situation at present; many of the studies focus on the effects of one or more of the physical parameters listed in the previous paragraph, and it is therefore difficult to compare results from different models.

Ilgner et al. (2004) have begun a series of studies to explore the effect on chemistry of mass transport through the disk. (This paper also provides a very useful summary of work on disk chemistry up to that date.) In their study, mass is injected continuously at the outer edge of the disk (at 10 AU from the central star), and the steady flow and associated chemistry are followed up to an inner radius of 1 AU (where the temperatures become too large for their chemistry to follow). Their paper also examines the effects on the chemistry of diffusion perpendicular to the plane of the disk. They conclude that the chemical evolution is strongly influenced by mass transport processes. Also, vertical mixing is shown to have an important effect on chemistry. Therefore, models limited to chemistry in the central plane are unlikely to be representative. However, the local conditions are not well-determined in any models and it is these conditions that determine the chemistry. Thus, this area is one in which molecular astrophysics has an important role to play but in which there is much yet to do.

Disk Interfaces

An interesting aspect of the study of stellar disks is to consider which parts of the disk may be the most readily detectable in molecular lines, rather than which regions are most chemically rich. Given that the temperature rises with distance from the central plane of the disk, it may be that warmer regions at or near the interface with the surrounding medium may be effective emitters. Disks around massive stars, and disks around low-mass stars but irradiated by a nearby OB association are eroded by photo-evaporation, as is clearly evident in images of nearby disks. This erosion is a constraint on planet formation. Johnstone et al. (1998) have shown that photo-evaporating material is gravitationally retained by the star and constitutes an envelope similar to a photon dominated region (a so-called PDR). PDRs are well-studied structures with intense characteristic emissions.

Nguyen et al. (2002) explored the chemistry of such a PDR as an interface between the disk and the surrounding medium. This was assumed to be created by radiation from a massive central star on a flared disk, or by radiation from a nearby OB association on the disk. Results for a 10 solar mass central star were compared with observations of a disk around the massive star GL2591 which is known to have hot gas around the disk and to have emissions in several molecular lines. The Nguyen et al. model produced results that were reasonably consistent with inferred molecular column densities, and also made predictions of a number of other molecules that should be detectable by virtue of their location in a PDR. PDRs associated with disks around low-mass stars, if irradiated by a nearby OB association, should also generate a rich chemistry with some differences from the predictions for a PDR/disk around a massive star. Nguyen et al. noted that detection of any PDR-like chemistry is a signature of disk erosion, and may imply limitations for planet formation in those systems.

4 Conclusions

Through a combination of observations and modelling, molecular astrophysics can explore may situations in astronomy where reasonably high densities exist. Interfaces between different media, especially those where winds impact on dense and dusty gas, can be a rich source of information. The techniques for the microphysics and chemistry are well established. However, our current ability to link dynamics and the chemistry are constrained in part by computer power, in part by lack of information about gas-phase and gas-solid interactions, and in part by our limited understanding of small scale gas dynamics. However, the prospects for continuing success are very bright.

5 Acknowledgments

John Dyson's work has encouraged me to widen my horizons in the type of astronomical problems that might be addressed in part through molecular astrophysics. I am grateful to him for his support in science and for his friendship over many years. Much of the work reported in this article has been carried out in conjunction with postdocs and students, and I am grateful to all of them for making my life so interesting. I am particularly grateful to Jonathan Rawlings and Serena Viti for many enjoyable collaborations and for reading and commenting on this article.

References

André, P., Ward-Thompson, D., Barsony, M. 1993 ApJ **406**, 122
Anglada, G., Sepulveda, I., Gomez, J.F. 1997 A&AS **121**, 255
Beichman, C.A., Myers, P.C., Emerson, J.P. et al. 1986 ApJ **307**, 337
Benson, P.J. and Myers, P.C. 1989 ApJSS **71**, 89
Bernasconi, P.A., Maeder, A. 1996 A&A **307**, 829
Choi, M., Evans, N.J.,II, Gregerson, E.M., Wang, Y. 1995, ApJ **448**, 742
Collings, M.P., Dever, J.W., Fraser, H.J., McCoustra, M.R.S., Williams, D.A., 2003 ApJ **583**, 1058
Collings, M.P., Anderson, M.A., Chen, R., Dever, J.W., Viti, S., Williams, D.A., McCoustra, M.R.S. 2004 MNRAS **354**, 1133
Dyson, J.E., Williams, D.A. "Physics of the Interstellar Medium" 1980 Manchester University Press, Manchester
Evans, N.J., Rawlings, J.M.C., Shirley, Y.L., Mundy, L.G. 2001 ApJ **557**, 193
Evans, N.J., II, Lee, J.-E., Rawlings, J.M.C., Choi, M. 2005 ApJ **626**, 919
Falle, S.A.E.G, Hartquist, T.W. 2002 MNRAS **329**, 195
Fraser, H.J., Collings, M.P., McCoustra, M.R.S., Williams, D.A. 2001 MNRAS **327**, 116
Garrod, R.T., Williams, D.A., Hartquist, T.W., Rawlings, J.M.C., Viti, S. 2005 MNRAS **356**, 654
Garrod, R.T., Williams, D.A., Rawlings, J.M.C. 2006 ApJ in press
Girart, J.M., Rodriguez, L.F., Anglada, G., et al. 1994 ApJ **435**, L145
Girart, J.M., Viti, S., Williams, D.A., Estalella, R., Ho, P.T.P 2002 A&A **388**, 1004
Girart, J.M., Viti, S., Estalella, R., Williams, D.A. 2005 A&A in press
Hartquist, T.W., Williams, D.A., Viti, S. 2001 A&A **369**, 605
Hirahara, Y., Suzuki, H., Yamamoto, S., Kawaguchi, K., Kaifu, N., Ohishi, M., Takano, S., Ishikawa, S.-I., Masuda, A. 1992 ApJ **394**, 539
Hogerheijde, M.R., van Dishoeck, E.F., Blake, G.A., van Langevelde, H.J. 1998 ApJ **502**, 315
Ilgner, M., Henning, Th., Markwick, A.J., Millar, T.J. 2004 A&A **415**, 643
Jessop, N.E., Ward-Thompson, D. 2001 MNRAS **323**, 1025
Jones, A.P., Williams, D.A. 1984 MNRAS **209**, 955
Jordan,C., Brueckner, G.E., Bartoe, J.-D.F., Sandlin, G.D., van Hoosier, M.E. 1977 Nature **270**, 326

Lintott, C.J., Viti, S., Rawlings, J.M.C., Williams, D.A., Hartquist, T.W., Caselli, P., Zinchenko, I., Myers, P. 2005a ApJ **620**, 795

Lintott, C.J., Viti, S., Williams, D.A., Rawlings, J.M.C., Ferreras, I. 2005b MNRAS **360**, 1527

Millar, T.J. 1993 in Dust and Chemistry in Astronomy, ed. Millar, T.J., Williams, D.A., IoP Publishing, Bristol, p259

Morata, O., Estalella, R., Lopez, R., Planesas, P. 1997 MNRAS **292**, 120

Morata, O., Girart, J.M., Estalella, R. 2003 A&A **397**, 181

Morata, O., Girart, J.M., Estalella, R. 2005 A&A **435**, 113

Nguyen, T.K., Viti, S., Williams, D.A. 2002 A&A **387**, 1083

Nomura, H., Millar, T.J., 2004 A&A **414**, 409

Pastor, O., Estalella, R., Lopez, R. et al. 1991, A&A **252**, 320

Peng, R., Langer, W.D., Velusamy, T., Kuiper, T.B.H., Levin, S. 1998 ApJ **497**, 482

Pirogov, L., Zinchenko, I., Caselli, P., Johansson, L.E.B., Myers, P.C. 2003 A&A **405**, 639

Raga, A.C., Williams, D.A. 2000 A&A **358**, 701

Rawlings, J.M.C. 2003 Astrophys Space Sci **285**, 777

Rawlings, J.M.C., Hartquist, T.W., Menten, K.M., Williams, D.A. 1992 MNRAS **255**, 471

Rawlings, J.M.C., Taylor, S.D., Williams, D.A. 2000 MNRAS **313**, 461

Rawlings, J.M.C., Yates, J.A. 2001 MNRAS **326**, 1423

Rawlings, J.M.C., Redman, M.P., Keto, E., Williams, D.A. 2004 MNRAS **351**, 1054

Redman, M.P., Rawlings, J.M.C., Nutter, D.J., Ward-Thompson, D., Williams, D.A. 2002 MNRAS **337**, L17

Redman, M.P., Keto, E., Rawlings, J.M.C., Williams, D.A. 2004 MNRAS **352**, 1365

Sonnabend, G., Wirtz, D., Schieder, R., Bernath, P.F. 2006 Solar Phys **233**, 205

Taylor, S.D., Williams, D.A. 1996 MNRAS **282**, 1343

Velusamy, T., Langer, W.D. 1998 Nature **392**, 685

Viti, S., Williams, D.A. 1999 MNRAS **310**, 517

Viti, S., Natarajan, S., Williams, D.A. 2002 MNRAS **336**, 797

Viti, S. Girart, J.M., Garrod, R., Williams, D.A., Estalella, R. 2003 A&A **399**, 187

Viti, S., Collings, M.P., Dever, J.W., McCoustra, M.R.S., Williams, D.A. 2004 MNRAS **354**, 1141

Walmsley, C.M., Schilke, P. 1993 in Dust andChemistry in astronomy, ed

Millar, T.J., Williams, D.A. IoP Publishing, Bristol p37

Whittet, D.C.B. "Dust in the Galactic Environment" Institute of Physics Publishing, Bristol, 2002

Williams, D.A. in "Solid State Astrochemistry", ed. V. Pirronello, Kluwer Academic Publishers, Dordrecht, 2003, p1

Zhou, S. 1992 ApJ **394**, 204

Zhou, S., Evans, N.J., II, Butner, H.M., Kutner, M.L., Leung, C.M., Mundy, L.G. 1990 ApJ **363**, 168

Zhou, S., Evans, N.J., II, Koempe, C., Walmsley, C.M. 1993 ApJ **404**, 232

Dusty Plasma Effects in Star Forming Regions

T.W. Hartquist[1] and O. Havnes[2]

[1] School of Physics and Astronomy, The University of Leeds, Leeds LS2 9JT, UK
twh@ast.leeds.ac.uk
[2] Department of Physics, University of Tromsø, N–9037 Tromsø, Norway
Ove.Havnes@phys.uit.no

1 Introduction

In his chapter on star formation, Sam Falle has demonstrated the importance of hydromagnetic effects in star formation. In his, David Williams has described some of the chemical consequences of the presence of dust in star forming regions. The dust grains carry charge at least a good fraction of the time and, consequently, are affected by the magnetic field. In this chapter, we show that the presence of dust in star forming regions has important dynamical consequences in cold (≈ 10 K) regions collapsing to produce stars and in shock-heated regions at temperatures of many hundreds to a few thousand degrees.

Section 2 contains an introduction to the charging of dust and the role that it plays in establishing the gas phase fractional ionization. Section 3 is a primer for multifluid descriptions of the dynamics of magnetized, dusty media. Section 4 concerns the effects of dust grains on the damping of Alfvén waves propagating parallel to the large scale magnetic field, while the role that dust plays in ion–neutral streaming (often referred to as ambipolar diffusion) is central to section 5. Section 6 is a summary of work on multifluid models of shocks in dusty star forming regions.

2 Dust Grain Charging and Ionization Balance

Throughout this chapter we focus on regions that are sufficiently dark that photoabsorption by grains plays no role in grain charging. Electrons move much more quickly than ions, if all gas phase species are at the same temperature. Assume that the dust grains have a sufficiently low number density that they carry a negligible fraction of the negative charge and that they are at rest with respect to the gas. Then a grain is more likely to encounter an electron than an ion as long as the absolute value of the negative charge on the grain is small enough that the absolute value of the electric potential at the grain surface, U, is small compared to the average thermal energy per gas phase particle. In equilibrium

$$U = \mathcal{K} \frac{k_B T}{e} \qquad (1)$$

where k_B, T, and e are the Boltzmann constant, the temperature, and the charge of a proton, respectively. $\mathcal{K}$ is a constant of order unity and depends on the mass of the ions; for protons $\mathcal{K} = -2.51$ (e.g. Spitzer 1978). For a grain radius, a, of 10^{-5} cm and $T = 10$ K, Eq. 1 implies that the average charge on the grains is roughly $-0.2\ e$.

Eq. 1 follows from an analysis in which the cross section of a neutral grain is equal to the geometric cross section. In fact, it is rather larger (e.g. Draine & Sutin 1987), and most interstellar grains in a dark star forming cloud with $T \approx 10$ K carry one negative charge each. When the magnitude of the average charge is so low, the probability distribution function for the charge carried by a grain must be calculated in order to determine the mean charge per dust particle accurately (Gail & Sedlmayr 1975).

The collisions of metallic ions, formed by charge transfer reactions involving molecular ions, with negatively charged grains constitute an important recombination process in interstellar cloud (Oppenheimer & Dalgarno 1974; Elmegreen 1979). In dark clouds, ionization is induced by cosmic rays. Figure 1 shows the number densities of gas phase ions (n_i), electrons (n_e), neutral grains (n_{g0}), grains carrying a charge of $+e$ each (n_{g+}), and grains carrying a charge of $-e$ each (n_{g-}) as functions of n_H, the number density of hydrogen nuclei. The assumed temperature, cosmic ray ionization rate, and the fractional abundances of CO and of gas phase heavy metals are 10 K, 10^{-17} s^{-1} per H_2 molecule, 1.5×10^{-5}, and 2×10^{-7}. Almost all hydrogen is in H_2. All grains are assumed to be spherical with radii of 10^{-5} cm. The number density of grains is $4 \times 10^{-12}\ n_H$. Typically, dark regions in interstellar clouds have 10^4 cm$^{-3} \lesssim n_H \lesssim 10^7$ cm^{-3}, but as collapse takes place n_H increases.

In shocks, temperatures are higher, and grains become the dominant charge carriers at much lower values of n_H than in cooler regions. In shocks, the average grain velocity can differ from those of the other charged species. Account of the systemic relative motion must be taken in the calculation of the rate at which ions recombine onto grains, as treated in detail by Havnes, Hartquist, & Pilipp (1987).

3 The Multifluid MHD Description

Mullan (1971) and Draine (1980) introduced the use of multifluid descriptions to model shocks in weakly ionized interstellar clouds. The papers by Draine, Roberge, & Dalgarno (1983) and by Draine (1986) are important for anyone who wishes to develop a code to construct such models. However, none of these

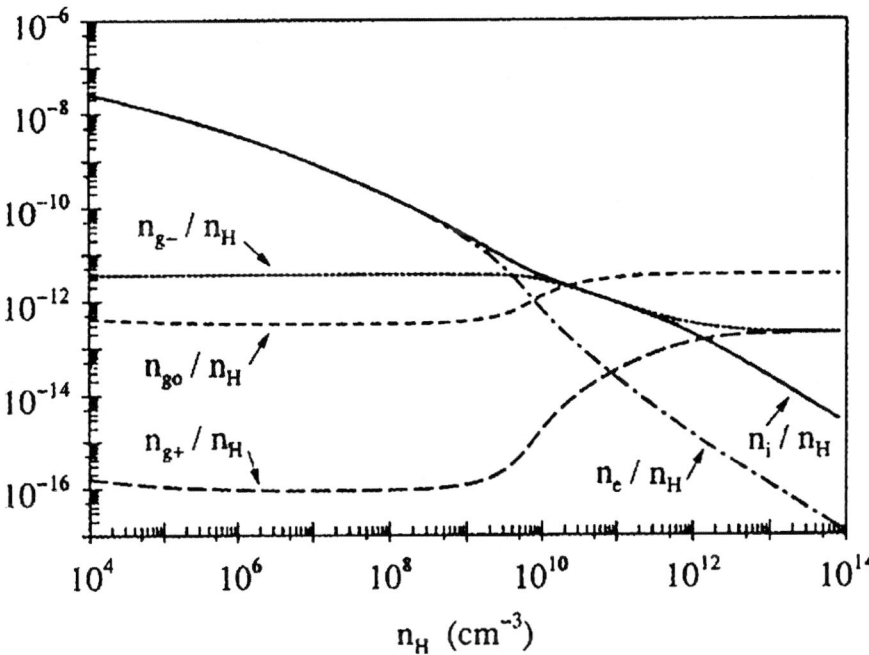

Fig. 1. Number densities of various species in dark regions. From Hartquist, Havnes, & Pilipp (1997).

four papers presents a treatment of grain dynamics that is valid in all parameter regimes relevant to shocks in star forming regions. Pilipp, Hartquist, & Havnes (1990) and Pilipp & Hartquist (1994) were the first to treat dust grains on an equal footing with other species in shock models by using a fluid description to follow their dynamics. The same approach is of utility when one addresses other dynamical problems concerning star forming regions. We summarize the approach here and refer the reader to the review by Hartquist, Havnes, & Pilipp (1997) for a more detailed introduction to it.

We assume that the medium consists of ion, electron, and neutral gas phase fluids and a number of grain fluids. Even if the grains are all of the same mass, shape, and size, different grain fluids must be included in the treatment if the magnitude of the mean grain charge is of order e or less; in this case, the different grain fluids consist of particles carrying different charges per particle. Of course, the presence of a distribution of grain sizes may necessitate the inclusion of more grain fluids in the calculations.

The conservation of mass of fluid α requires that

$$\frac{\partial \rho_\alpha}{\partial t} + \nabla \cdot (\rho_\alpha \mathbf{u}_\alpha) = S_\alpha \qquad (2)$$

ρ_α and $\mathbf{u}_\alpha$ are the mass density and velocity of the αth fluid. t is time, and S_α is a source term due to reactions. For instance, if the αth fluid were the fluid of neutral grains of a particular size, contributions to S_α would include the rate per unit volume per unit time at which grains of the same size but carrying a charge of $-e$ remove singly charged positive gas phase ions. Draine (1986) and Hartquist, Havnes, & Pilipp (1997) give fuller expositions on all of the source terms mentioned in this section.

Newton's second law requires that

$$\frac{\partial(\rho_\alpha \mathbf{u}_\alpha)}{\partial t} + \sum_j \frac{\partial}{\partial x_j}(u_{\alpha j} \rho_\alpha \mathbf{u}_\alpha) + \nabla P_\alpha = \frac{\rho_\alpha}{m_\alpha} Z_\alpha e \left(\mathbf{E} + \frac{\mathbf{u}_\alpha}{c} \times \mathbf{B}\right) + \rho_\alpha \mathbf{a}'_\alpha + \mathbf{F}_\alpha \quad (3)$$

$Z_\alpha e$ and m_α are the charge and mass, respectively, of each particle in the αth fluid. P_α is the thermal pressure of the αth fluid, which is set to zero for a grain fluid. $\mathbf{E}$, $\mathbf{B}$, and c are the electric field, the magnetic field, and the speed of light, respectively. The source term $\mathbf{F}_\alpha$ is the net rate per unit volume per unit time at which collisions and reactions transfer momentum to the αth fluid from other fluids. $\rho_\alpha \mathbf{a}'_\alpha$ is the force per unit volume due to other nonelectromagnetic forces, including gravity and radiation pressure.

The conservation of energy implies that

$$\frac{\partial}{\partial t}\left(\frac{3}{2}P_\alpha + n_\alpha U_\alpha + \frac{1}{2}\rho_\alpha u_\alpha^2\right) + \nabla \cdot \left[\mathbf{u}_\alpha \left(\frac{5}{2}P_\alpha + n_\alpha U_\alpha + \frac{1}{2}\rho_\alpha u_\alpha^2\right)\right] \quad (4)$$
$$= \rho_\alpha \mathbf{u}_\alpha \cdot \mathbf{a}'_\alpha + n_\alpha Z_\alpha e \mathbf{u}_\alpha \cdot \mathbf{E} + \mathbf{u}_\alpha \cdot \mathbf{F}_\alpha + G_\alpha + H_\alpha$$

n_α is ρ_α/m_α. $\mathbf{u}_\alpha \cdot \mathbf{F}_\alpha + G_\alpha$ is the difference between the rate per unit volume per unit time at which energy is put into particles of species α and the rate per unit volume per unit time at which energy is removed from them due to collisions of particles of species α with particles of other species. Radiative losses due to collisions are included in G_α. H_α is the rate per unit volume per unit time at which energy is transferred to species α from other external sources of which account has not been taken in other terms. For instance, absorption of photons emitted by a nearby star would contribute to H_α. U_α is the mean internal energy per particle of species α.

As $P_\alpha = 0$ if the αth fluid is a grain fluid, Eq. 4 need not be considered for any grain fluid. In practice, in the study of shocks in very weakly ionized clouds, the left hand sides of Eq. 3 and 4 may be set to zero for the ion and electron fluids. The left hand side of Eq. 3 is usually set to zero for grain fluids, but grain inertial terms are important for shock speeds above about 40 km s^{-1}.

Typically, in the study of shocks in weakly ionized clouds, quasineutrality is assumed. This requires that

$$\sum_\alpha Z_\alpha n_\alpha = 0 \quad (5)$$

Also

$$\nabla \cdot \mathbf{B} = 0 \quad (6)$$

$$\nabla \times \mathbf{B} = \frac{4\pi}{c} \mathbf{J} \tag{7}$$

$$\nabla \times \mathbf{E} = -\frac{1}{c} \frac{\partial \mathbf{B}}{\partial t} \tag{8}$$

and

$$\mathbf{J} = \sum_\alpha n_\alpha Z_\alpha e \mathbf{u}_\alpha \tag{9}$$

$\mathbf{J}$ is the current density.

Falle (2003) has addressed the issue of the numerical solution of the time-dependent equations for the various assumptions mentioned previously in this section.

4 Dust and the Damping of Linear Alfvén Waves Propagating Parallel to the Large Scale Magnetic Field

In Section 3 of his chapter, Sam Falle has summarized the properties of translucent clumps and dense cores. Observations of translucent clumps (e.g. Williams, Blitz, & Stark 1995) and of dense cores (e.g. Caselli & Myers 1995) in star forming giant molecular clouds show that spectral lines are broadened by nonthermal motions. The nonthermal broadening is often attributed to the presence of waves with sub-Alfvénic velocity amplitudes (Arons & Max 1975). In translucent clumps the velocity amplitudes are typically 1 to 3 km s^{-1}. The waves may dissipate through nonlinear effects, but a lower bound to the rate at which they damp is the rate at which frictional dissipation due to the motions of charged species relative to neutrals causes linear waves to damp.

Arons & Max (1975) argued that nonlinear effects may be less important for Alfvén waves than other types of waves. Thus, we focus on the frictional damping of Alfvén waves propagating in the direction of the large-scale magnetic field. Such waves are noncompressive. Consequently, no thermal pressure term appears in the appropriate linearized version of Eq. 3 if we assume that the medium is isothermal. It may be written

$$\rho_\alpha \frac{\partial \mathbf{u}_\alpha}{\partial t} = Z_\alpha n_\alpha e \left(\mathbf{E} + \frac{\mathbf{u}_\alpha}{c} \times B_0 \hat{z} \right) - \rho_\alpha \sum_\beta \nu_{\alpha\beta} (\mathbf{u}_\alpha - \mathbf{u}_\beta) \tag{10}$$

We have assumed that $\mathbf{a}'_\alpha = 0$. The second term on the right hand side is the linearized version of $\mathbf{F}_\alpha$. Each $\nu_{\alpha\beta}$ is a constant proportional to $\rho_{\beta 0}$ and depending on $T_{\alpha 0}$ and $T_{\beta 0}$. The subscript "0" indicates the value of the quantity appropriate for the unperturbed background. The large-scale magnetic field is in the z direction.

Hartquist, Havnes & Pilipp (1997) have given values for the $\nu_{\alpha\beta}$'s:

$$\nu_{ni} = 1.8 \times 10^{-12} \text{ s}^{-1} \ (n_i/10^{-3} \text{ cm}^{-3}) \tag{11}$$

$$\nu_{in} = 1.3 \times 10^{-6} \text{ s}^{-1} \left(n_n/10^4 \text{ cm}^{-3} \right) \cdot (29 m_n/2m_i) \tag{12}$$

$$\nu_{nc} = \frac{\rho_c}{\rho_n} \nu_{nu} = 1.3 \times 10^{-12} \text{ s}^{-1} \left(\frac{\rho_c}{4 \times 10^{-22} \text{ g cm}^{-3}} \right) \left(\frac{a}{10^{-5} \text{ cm}} \right)^{-1} \left(\frac{T_n}{10 \text{ K}} \right)^{1/2} \tag{13}$$

$$\nu_{cn} = \nu_{un} = 1.3 \times 10^{-10} \text{ s}^{-1} \left(\frac{n_n}{10^4 \text{ cm}^{-3}} \right) \left(\frac{a}{10^{-5} \text{ cm}} \right)^{-1} \left(\frac{T_n}{10 \text{ K}} \right)^{1/2} \tag{14}$$

$$\nu_{cu} = 4.5 \times 10^{-8} \text{ s}^{-1} \left(\frac{a}{10^{-5} \text{ cm}} \right) \left(\frac{n_i}{10^{-3} \text{ cm}^{-3}} \right) \left(\frac{m_i}{29 \text{ a.m.u.}} \right)^{-1/2} \left(\frac{T_i}{10 \text{ K}} \right)^{-1/2} \tag{15}$$

$$\nu_{uc} = 6.2 \times 10^{-7} \text{ s}^{-1} \left(\frac{a}{10^{-5} \text{ cm}} \right) \left(\frac{n_e}{10^{-3} \text{ cm}^{-3}} \right) \left(\frac{T_e}{10 \text{ K}} \right)^{1/2} \tag{16}$$

All other $\nu_{\alpha\beta}$'s may be set to zero in the study of wave damping. The subscripts u and c indicate uncharged and charged grains. All charged grains are assumed to carry a charge of $-e$ each. Eq. 14 is applicable only if $aT_i \ll 10^{-3}$ cm K. Eq. 16 is based on the assumption that the cross section for electron–neutral grain collisions is the geometric cross section and, thus, gives a lower bound to ν_{uc}.

The gyrofrequencies of gas phase ions and of grains carrying a charge $-e$ each are also relevant. They are

$$\Omega_i = 3.3 \times 10^{-2} \text{ s}^{-1} \left(\frac{B_0}{10^{-4} \text{ G}} \right) \left(\frac{m_i}{29 \text{ a.m.u.}} \right)^{-1} \tag{17}$$

$$|\Omega_c| = 3.8 \times 10^{-9} \text{ s}^{-1} \left(\frac{B_0}{10^{-4} \text{ G}} \right) \left(\frac{a}{10^{-5} \text{ cm}} \right)^{-3} \tag{18}$$

For Eqs 13–16 and 18, we have assumed that the grains are not porous and are composed of material with a density of 1 g cm^{-3}. For given values of the speed of low frequency or "coupled" Alfvén waves, v_{AC}, and of ρ_n, one finds

$$B_0 = 2.1 \times 10^{-4} \text{ G} \left(\frac{v_{AC}}{3 \text{ km s}^{-1}} \right) \left(\frac{\rho_n}{4 \times 10^{-20} \text{ g cm}^{-3}} \right)^{1/2} \tag{19}$$

The Hall parameter for charged grains is

$$\beta_{HC} = \frac{|\Omega_c|}{\nu_{cn}} \tag{20}$$

and is of order unity or less for some realistic choices of interstellar dark region properties. The Hall parameters for the gas phase charged species are very large.

The angular frequency, ω, of the waves is also relevant. The sizes of dense cores and translucent clumps range from about 0.1 pc. to several parsecs. The angular frequency of a coupled wave is

$$\omega = 6 \times 10^{-13} \text{ s}^{-1} \left(\frac{v_{AC}}{3 \text{ km s}^{-1}}\right)\left(\frac{\lambda}{\text{pc}}\right)^{-1} \qquad (21)$$

We assume that perturbation quantities vary as $\exp(ikz - i\omega t)$ where k is the wavenumber. Then Eq. 10 yields

$$-i\omega(1 - \delta_{\alpha e})\rho_\alpha u_\alpha = \rho_\alpha \Omega_\alpha \left(\frac{cE}{B_0} \pm i u_\alpha\right) - \rho_\alpha \sum_\beta \nu_{\alpha\beta}(u_\alpha - u_\beta) \qquad (22)$$

with

$$\Omega_\alpha = \frac{Z_\alpha e B_0}{m_\alpha c} \qquad (23)$$

and where $\delta_{\alpha e}$ is the Kronecker symbol. The appropriateness of $+$ or $-$ in the $i u_\alpha$ term in Eq. 22 depends on the circular polarization of the wave.

Eqs. 7 through 9 and the assumed dependence of perturbation quantities on t and z yield

$$\frac{c^2}{4\pi e} k^2 E = i\omega \sum_\alpha Z_\alpha n_\alpha e u_\alpha \qquad (24)$$

Pilipp et al. (1987) have solved Eqs. 22 and 24 numerically for a range of parameters. ω was specified and k_r and k_i were calculated.

$$k = k_r + i k_i \qquad (25)$$

where k_r and k_i are both real constants. They found that for some realistic interstellar conditions, charge fluctuations (i.e. $\nu_{cu} \neq 0$ and $\nu_{uc} \neq 0$) affect the damping.

Hartquist, Havnes, & Pilipp (1997) have given analytic results for cases in which

$$|\Omega_i| \gg \nu_{in} \gg \nu_{ni} \qquad (26)$$

$$\left|\Omega_c \frac{\rho_c}{\rho_u + \rho_c}\right| \gg \nu_{cn} = \nu_{un} \gg \nu_{nc} + \nu_{nu} \qquad (27)$$

$$\omega \ll \nu_{ni} + \nu_{nc} + \nu_{nu} \qquad (28)$$

and

$$|\Omega_i| \gg \nu_{ni} + \nu_{nc} + \nu_{nu} \qquad (29)$$

If these conditions are met, then

$$k_r \approx \omega \left(\frac{4\pi \rho_n}{B_0^2}\right)^{1/2} \qquad (30)$$

and

$$k_i \approx \frac{1}{2} \frac{\omega^2}{\nu_{ni} + \nu_{nu} + \nu_{nc}} \left(\frac{4\pi \rho_n}{B_0^2}\right)^{1/2} \qquad (31)$$

As Eq. 31 indicates, the presence of dust in dark interstellar clouds decreases the damping rate of linear waves in many situations.

5 Dust and Ion–Neutral Streaming in Dense Cores

Dense cores in which low-mass stars form may be magnetically subcritical in the sense that the mass of one is too low for gravity to induce its further collapse against the magnetic support on a timescale comparable to the free–full timescale. However, the direct force of the magnetic field on charged particles causes them to stream relative to the neutrals at a rate that is limited by the collisions between the charged particles and the neutral gas. This streaming is often called ambipolar diffusion, and its role in star formation has been studied widely. Numerical studies of star formation including ambipolar diffusion and the effects of dust on it include those described by Tassis & Mouschovias (2005a,b) and in a number of the references contained therein.

Charge fluctuations can make important differences and have been included by Mouschovias and his collaborators. However, here we mostly assume that all grains are the same size and carry a charge of $-e$ each. Following Hartquist, Havnes, & Pilipp (1997), who were influenced by Elmegreen (1979) and Nakano & Umebayashi (1980), we consider a dusty medium in which $\mathbf{B} = |B_0|\hat{x}$ and the gravitational field is $-g\hat{z}$. We assume that the neutrals remain static and that the inertial and gravitational terms in the electron, ion, and grain equations of motion to be negligible. We will use g to denote the grain fluid and assume that collisions with neutrals dominate the drag on ions and grains. We assume the electron fluid to be perfectly conducting and neglect thermal pressure.

Thus, we approximate Eq. 3 for the different fluids by

$$\nu_{ni}\mathbf{u}_i + \nu_{ng}\mathbf{u}_g - g\hat{z} = 0 \tag{32}$$

$$-\rho_i\nu_{in}\mathbf{u}_i + n_i e\left(\mathbf{E} + \frac{\mathbf{u}_i}{c} \times \mathbf{B}\right) = 0 \tag{33}$$

$$n_e e\left(\mathbf{E} + \frac{\mathbf{u}_e}{c} \times \mathbf{B}\right) = 0 \tag{34}$$

$$-\rho_g\nu_{gn}\mathbf{u}_g - n_g e\left(\mathbf{E} + \frac{\mathbf{u}_g}{c} \times \mathbf{B}\right) = 0 \tag{35}$$

We assume quasineutrality (Eq. 5), add Eq. 32 through Eq. 35 and use the definition of current (Eq. 9) to find that

$$\rho_i\nu_{in}\mathbf{u}_i + \rho_g\nu_{gn}\mathbf{u}_g = \frac{1}{c}(\mathbf{J} \times \mathbf{B}) \tag{36}$$

From Eq. 32 and Eq. 36 and the relations between ν_{in} and ν_{ni} and ν_{gn} and ν_{ng}, we see that

$$\frac{1}{c}(\mathbf{J} \times \mathbf{B}) = g\rho_n\hat{z} \tag{37}$$

From Eq. 34 we find that

$$\mathbf{E} = -\frac{\mathbf{u}_e}{c} \times \mathbf{B} \tag{38}$$

We use that $\mathbf{E}\cdot\mathbf{B} = 0$ from Eq. 38 when finding solutions to Eq. 33 and Eq. 35 for $\mathbf{u}_i$ and $\mathbf{u}_g$ and then eliminate $\mathbf{E}$ from those solutions by using Eq. 38 to obtain

$$\mathbf{u}_\alpha = \frac{\Omega_\alpha}{\nu_{\alpha n}} \frac{\left(\hat{\mathbf{x}} \times \mathbf{u}_e + \frac{\Omega_\alpha}{\nu_{\alpha n}}\mathbf{u}_e\right)}{\left(1 + \frac{\Omega_\alpha^2}{\nu_{\alpha n}^2}\right)} \qquad \alpha = i, g \qquad (39)$$

with

$$\mathbf{u}_e = c(\mathbf{E} \times \mathbf{B})/B^2 \qquad (40)$$

Use of Eq. 39 to substitute for $\mathbf{u}_i$ and $\mathbf{u}_g$ in Eq. 36 yields an equation for $\mathbf{u}_e$ which can be solved to find that

$$\mathbf{u}_e = \frac{A_0 \frac{1}{c}(\mathbf{J} \times \mathbf{B}) \times \hat{\mathbf{x}} + A_1 \frac{1}{c}(\mathbf{J} \times \mathbf{B})}{A_0^2 + A_1^2} \qquad (41)$$

with

$$A_0 = \sum_{\alpha=i,g} \rho_\alpha \nu_{\alpha n} \frac{\Omega_\alpha}{\nu_{\alpha n}} \frac{1}{\left(1 + \frac{\Omega_\alpha^2}{\nu_{\alpha n}^2}\right)} \qquad (42)$$

and

$$A_1 = \sum_{\alpha=i,g} \rho_\alpha \nu_{\alpha n} \frac{\Omega_\alpha^2}{\nu_{\alpha n}^2} \frac{1}{\left(1 + \frac{\Omega_\alpha^2}{\nu_{\alpha n}^2}\right)} \qquad (43)$$

By taking the curl of Eq. 35 and using Eq. 8 one sees that the evolution of the magnetic field depends only on $\mathbf{B}$ and $\mathbf{u}_e$. Hence, the timescale on which the magnetic field evolves will be significantly affected by the presence of grains if their presence substantially affects the magnitude of u_{ez} which from Eq. 41 and Eq. 37 is given by

$$u_{ez} = \frac{A_1}{A_0^2 + A_1^2} \rho_n g \qquad (44)$$

From Eq. 42 and Eq. 43 and the relation between ν_{ni} and ν_{in}, it follows that in the absence of dust

$$\frac{A_1}{A_0^2 + A_1^2} = \frac{1}{\rho_n \nu_{ni}} \qquad (45)$$

To a good approximation in interstellar clouds with dust

$$A_0 \approx \rho_n \nu_{ni} \frac{\nu_{in}}{\Omega_i} + \rho_n \nu_{ng} \frac{\nu_{gn}\Omega_g}{\nu_{gn}^2 + \Omega_g^2} \qquad (46)$$

$$A_1 \approx \rho_n \nu_{ni} + \rho_n \nu_{ng} \frac{\Omega_g^2}{\nu_{gn}^2 + \Omega_g^2} \qquad (47)$$

From Eq. 44 and Eq. 45 it follows that the presence of grains greatly lowers $|u_{ez}|$ if

$$\frac{A_1}{A_0^2 + A_1^2} \rho_n \nu_{ni} \ll 1 \qquad (48)$$

If $|\nu_{gn}/\Omega_g| \ll 1$, condition 48 is satisfied whenever

$$\nu_{ng} \gg \nu_{ni} \qquad (49)$$

If $|\nu_{ng}/\Omega_g| \gg 1$, condition 48 is satisfied if

$$\frac{\nu_{ni}^2}{\nu_{ng}^2} \frac{\nu_{gn}^2}{\Omega_g^2} \ll 1 \qquad (50)$$

We now consider the necessary and sufficient condition that must be met if charged grains are to remain coupled to the magnetic field

$$\frac{|u_{ez} - u_{gz}|}{|u_{ez}|} \ll 1 \qquad (51)$$

From Eq. 39 and Eq. 41 and the fact that **B** is in the $\hat{\mathbf{x}}$ direction and $\mathbf{J} \times \mathbf{B}$ is in the $\hat{\mathbf{z}}$ direction it follows that

$$u_{gz} = \frac{\nu_{gn}\Omega_g}{\nu_{gn}^2 + \Omega_g^2}\left(\frac{A_0}{A_1} + \frac{\Omega_g}{\nu_{gn}}\right)u_{ez} \qquad (52)$$

From use of Eq. 46 and 47 in 52, it follows that if $|\nu_{gn}/\Omega_g|^2 \ll 1$ condition 51 is satisfied. It also follows that if $|\nu_{gn}/\Omega_g|^2 \gg 1$, condition 51 is met whenever

$$\frac{\nu_{ng}}{\nu_{ni}} \frac{\Omega_g^2}{\nu_{gn}^2} \gg 1 \qquad (53)$$

In the above, we have taken the neutral fluid to be static. In reality it collapses to form stars, and a full dynamical, multidimensional study is necessary.

In multifluid models of oblique shocks propagating in dense, dusty, weakly ionized star forming regions, the magnetic field tends to rotate around the direction in which the shock propagates (Pilipp & Hartquist 1994; the next section of this review). Knowing this, Hartquist, Havnes, & Pilipp (1997) suggested that at high densities in a collapsing protostar, rotation of the magnetic field around the direction of the collapse velocity may have dynamical consequences. Independently, Wardle & Ng (1999) and Wardle (2004) noted that certain grain size distributions would lead to Hall conductivity being important at $T = 10$ K and m_H as low as 10^7 cm^{-3}. If the Hall conductivity is important, the simple geometries assumed in many studies of ambipolar diffusion in star formation are inappropriate.

The current is related to the electric field by

$$\mathbf{J} = \sigma_\| \mathbf{E}'_\| + \sigma_1 \hat{\mathbf{B}} \times \mathbf{E}'_\perp + \sigma_2 \mathbf{E}'_\perp \qquad (54)$$

where the prime indicates that the electric field is measured in the frame comoving with the fluid. $\mathbf{E}'_\|$ is the component of $\mathbf{E}'$ parallel to the magnetic field, and $\mathbf{E}_\perp$ is the component perpendicular to it.

$$\sigma_\| = \frac{ec}{B} \sum_\alpha \frac{n_\alpha Z_\alpha \beta_\alpha}{1+\beta_\alpha} \tag{55}$$

where

$$\beta_\alpha = \frac{|\Omega_\alpha|}{\nu_{\alpha n}\rho_n} \tag{56}$$

is the Hall parameter for the αth fluid and we have assumed that collisions with neutrals dominate the resistivity. The Hall conductivity is

$$\sigma_1 = \frac{ec}{B} \sum_\alpha \frac{n_\alpha Z_\alpha}{1+\beta_\alpha^2} \tag{57}$$

and the Pedersen conductivity is

$$\sigma_2 = \frac{ec}{B} \sum_\alpha \frac{n_\alpha Z_\alpha \beta_\alpha}{1+\beta_\alpha^2} \tag{58}$$

Finally,

$$\sigma_\perp = \sqrt{\sigma_1^2 + \sigma_2^2} \tag{59}$$

Like Hartquist, Havnes, & Pilipp (1997), Wardle & Ng (1999) have noted that in some situations in which the Hall conductivity is important, it can give rise to the production of a magnetic field component perpendicular to the original magnetic field. This component would not arise in analogous situations in which the Hall conductivity is unimportant.

6 Multifluid Models of Shocks in Dusty Weakly Ionized Regions

As mentioned in section 3, Mullan (1971), Draine (1980), and Draine, Roberge, & Dalgarno (1983) introduced and developed multifluid models of shocks in magnetized, weakly ionized interstellar clouds. Peter Brand has considered them thoroughly in his chapter and discussed their relevance to the interpretation of observations. The simplest such models are of shocks propagating perpendicular to the magnetic field and are based on the assumption that all charged species have such large Hall parameters that they move with the same velocity component in the shock propagating direction. Work has been focussed on regions in which the upstream magnetic pressure is very large compared to the upstream thermal pressure; consequently, the upstream Alfvén and fast-mode speeds may be taken as approximately equal.

The upstream speed of Alfvén waves in which the motions of all species are well coupled is

$$v_{AC} = B_0 / \sqrt{\sum_j (4\pi\rho_j)} \tag{60}$$

The speed of high frequency waves in which the motions of charged species are well coupled to one another but uncoupled to the motion of the neutrals is

$$v_{AU} = B_0 / \sqrt{\sum_{j \neq n}(4\pi\rho_j)} \tag{61}$$

B_0 is the magnitude of the upstream magnetic field.

For perpendicular shocks in which the Hall parameters of all charged species are large, there is a range of values for the shock speed, U_S, greater than v_{AC} but still less than v_{AU}, for which a so-called "precursor" exists. Throughout much of this precursor, the mean velocity of the charged particles is closer to the distant downstream mean flow velocity than to the distant upstream mean flow velocity, whereas the mean velocity of the neutrals is closer to the mean velocity of the distant upstream flow. Figure 2 shows a schematic plot of the velocity components of ions and of neutrals in the shock propagation direction. If ν_{ni} is much greater than ν_{nj} for all j's other than i and n, then the precursor length or thickness is given roughly by

$$\Delta \approx \frac{v_{AC}^2}{\nu_{ni} U_S} \tag{62}$$

(Draine 1980). Here ν_{ni} is evaluated for a far upstream position and chemistry is assumed to have little effect on the flux in the precursor.

From the previous section, it is clear that ν_{ng} can be comparable to or greater than ν_{ni} in dark star forming regions. Consequently, the effects of dust on shock structure cannot be neglected. Draine and his collaborators included these effects in an approximate fashion that is suitable for a broad region of parameter space. However, shocks for some interesting ranges of parameters can be modelled reliably only if a somewhat more sophisticated treatment of grain dynamics is employed.

Pilipp, Hartquist, & Havnes (1990) constructed steady, plane-parallel multifluid models of perpendicular shocks in which the grains were treated as a fluid. They assumed that all grains are of the same size and are spherical. They followed the average grain charge and evolution of the gas phase electron and ion number densities with the appropriate differential equations including source terms due to collisions of gas phase charged particles with grains and gas phase chemical reactions. Significant differences between models incorporating the Pilipp et al. (1990) treatment of grains and that employed by Draine and his collaborators exist for cases in which the preshock value of ρ_n/m_n is greater than roughly 10^6 cm^{-3}.

In a model of a 15 km s^{-1} perpendicular shock propagating into a medium with B_0 of 4.74×10^{-3} G and a neutral number density of 10^7 cm^{-3}, the fractional ionization drops about two orders of magnitude relative to its upstream value. Furthermore, the grains become the dominant carriers of negative charge. As a consequence, a runaway effect, not seen in previous models, occurs.

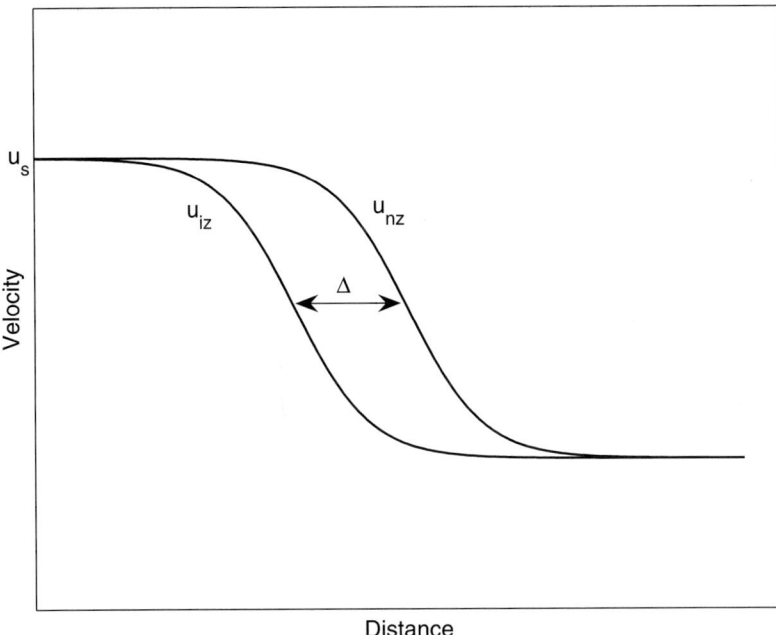

Fig. 2. Ion and neutral velocity components in the shock propagation direction for a two-fluid model of a C-type shock. U_S is the shock speed. Δ is the precursor width. A C-type shock is a shock in which there are no discontinuities in any flow parameters.

To understand the runaway effect, assume that the shock propagates in the z direction and that the magnetic field is in the x direction. As $n_e/|Z_g|n_g$ drops below unity, $|Z_g|$ also starts to drop since fewer gas phase electrons are able to charge the grains. For these conditions, the grain Hall parameter is comparable to unity or smaller, while the ion Hall parameter is large; consequently a charge separation electric field component in the z direction balances most of the neutral drag on grains. As $|Z_g|$ drops, $|E_z|$ increases. As $|E_z|$ increases, the $\mathbf{E} \times \mathbf{B}$ drift speed of the ions increases. The increase in this speed causes collisions of ions with grains to be more rapid which decreases $|Z_g|$ further, which increases $|E_z|$, which increases the drift speed further. The runaway has significant consequences for the ionization structure and heating of the gas. However, it is, in part, due to the assumption that the shock velocity is exactly perpendicular to the magnetic field.

Pilipp & Hartquist (1994) extended the approach of Pilipp et al. (1990) to oblique shocks. Their most notable result was the finding that the magnetic field rotates around the shock propagation direction. To understand this result, consider an upstream fluid moving with the velocity $U_S\hat{z}$ with a magnetic field $\mathbf{B}_0 = B_{0x}\hat{x} + B_{0z}\hat{z}$ where B_{0x} and B_{0z} are constant. In the shock frame, $E_x = 0$ and $E_y = -U_S B_{x0}/c$. For a perfectly conducting fluid (i.e.

large Hall parameter), $\mathbf{E} + (\mathbf{u}_j \times \mathbf{B})/c = 0$, which implies that $\mathbf{u}_{j\perp} \propto \mathbf{E} \times \mathbf{B}$ and $\mathbf{u}_{j\perp z} \propto E_y B_x$. Here $\mathbf{u}_{j\perp}$ is the component of the mean velocity of the jth fluid that is perpendicular to $\mathbf{B}$, and $\mathbf{u}_{j\perp z}$ is the projection of $\mathbf{u}_{j\perp}$ onto the z axis. Grains have a Hall parameter that is not small, implying that $\mathbf{u}_{g\perp z}$ differs from $\mathbf{u}_{e\perp z}$ and $\mathbf{u}_{i\perp z}$. Thus, the z component of the "Hall" current is nonzero. However, if we restrict attention to steady flow depending on z only, charge conservation require that $J_z = 0$. A component of the current parallel to $\mathbf{B}$ having a projection onto the z axis that cancels the z component of the Hall current must exist. That "parallel" current has a finite projection on the x axis, and J_x is nonzero. From Ampere's law $dB_y/dz = -4\pi J_x/c$, which induces the rotation of the magnetic field around the shock propagation direction.

Pilipp & Hartquist (1994) integrated the relevant ordinary differential equations in the downstream diretion. They were able to obtain intermediate–mode shock solutions. Such solutions are unphysical (e.g. Falle & Komissarov 2001). They were unable to find fast-mode shock solutions, a result explained by Wardle (1998). He showed that the downstream state of a fast-mode shock corresponds to a saddle point with the consequence that integration of time-independent equations in the downstream direction will not lead to a solution with the downstream state of a fast-mode shock. Making a number of simplifying assumptions (e.g. the ion flux is constant.), Wardle (1998) was able to obtain fast-mode solutions by integrating in the upstream direction. Such an approach is invalid in cases in which local thermal, ionization, and chemical equilibrium does not obtain.

Consequently, Falle (2003) has shown that the numerical scheme he has developed for multifluid time-dependent magnetohydrodynamics is suitable for the construction of shock models. A treatment of nonequilibrium conditions is possible with his approach.

Wardle (1998), Falle (2003), and Chapman & Wardle (2006) have found fast-mode shock solutions for which the rotation of the magnetic field is substantial. The grain Hall parameter is not large in these cases.

Caselli, Hartquist, & Havnes (1997) argued that the use of fast-mode oblique shock solutions in which the grain dynamics are described as in the work by Pilipp and Hartquist (1994) would give much lower minimum shock speeds for substantial sputtering of grains in dense regions than obtained on the basis of perpendicular shock models in which less sophisticated grain dynamical treatments are employed. Their work can now be extended given the availability of the methods due to Falle (2003).

The work of Chapman & Wardle (2006) underscores the sensitivity that shock models have to the assumed grain size distribution. Even so, the sensitivity may be even greater than their results indicate, because they did not follow the fractional ionization, which can be affected greatly if many grains much smaller than 0.1 micron are present.

A considerable amount of work on shocks in dusty star forming regions remains to be completed.

References

Arons, J., Max, C.E. 1975 Ap J **196**, L77
Caselli, P., Myers, P.C. 1995 Ap J **446**, 665
Chapman, J.F., Wardle, M. 2006 MNRAS, submitted
Draine, B.T. 1980 Ap J **241**, 1021
Draine, B.T. 1986 MNRAS **220**, 133
Draine, B.T., Sutin, B. 1987 Ap J **320**, 803
Draine, B.T., Roberge, W.G., Dalgarno, A. 1983 Ap J **264**, 485
Elmegreen, B.G. 1979 Ap J **232**, 729
Falle, S.A.E.G. 2003 MNRAS **344**, 1210
Falle, S.A.E.G., Komissarov, S.S. 2001 J Plasma Phys **65**, 29
Gail, H.-P., Sedlmayr, E. 1975 A & A **41**, 359
Hartquist, T.W., Havnes, O., Pilipp, W. 1997 Ap & SS **246**, 243
Havnes, O., Hartquist, T.W., Pilipp, W. 1987 in Morfill, G.E. and Scholer, M. (eds.), Physical Processes in Interstellar Clouds, Reidel, D., Dordrecht
Mullan, D.J. 1971 MNRAS **153**, 145
Nakano, T., Umebayashi, T. 1980 Publ Astron Soc Jpn **32**, 405
Oppenheimer, M., Dalgarno, A. 1974 Ap J **192**, 29
Pilipp, W., Hartquist, T.W. 1994 MNRAS **267**, 801
Pilipp, W., Hartquist, T.W. 1994 MNRAS **267**, 801
Pilipp, W., Hartquist, T.W., Havnes, O. 1990 MNRAS **243**, 685
Spitzer, L., Jr. 1978 Physical Processes in the Interstellar Medium, John Wiley & Sons, New York, p. 199
Tassis, K., Mouschovias, T.Ch. 2005a Ap J **618**, 769
Tassis, K., Monschovias, T.Ch. 2005b Ap J **618**, 783
Wardle, M. 1998 MNRAS **298**, 507
Wardle, M. 2004 Ap & SS **292**, 317
Wardle, M., Ng, C. 1999 MNRAS **303**, 239
Williams, J.P., Blitz, L., Stark, A.A. 1995 Ap J **451**, 252

Massive Star Formation

M.G. Hoare[1] and J. Franco[2]

[1] School of Physics and Astronomy, University of Leeds, Leeds, LS2 9JT, UK
   mgh@ast.leeds.ac.uk
[2] Instituto de Astronomía-UNAM pepe@astroscu.unam.mx

1 Introduction

The formation and evolution of massive ($> 10 M_\odot$) stars plays a key role in the final fate of their parental molecular clouds, and in the appearance and evolution of their host galaxies. They inject large amounts of mechanical and radiative energy creating, either by a single star or a stellar cluster, the most spectacular gaseous nebulae in the Cosmos. Also, they generate fast shocks which heat up the surrounding plasma to temperatures above 10^6 K, and carve large interstellar "holes" that continuously stir the general interstellar medium.

From the moment they are born their powerful outflows begin to plough into the surrounding molecular material. Similarly, their emission of Lyman continuum radiation, which sets them apart from lower mass stars, ionizes the clouds and creates dense and hot H II regions (cf. the chapter by Will Henney). The expansion of such a photoionized nebula drives a shock wave that both compresses and may trigger further star formation on the one hand, and disperse a large portion of the molecular cloud on the other. As a new OB star enters the field population its wind and ultra-violet radiation continue to influence the general interstellar medium. The wind and radiation field strengthen after the short main sequence lifetime, through the supergiant and Wolf-Rayet phases. The evolution culminates in the most energetic of stellar events, a supernova explosion. Thus, the energy input from massive stars, via the combined effects of expanding H II regions and supernova remnants, can shape the interstellar medium of gaseous galaxies, creating large, expanding structures that may even vent mass and energy into the halo (cf. the chapter by Alan Pedlar and Karen Wills on the M82 starburst galaxy). It is this litany of energetic phenomena that gives massive stars such a pivotal role in astrophysics. Hence, it is not surprising to find that the dynamics of these phenomena have figured strongly in John's work over the years.

2 Accretion

Larson & Starrfield (1971) and John's mentor, Kahn (1974), first investigated one of the intriguing questions in massive star formation - how can accretion continue in the face of extreme radiation pressure? This question highlights another key difference between low and high mass star formation. The timescale for the evolution of a high mass protostar to a main sequence configuration is shorter than the timescale for dense core material to collapse due to its gravity. Hence, a high mass star is probably still accreting at the surface when hydrogen burning begins in the core. As the mass increases the luminosity from fusion and accretion exerts a high radiation pressure on the dust grains in the infalling cloud. In a spherically symmetric treatment Kahn (1974) deduced that this effect would limit the mass of a star that could be formed by accretion to about 40 $M_\odot$. The adoption of more appropriate dust parameters removes this strict limitation, but does require very high infall rates for the ram pressure to overcome the radiation pressure. Wolfire & Cassinelli (1987) updated this with a full radiative transfer solution and concluded that normal interstellar dust opacities would not allow inflow to occur. Even with depleted dust models they found that high accretion rates ($\sim 10^{-3}$ $M_\odot$ yr^{-1}) would be necessary to form the most massive stars.

Such accretion rates were thought to be unreasonably high compared to the $\sim 10^{-5}$ $M_\odot$ yr^{-1} expected from the collapse of a cloud initially close to equilibrium and held up by thermal pressure alone ($\dot{M}_{\rm acc} \sim c^3/G$, where c and G are the sound speed and the gravitational constant); as is thought to be the case for low-mass star formation (e.g., Shu 1977). However, several arguments have been put forward for higher and time variable accretion rates, which can also overcome the radiation pressure problem. In any case, one might intuitively expect more massive objects to accrete faster.

Norberg & Maeder (2000) and Behrend & Maeder (2001) have suggested that the accretion rate increases as the mass of the star grows. There is a region of the mass-luminosity plane where spherical accretion rates can grow and keep above that needed to overcome radiation pressure on dust in the cocoon, but below the Eddington limit due to radiation pressure on electrons in the stellar atmosphere. They attempted to put this on a physical basis by using the observed relation between outflow rates and bolometric luminosity (e.g. Churchwell 1998) and then making the accretion rate a fraction of the outflow rate. This invokes the often found result from outflow models that the outflow rate is a fixed fraction of the accretion rate. Their prescription gave very high accretion rates ($\sim 10^{-2}$ $M_\odot$ yr^{-1}) at the upper end of the mass range. However, the observed outflow rate versus luminosity relation is likely to be severely affected by selection effects for massive objects. Firstly, the objects observed are likely to be at a wide range of evolutionary stages. More importantly, Ridge & Moore (2001) showed that when a constant distance sample is studied there is much less of a correlation between outflow rate and luminosity and that previous studies were affected by Malmquist bias.

Another approach has been based on the fact that the cores that form massive stars are not supported by thermal pressure, but by a combination of turbulence and MHD waves (see the book edited by Franco & Carramiñana 1999). Bernasconi & Maeder (1996) used the empirically derived Larson (1981) relations, which show the non-thermal linewidth increasing with increasing size of cloud. This is thought to be a natural consequence of turbulent or magnetic support. In the simple picture in which $\dot{M}_{acc} \propto c^3$, this gives a physical basis for the accretion rate increasing with time as the inside-out collapse proceeds if c is taken to be related to the sum of the thermal, magnetic and kinetic pressures in the same way as the ordinary sound speed is to the thermal pressure. Bernasconi & Maeder (1996) found accretion rates up to $\sim 10^{-4}$ $M_\odot$ yr^{-1} for the most massive stars.

McKee & Tan (2003) have developed a turbulent core model, which results in accretion rates about an order of magnitude higher than this ($\sim 10^{-3}$ $M_\odot$ yr^{-1} for the most massive stars). This they justify from the observed high pressures of massive star forming cores that require higher linewidths than given by the usual Larson relations. They still appeal to the form of the Larson relations to get accretion rates increasing with time, although there is little direct evidence that they apply to massive star forming clumps at present. Indeed, in the Plume et al. (1997) study, upon which McKee & Tan draw, they found no correlation between linewidth and size. Most of these cores also already have plenty of star formation activity in them, which will inevitably affect the observed linewidths.

The spherical accretion rates in the turbulent core model are sufficient to overcome radiation pressure. The time variable rates also make the star formation timescale rather independent of the mass which helps to produce apparently coeval clusters containing a range of masses. Their fiducial model also has a radial density distribution on large scales that is close to r$^{-1.5}$, similar to that seen in several studies of massive star forming regions (e.g. Hatchell & van der Tak 2003). This type of gradient has usually been interpreted as being consistent with infall in a rapid star formation scenario in which the whole region is collapsing. In the McKee & Tan model it is seen as a longer lived quasi-equilibrium structure supported by turbulence.

It has been fairly clear ever since the first examples of massive young stars were found that they form from accretion via a disc rather than spherical infall. That is because the main manifestations of luminous embedded sources are their ubiquitous bipolar outflows (Lada 1985). Since these most naturally arise from discs in most viable models this has always been good indirect evidence for discs. Accretion through a disc is also the key to overcoming the radiation pressure in a number of ways. Firstly, the stellar radiation is isotropic and so a disc only intercepts a small fraction of the total luminosity. Secondly, if the infall is concentrated through a thin disc then the effective accretion rate is amplified. Finally, there is likely to be a large self-shielding effect, whereby material can accrete through the mid-plane, whilst the upper layers of the disc take the brunt of the radiative effects.

This was investigated by Nakano (1989) and Jijina & Adams (1996) who showed analytically that non-spherical accretion caused by magnetised collapse or rotation can overcome the radiation pressure for a range of reasonable starting conditions and accretion rates. Yorke & Sonnhalter (2002) performed 2D axisymmetric radiation hydrodynamical simulations on a collapsing rotating cloud and came to similar conclusions concerning the upper mass limit. They found accretion rates of about $\sim 10^{-3}$ $M_\odot$ yr^{-1} for a very overdense, non-turbulent core. Their treatment of radiative transfer also showed that a substantial amount of radiation escapes along the polar axis of the structure where it does drive away some material, but it is along the equatorial axis that the accretion occurs. Numerical simulations have now moved on to three dimensions with the work of Krumholz et al. (2005a). The application of adaptive mesh refinement (AMR) hydrodynamical codes (see the chapter by Falle) to this problem is finally demonstrating conclusively that radiation pressure is not a barrier to massive star formation. As material rains down on the accretion disc, radiation bubbles repeatedly blow out perpendicular to the disc plane and then collapse down again, but overall the accretion continues. They also find accretion rates of order $\sim 10^{-3}$ $M_\odot$ yr^{-1} and it would appear that a consensus may be emerging over these kind of rates.

The magnetic field is the next major ingredient that needs to be added to these simulations as computer power increases. It is possible that magnetic flows set in very early in the accretion process as in the simulations of Tomisaka (2002) for the collapse of magnetised, rotating cores. It is unlikely that the MHD outflow mechanisms developed mostly for low-mass protostars can be simply scaled up to high mass accretors. Again the high output of UV radiation will increase the ionization fraction and hence, the coupling of the magnetic field with the material. The onset of a radiative envelope (e.g. Palla & Stahler 1992) in the star as it adopts a more main sequence structure removes the usual method for the generation of strong stellar surface magnetic fields. This may take away the driving force in the widely used X-wind model for protostellar outflows. Outflow mechanisms driven by magnetic fields generated or dragged in by the disc could still operate if a strong stellar field is absent (Banerjee & Pudritz 2006). Krumholz et al. (2005b) have made an initial investigation of how outflows affect accretion onto massive stars As expected, the punching of holes along the axis by these flows allows the radiation pressure to be relieved in the centre, and accretion to continue to higher masses still.

The direct observational evidence of accretion discs around massive young stars is now beginning to arrive. Millimetre interferometry had delivered convincing observations of compact flattened structures around the more luminous intermediate mass young stellar objects such as GL490 (Mundy & Adelman 1988) and G192.16-3.82 (Shepherd & Kurtz 1999). Now sub-millimetre interferometry has heralded the best example of a disc around a genuinely luminous object. Patel et al. (2005) have used the SMA to resolve an elongated dust and molecular structure 1000 AU in size that is perpendicular to

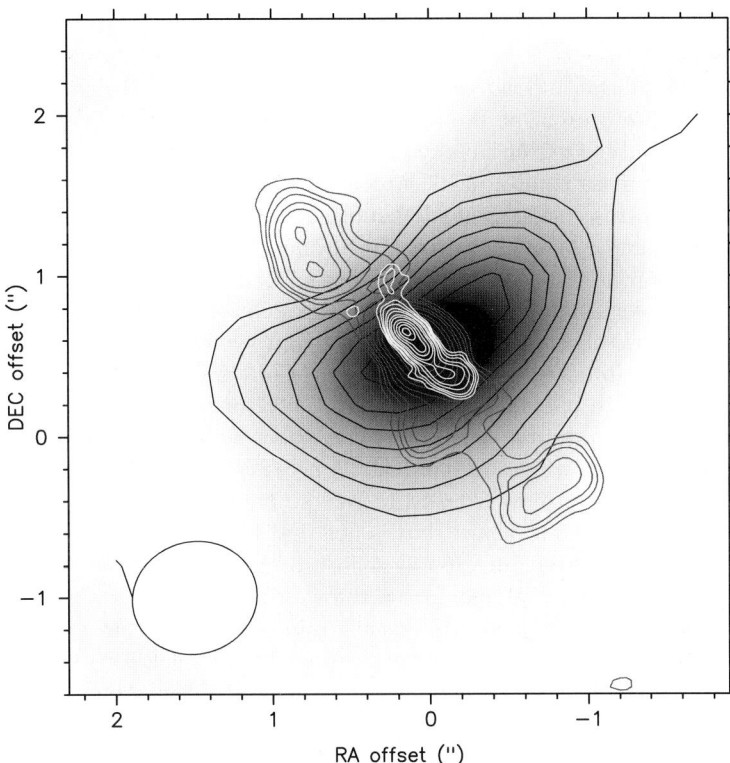

Fig. 1. SMA image of the 325 GHz dust emission from the massive YSO Cep A2. Dark contours show the integrated emission from the CH_3CN J=18-17 transition that shows a rotational kinematic signature consistent with a disc origin. Grey and white contours show the 3.6 and 1.3 cm radio continuum emission from the ionized jet in this source. Note the elongated structure (SE-NW) in the dust and molecular gas perpendicular to the radio jet (SW- NE). From Patel et al. (2005).

the radio jet in Cep A2 (Fig 1). The molecular line has a rotational kinematic signature which helps to confirm the disc interpretation. This is a taster of what ALMA will do for massive star formation studies. The use of spectropolarimetry to investigate discs is addressed in the chapter by René Oudmaijer, who has also discussed some of the data mentioned in this and the following section of this chapter.

The alternative coalescence model by Bonnell et al. (1998) seems to be unnecessary, since there was never really a fundamental problem with the accretion picture as described above. Also, the stellar densities required for it to work are many orders of magnitude above those seen or inferred, and it seems inconceivable that energetic events such as stellar collisions are common without any observational sign of them.

3 Outflows

As mentioned above, bipolar molecular outflows have long indicated that accretion discs are present at the heart of massive star formation. It has been thought that outflows from massive YSOs are not as well collimated as those from their low-mass counterparts. Much of this impression may have arisen from low spatial resolution single-dish molecular line observations of high-mass systems that are on average at least ten times further away (Beuther et al. 2002a). The strong clustering in massive star forming regions can also cause multiple outflows to appear to merge at low resolution. Again, the application of interferometric observations has revealed that some regions are made up of multiple, well-collimated outflows (Beuther et al. 2002b), although these are not associated with particularly high luminosity sources.

However, other systems still appear to show little evidence for high degrees of collimation when observed at high resolution. Fig 2 shows OVRO observations of the outflow from S140 IRS 1. The blueshifted lobe does display an approximate bow shape, but is not that highly collimated, whilst the redshifted lobe is severely affected by self-absorption. If these type of outflows are driven by jets then we should see the shock-excited emission lines characteristic of low-mass objects. However, ground-based searches for molecular hydrogen emission at 2μm have often turned up very little to support this picture (Davis et al. 1998), although see Davis et al. (2004). Optical searches for shocked emission in the outer reaches of molecular clouds where the flows terminate and extinction is expected to be less have also not detected anything significant (Alvarez & Hoare 2005). Images from the SPITZER satellite in the 4.5μm filter are turning up numerous examples of outflow lobe emission, most likely from rotational lines of molecular hydrogen (Noriega-Crespo et al. 2004). This band occurs close to where there is a minimum in the extinction curve (Indebetouw et al. 2005) and should be a useful probe of massive outflows.

As well as molecular outflows, the massive YSOs also have compact, ionized winds. These manifest themselves through thermal radio spectra with spectral indices close to +0.6 which is expected for a constant velocity wind (Wright & Barlow 1975). They also give rise to broad, single-peaked H I emission lines in the IR (Bunn et al. 1995). These typically have FWHM of about 100 km s^{-1} and FWZI up to several hundred km s^{-1} which is a lower limit on the wind's terminal velocity. NLTE modelling shows that there is a common origin for the radio continuum and IR line emission (e.g. Höflich & Wehrse 1987). The typical mass-loss rates estimated for these ionized winds are of order $\sim 10^{-6} M_\odot$ yr^{-1}.

Clues to what is driving the ionized winds comes from resolving their spatial structure. Here the picture is also mixed as it is for the molecular outflows (see review by Hoare 2002). In some cases radio jets are seen as in their low-mass counterparts (see the chapter by Tom Ray). By far the most spectacular example is the 2.6 pc long wiggling jet seen in the GGD27 system

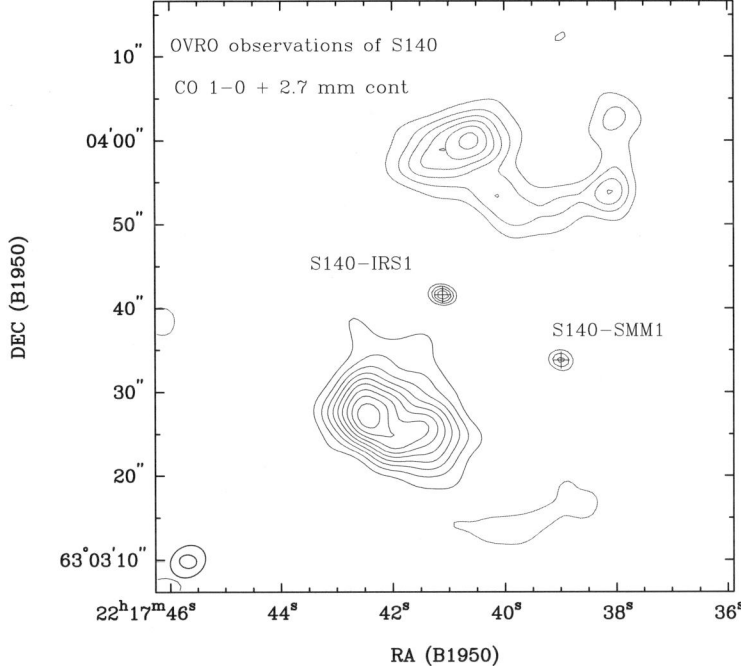

Fig. 2. OVRO map of the bipolar molecular outflow from S140 IRS 1. The ^{12}CO 1-0 observation has a resolution of about $4''$ whilst the 2.7 mm continuum emission showing the two point sources was made at a resolution of $2''$. From Gibb, Hoare & Shepherd, in prep.

(Martí et al. 1993). This powers radio and optical Herbig-Haro objects at its termination and looks just like a scaled up low-mass system. Proper motion studies reveal velocities of 500 km s^{-1} (Martí et al. 1998). Other examples are G35.2N (Gibb et al. 2003) and IRAS (Rodríguez et al. 2005) and possibly W3 IRS 5d2 (Wilson et al 2003), which are mostly much more knotty and have more exaggerated point symmetry. In other cases we just resolve the base of the jet; the best example being Cep A2 as in Fig 1, and GL 2591 also falls in this category (Trinidad et al. 2003). The fast motion in the Cep A2 jet has recently been confirmed by proper motion studies (Curiel et al. 2006) (Figure 3). One expects that such highly collimated flows are driven by MHD mechanisms in the star-disc system as for low-mass outflows.

However, jets are not the only radio morphology seen in massive YSOs. The exciting source of the peculiar bipolar H II region S106, which otherwise shows all the characteristics of a massive YSO (Drew et al. 1993), has radio emission elongated perpendicular to the outflow axis (Hoare et al. 1994; Hoare & Muxlow 1996; Hoare 2002). S140 IRS 1 also has such an equatorial wind of about 500 AU in size (Hoare & Muxlow 1996; Hoare 2002). Originally interpreted as a jet (Schwartz 1989; Tofani et al. 1995) it is clearly perpendicular to

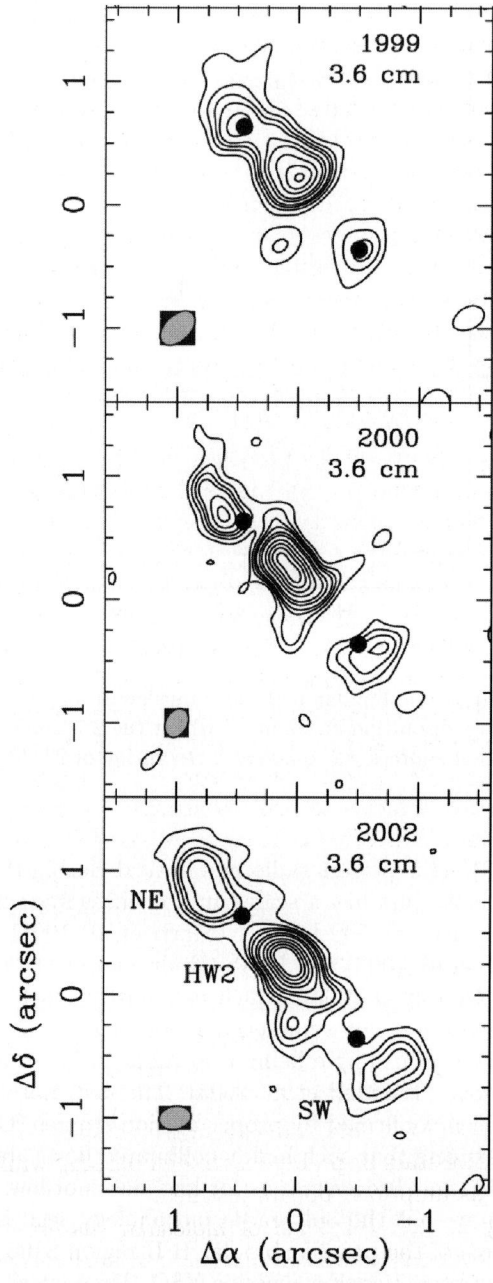

Fig. 3. Series of VLA 3.6 cm continuum maps of the jet from the massive YSO Cep A2. Note the outward motion of the knots in the jet which corresponds to a speed of 480 km s^{-1} on the sky and nearly 600 km s^{-1} after correction of inclination. The filled circles mark the position of the knots in the 1999 epoch. From Curiel et al. (2006).

the large scale bipolar molecular outflow and to the monopolar IR reflection nebula at the base of the blueshifted outflow cavity revealed in speckle observations (Schertl et al. 2000; Alvarez et al. 2004). Proper motions confirm that this structure is not moving outwards as would be expected for a jet (Hoare 2002). Another possible example of an equatorial wind is in GL490 (Campbell et al. 1986), although the extension in the radio emission along the disc plane (Mundy & Adelman 1988) is somewhat noisy.

Drew et al. (1998) have proposed that the radiation pressure due to central star and inner disc act on the ionized gas at the surface of the disc to drive such equatorial winds. The gas is pushed sideways, across the surface of the disc forming an equatorial wind. This is the same mechanism that drives the stellar winds in field main sequence stars where the terminal velocity is a few thousand km s^{-1}, i.e. of order the escape speed at the stellar surface (Prinja et al. 1990). In the massive YSO case, the surface gravity in the disc atmosphere a few stellar radii out from the star is lower and so the typical speed of the equatorial wind is a few hundred km s^{-1}. Hence, this mechanism has the potential to explain both the morphology, speed and mass-loss rate of the observed equatorial wind systems. Simulated line profiles arising from such winds have too high a rotational component and result in double-peaked line profiles, whereas only single-peaks are usually seen (Sim et al. 2005).

Note that this radiatively driven flow is very different from a flow induced by the photo-evaporation mechanism discussed by Franco et al. (1989) and Hollenbach et al. (1994). That drives a much slower, thermal flow from the outer regions of the disc, and they did not consider the effect of radiation pressure on the inner disc. Both mechanisms could operate simultaneously, but if the accretion disc does extend down close to the star there is little doubt that the radiation pressure acting on the gas will drive such a wind and is likely to dominate the dynamics.

4 The Transition to UCHII Region

The fact that the massive YSOs do not ionize their surrounding molecular gas as soon as they become luminous enough has long been a puzzle. All objects of above about $10^4 L_\odot$ should emit copious amounts of Lyman continuum radiation if they are fully on the zero age main sequence (ZAMS). It is commonly accepted that massive stars begin core hydrogen burning whilst still accreting and so the ZAMS assumption appears justified.

One possibility is that the infall of molecular material "quenches" the H II region (Walmsley 1995). The critical accretion rate for this to occur is obtained from inserting an infalling spherically symmetric density distribution into the Strömgren radius equation. The rates are high, but not as high as the $\sim 10^{-3} M_\odot$ yr^{-1} needed to overcome radiation pressure on the dust in the same spherical treatment. In this scenario, the H II region is not absent, but just very dense ($\sim 10^{12}$ cm^{-3} with a Strömgren radius close to the star. One problem with this approach is that the infall would have to be approximately spherically symmetric in order to stop the H II region breaking out in low density directions. In any accretion scenario, most of the infalling material will arrive on the disc rather than the star and a bipolar H II region would ensue with ionized lobes above and below the disc (Franco et al. 1989). It also neglects the effect of winds and outflows which are likely to push the inner radius of the infalling envelope some distance away from the star, again probably mostly in the polar directions.

Another picture was developed by Tan & McKee (2003). In it strong outflow rather than infall confines the H II region. They used an approximate X-wind outflow density distribution where the H II region propagates along the cavity along the axis. These models are basically X-wind jets with extra photo-ionisation. They predict ionized zones that are very narrow (few AU), whereas most high-mass radio jets are resolved across the minor axis (~ 50 AU) (Martí et al. 1999; Curiel et al. 2006). Jets from high-mass objects are also likely to be significantly ionized as they are launched. Low-mass jets are partially ionized, and it is difficult to predict at this moment how the ionization fraction of MHD driven jets would change for higher mass stars. If they are significantly ionized then their ability to confine the H II regions would be dramatically reduced.

Again a difficulty is that as soon as the central star becomes at all luminous it will start to drive strong stellar winds due to radiation pressure. These will open up cavities further exposing more material to the ionizing radiation. The equatorial winds discussed above would also not be able to confine the H II region since they do not cover the polar regions. Indeed, in the models by Drew et al. (1998) the polar regions are occupied by a normal O star stellar wind flowing at a few thousand km s^{-1}.

One possible solution is to tackle the problem at the source through an examination of whether the star really is emitting large amounts of Lyman continuum radiation during the massive YSO phase. Although the core is very likely to be already on the main sequence, it is unclear that the outer layers of the star have contracted fully to a main sequence configuration. It is well known that accreting stars swell up well beyond their ZAMS radius, mainly due to shell deuterium burning (Palla & Stahler 1992). Furthermore, the higher the accretion rate the more the radius increases and extends to higher masses before the rapid contraction onto the main sequence occurs. The calculations by Palla & Stahler only consider accretion rates up to $\sim 10^{-4}$ M$_\odot$ yr^{-1} and masses up to 15 M$_\odot$ (see Figure 4), but the current thinking points towards accretion rates as high as $\sim 10^{-3}$ M$_\odot$ yr^{-1}. Figure 4 shows that we would then expect a much greater swelling, up to around 30 R$_\odot$, before the contraction to the main sequence at around 30 M$_\odot$ (L$\sim 10^5$ L$_\odot$) for these rates of accretion (Palla, private communication).

The key point here is that a larger radius means a lower effective temperature. If ongoing accretion keeps the effective temperature below about 30 000 K, then there is no need to invoke any mechanism to quench the H II

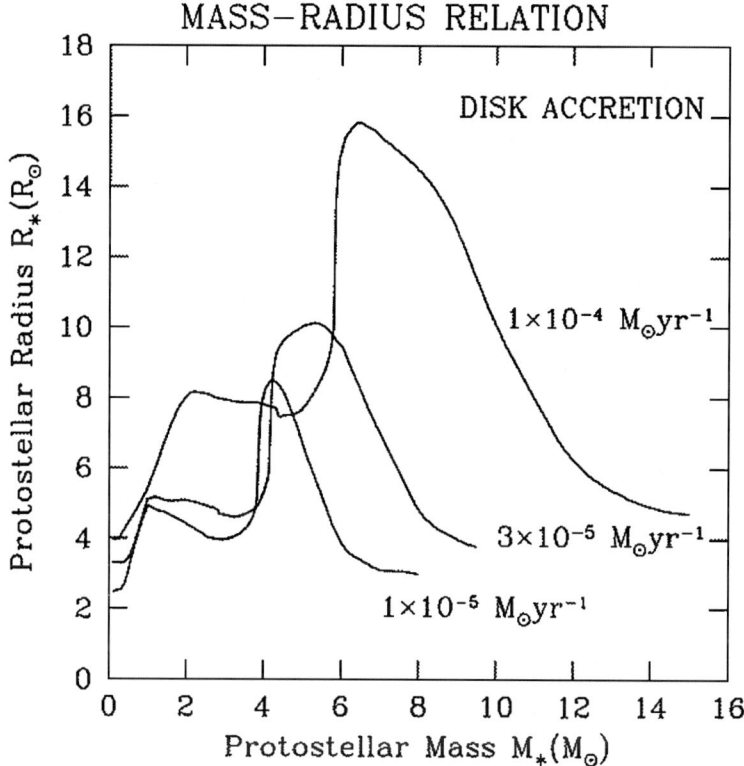

Fig. 4. Variation of the radius of an accreting star as it grows in mass with the steady accretion rates shown. From Palla & Stahler (1992).

region since there will not be a significant Lyman continuum output from the star in any case. Future accretion models need to be calculated with a fully self-consistent treatment of the stellar mass-radius relation to investigate this aspect. The McKee & Tan (2003) models, in which the stars accrete at rates significantly above $\sim 10^{-4}$ $M_\odot$ yr^{-1}, implement an approximate treatment of the Palla & Stahler (1992) results and their stars join the main sequence at masses of around 20 $M_\odot$ as expected. However, their radii never get above the 15 $R_\odot$ for Palla & Stahler's $\sim 10^{-4}$ $M_\odot$ yr^{-1} case suggesting that these have been under-estimated for the highest accretion rates. The Behrend & Maeder (2001) mass-radius relation shows a similar pattern of swelling towards larger radii (but again not as much as one would expect based on Figure 4), since their accretion rates exceed $\sim 10^{-3}$ $M_\odot$ yr^{-1} through the relevant mass range.

Nakano et al. (2000) used a polytropic model to estimate that a young star accreting at $\sim 10^{-2}$ $M_\odot$ yr^{-1} would have a maximum radius of about 30 $R_\odot$. In a different context Kippenhahn & Meyer-Hofmeister (1977) calculated the effect on the radii of stars already fully on the main sequence of accretion at rates in the $\sim 10^{-4}$ $M_\odot$ yr^{-1} to $\sim 10^{-2}$ $M_\odot$ yr^{-1} range of interest here. They found that the stars can expand greatly, e.g. about 100 $R_\odot$, occupying the supergiant part of the Hertzsprung-Russell diagram before returning to the main sequence at higher mass. This was not because of deuterium burning, but simply because the accretion timescale ($M/\dot{M}$) is shorter than the thermal adjustment or Kelvin-Helmholtz timescale. The expanded phase lasts for the accretion timescale if the accretion rate is constant. It is 10^4 years for a 10 $M_\odot$ star accreting at $\sim 10^{-3}$ $M_\odot$ yr^{-1}. If the accretion rate is increasing with time this phase would be longer still.

Overall, if the accretion rates during the hot core/massive YSO phases are as high as $\sim 10^{-3}$ $M_\odot$ yr^{-1}, then it is plausible that the reason for the lack of an H II region is simply that ongoing accretion keeps the effective temperature of the star too low. This scenario would also explain several other features of massive YSOs. If the ongoing accretion keeps the stellar radius high then it also lowers the surface gravity. It has been noted several times in the literature how the spectroscopic characteristics of massive YSOs resemble those of evolved OB stars (Simon & Cassar 1984). Most of the massive YSOs amenable to near-IR spectroscopy have luminosities around a few 10^4 $L_\odot$ and thus are about 12 $M_\odot$, just where accretion could easily result in a low surface gravity. The emission line spectra with broad profiles of a few hundred km s^{-1} are reminiscent of those of B supergiants. The low surface gravity leads directly to low escape speeds and increased mass-loss rates.

Large central stars may be central to the explanation of the near- IR line profiles in equatorial wind sources like S106IR. The current attempts to model the H I lines result in double-peaked rather than single-peaked lines (Sim et al. 2005). If the central star is larger than the ZAMS ones assumed, then the rotation of the inner disc is much slower, reducing the rotational splitting of the disc wind. The slower, denser stellar wind from the poles may also help fill in the profile to improve agreement with observations. Such a wind could also help trap the Lyman continuum radiation before it can ionize the surroundings.

Whichever of the above mechanisms (quenching by infall, outflows or accretion swelling the star) prevents the formation of an H II region during the massive YSO phase, all are associated with ongoing accretion. This would tend to indicate that the onset of the H II region phase is not to do with the star evolving to a stage where it is hot enough to ionize hydrogen, but more to do with the cessation of accretion. If accretion swelling the star is the relevant mechanism, then the end of the high accretion rate phase will then allow the star to contract down onto the ZAMS. Then, the Lyman continuum output will increase and the H II region phase will begin.

When ionization of the surrounding material does start the ionization front will move faster in the low density directions. These are most likely to be the polar lobes due to the flattening effect of the centrifugal barrier during the infall phase and bipolar outflows during the YSO phase. In this scenario, one would expect the youngest, most compact H II regions to be bipolar. Within the class of so-called hyper-compact H II regions (HCHII, see Kurtz

2005) there are examples of bipolar objects. These also tend to display broad recombination lines ($\gtrsim 40$ km s^{-1}), which has been used as another criterion to set some of the hyper-compact objects apart as a new class (Jaffe & Martín-Pintado 1999; Sewilo et al 2004).

Figure 5 shows a plot of radio luminosity versus linewidth for UCHII regions, HCHII regions, young stellar wind sources, and jets (see Hoare et al. 2006 source of these data). Unlike in Hoare et al. (2006), here we have plotted the half-width-zero-intensity (HWZI) for the recombination lines rather than FWHM. For the UCHII and hyper-compact H II regions this does not make a great deal of difference, but some stellar wind sources appear to have a narrow optically thin component which dominates the FWHM whilst a broader optically thick component dominates the HWZI. In these cases, the HWZI gives a better indication of the kind of speeds the wind is attaining. However, it is a more difficult parameter to work with since where the line returns to the continuum is dependent on the signal-to- noise of the spectrum. We have also added the two massive YSO jet sources that have measured proper motions to this plot: GGD 27 and Cep A2. These have been corrected for the inclination where known and thus measures the full speed of the gas. It is unfortunate at present that the jet sources are not amenable to IR spectroscopy to probe their line profiles as they are deeply embedded and not directly visible.

In Figure 5, NGC 7538 IRS 1 appears to stand out relative to the UCHII regions and stellar wind/jet sources. There is even a question mark over this source since the Sewilo et al. (2004) measurement plotted is for a low frequency line that may be affected by pressure broadening and NLTE effects, whilst the mm-wave recombination lines show HWZIs nearer 80 km s^{-1} (Jaffe & Martín-Pintado 1999). Other designated hyper-compacts appear to form more of a continuum with the UCHII regions in this plot and in the size-linewidth relation (Hoare et al. 2006). There is no real break around 40-60 km s^{-1}. NGC 7538 IRS 1 is distinctive in that the bipolar lobes have velocity widths in excess of 100 km s^{-1}, rather like the massive YSO stellar wind sources, but it is much more luminous in the radio continuum. The brightness of the radio lobes has recently been shown to be getting significantly fainter on a timescale of 10 years (Franco-Hernández et al. 2004). The short lifetime for this phase would be consistent with the rarity of objects like NGC 7538 IRS 1.

Another object commonly placed in this category is MWC 349A. However, its linewidths are somewhat narrower and the mm-wave recombination lines are masing which is not seen in other hyper-compacts (Martín-Pintado et al. 1989). There are still question marks over the evolutionary stage of MWC 349A and even whether it is a young or post-main sequence object (e.g. Meyer et al. 2002).

It is tempting to associate these bipolar hyper-compacts with the turn-on phase of the H II regions as they expand down the previously excavated bipolar outflow cavities. For simple H II region expansion, the gas can reach speeds of a few times the ionized gas sound speed (10 km s^{-1}) when travelling down the steep density gradients expected away from the disk plane. Franco

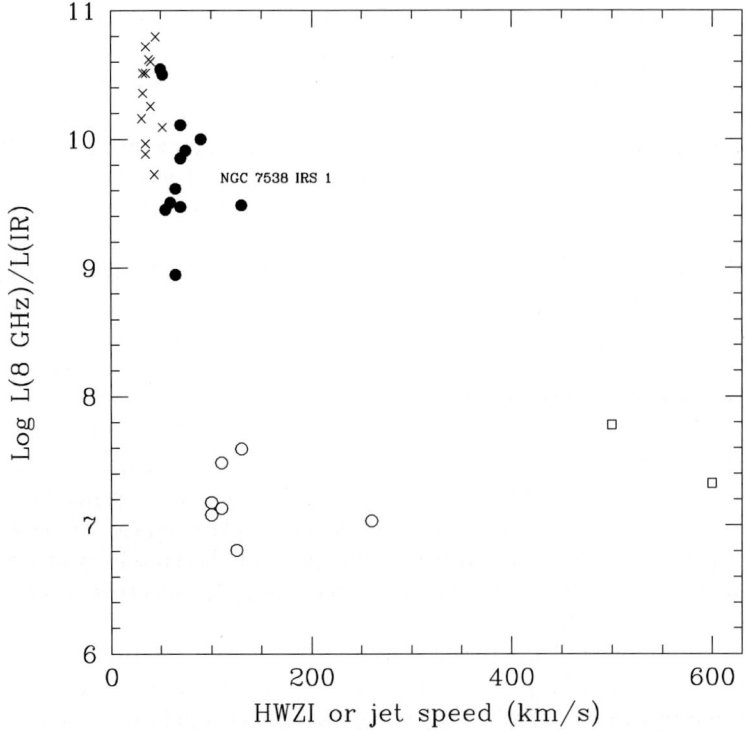

Fig. 5. Ratio of the radio luminosity at 8 GHz (W Hz^{-1}) to the bolometric luminosity from the IR (L$_\odot$) for UCHII regions (crosses), HCHII regions (solid circles) and massive young stellar object wind sources (open circles) and jets (open squares). Line widths are HWZI.

et al. (1989, 1990) found speeds of up to around 30 km s^{-1} for steep power-law density gradients and for some disk-like density gradients.

In particular, the bipolar H II regions resulting from disk-like density distributions have some interesting features that may be relevant to features appearing at the density distributions expected at the turn-on phase, and need further exploration. For instance, as pointed out by Franco et al. (1989), the dynamics of the H II regions depend on the details of the density distribution, and its appearance will be affected by recombination fronts that can result in bipolar molecular outflows, originally generated by the photoionized plasma. This type of phenomenon and the fragmentation created by instabilities in the ionization-shock fronts (Garcia-Segura and Franco 1996), along with the complex structure of combined ionization-photo-dissociation fronts (Diaz-Miller et al 1998), may provide some hints to the origin of the multiple flows with knotty structures discussed above.

Another model which is commonly invoked to explain bipolar hypercompacts, is the photo-evaporating disk model by Hollenbach et al. (1994)

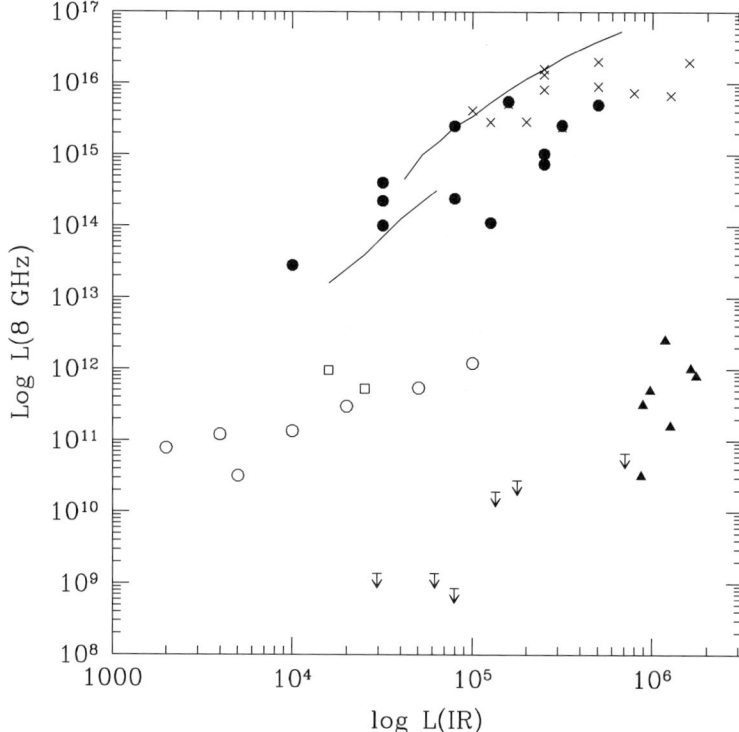

Fig. 6. Plot of the radio luminosity at 8 GHz (W Hz^{-1}) versus the bolometric luminosity from the IR (L$_\odot$) for UCHII regions (crosses), HCHII regions (solid circles) and massive YSO wind sources (open circles) and jets (open squares), evolved OB stars (filled triangles) and MS OB stars (upper limits). OB star data from thermal emitters in Bieging et al. (1989). Solid lines represent the expected optically thin radio luminosity at 8 GHz from MS stars at the given luminosity using O star parameters from Martins et al. (2005) and B star parameters from Smith et al. (2002).

mentioned above. This is a variation of the Franco et al (1989) model, in which the disk is now supposed to be formed from the gas accreted into the massive protostar. Lugo et al. (2005) show that such a picture can match the radio continuum spectrum for MWC 349A and NGC 7538 IRS 1. Such a thermal expansion model is capable of explaining the relatively low velocities in MWC 349A, but in no way can it account for the much faster flows in NGC 7538 IRS 1. MWC 349A does not appear to be deeply embedded in a molecular cloud and so a circumstellar disc origin is more likely. However, NGC 7538 IRS 1 is deeply embedded with evidence of large and small scale outflow activity (Davies et al. 1998; Kraus et al. 2006).

To explain the velocities of the order of a hundred km s^{-1} in NGC 7538 IRS 1 and its intermediate nature between wind/jet sources and UCHII regions, it is natural to think of some kind of transition object. One scenario is that a fast stellar wind, perhaps more like a main sequence radiatively driven wind travelling at around a thousand km s^{-1} is beginning to blow down the bipolar cavity and entrain material. As the star contracts onto the main sequence, as it must do in order to produce the Lyman continuum in the first place, such a wind will inevitably come with the increase in extreme UV radiation. It is certainly within the realms of possibility that such a wind will get mass-loaded with material from the walls of the bipolar cavity (see the chapter by Julian Pittard).

In this picture one would also expect to find much larger bipolar objects as the wind and ionization break out the axes. One such object is the bipolar H II region S106. The ionized lobes of this object have broad lines with HWZI=95 km s^{-1} (Solf & Carsenty 1982; Jaffe & Martín-Pintado 1999) and the full extent is about 0.7 pc. It has a limb-brightened appearance (Felli et al. 1984), which together with the presence of actual line-splitting indicates a swept-up bipolar shell structure. With his usual insight, Dyson (1983) developed a model for S106 whereby a star turns on in a plane-stratified medium where the H II region quickly breaks out in the polar direction and is followed by the stellar wind sweeping the nebula into a thin bi-cylindrical structure. At the ends of the nebula the shell is predicted to be Rayleigh-Taylor unstable and break up. Such a picture is consistent with the latest dramatic near-IR pictures of the nebula obtained with 8 m telescopes (Oasa et al. 2006). To get the right expansion velocity of the shell, Dyson used main sequence parameters for the stellar wind which are consistent with the overall luminosity and spectral type. However, the spectral characteristics of the exciting source are more like those of a massive YSO (Drew et al. 1993). It is also one of the equatorial wind sources (Hoare et al. 1994), although this does not preclude a main sequence wind blowing from the poles of the star.

Objects like S106 are rare. In fact, as an optically visible example it is currently unique. Dyson (1983) rightly pointed out that these objects would be rare with only a few in the whole galaxy. K3-50A (Turner & Welch 1984; De Pree et al. 1994) and W49N A2 (De Pree et al. 2004) appear to be similar objects, although they have strong ionized emission in the disc plane, whereas S106 has a lack of emission in the disc plane. NGC 6334 A also shows bipolar lobes with a significant velocity gradient, but this time breaking out from a more shell-like central source (De Pree et al. 1995). G5.89-0.39 may be another example of this structure (Acord et al. 1998). The dynamical ages of these objects also appear very short, again attesting to their rarity, but it also seems that the vast majority of ultra-compact H II regions do not have a bipolar morphology.

5 Ultra-compact H II Regions

A global review of the most important features for the evolution of HII regions, from the ultra-compact stages to the extended phases, is given by Dyson and Franco (2001). The usual definition of UCHII regions is that they have sizes smaller than 0.1 pc and electron densities above 10^4 cm^{-3}, and are located in the inner, high-pressure, parts of the parental molecular clouds (see Kurtz and Franco 2002, Kurtz et al. 2000 and Hoare et al. 2006 for reviews). As stated above, there is also an even more compact stage referred to as the hyper-compact phase.

About a thousand candidates have been identified, and they are classified into a variety of morphologies, ranging from spherical to irregular. Their internal density structure can be derived with the method of spectral index analysis, and power-law decreasing distributions with steep slopes (exponents below -2) have been obtained for some objects (Franco et al. 2000). The large number of objects at this UC stage may be due to two independent factors; one is that some of them have extended emission that was missed due to selection effects in the early studies (Kurtz et al. 1999; Kim and Koo 1996, 2001), and the other one is that they do not seem to be the short-lived phase predicted by simple models of expanding H II regions. The first factor simply implies that the actual number of UCHII regions has to be revised, and is probably smaller than initially stated. In addition, the possibility of having UCHII regions with a compact core and a large, more diffuse, envelope seems to be a logical consequence of the steep density gradients found by Franco et al. (2000). The second factor implies that the simple model is not directly applicable at this UC stage, and there may be several mechanisms that would confine them and delay their expansion. Indeed, there are a number of models proposed to explain the apparently slower growth rates, including the possibility of reaching pressure equilibrium inside cloud cores.

The cores of massive molecular clouds are clumpy, turbulent, and highly pressurised regions. Their total central pressures are above 10^6 dyn cm^{-2} (see Garcia-Segura and Franco 1996), and some H II regions can reach pressure equilibrium within the central uniform-density core. If this occurs, the resulting pressure-confined regions will have sizes of $2.9 \times 10^{-2} F_{48}^{1/3} T_4^{2/3} P_7^{-2/3}$ pc, and densities around $3.6 \times 10^4 P_7 T_4^{-1}$ cm^{-3}, where the UV photon flux is $F_{48} = F/10^{48}$ photons s^{-1}, $P_7 = P/10^7$ dyn cm^{-2}, and the photoionized plasma temperature is $T_4 = T/10^4$ K (the sizes are similar for the case of massive star wind-driven bubbles; see Garcia-Segura and Franco 1996, and Kurtz et al 2001). As indicated above, these sizes and densities are already typical of UCHII regions. If one includes dust absorption, the sizes are even smaller (see Diaz-Miller et al 1998 and Arthur et al. 2004), making this mechanism even more attractive as a means of confining the growth of UCHIIs. These, however, are only a few parts of a much more complicated story; the simplified pressure equilibrium scheme does not include the presence of clumps in highly dynamical cloud cores, nor the motion of stars inside the cloud. In addition, it does not explain the variety of morphologies that are already known for these objects.

Again, John's physical intuition was used to explore the effect of these cloud core clumps in the expansion of the ionization front. The net result is that, as the clumps are photo-evaporated inside the H II region, the growth of the ionized region stalls because the recombination rate increases as the mass and density of the photoionized gas is increased (Dyson et al. 1995; Williams et al 1996; Redman et al. 1996; Redman et al. 1998). Thus, the destruction of clumps is an effective mass-loading mechanism that reduces the expansion rate of the H II region, leading to a substantially longer lifetime of the ultra compact stage.

Another issue that has been considered in some detail is the origin of cometary UCHIIs. Their structure, with a bright head and a more diffuse extension on one side, resembles the morphological features of comets. Also, cometary UCHII regions are among the most common objects (they represent at least 20 % of them and maybe many more (Hoare et al. 2006)), and their shapes have been ascribed to either bow-shocks generated by the motion of the exciting star (see Van Buren and McCray 1988 and Van Buren and Mac Low 1992) or by the presence of a density gradient on one side of the parental cloud, that creates a 'champagne flow' (e.g., Tenorio-Tagle 1979). A recent study by Arthur and Hoare (2006), who modelled the structuring and emission of the objects created by both types of models, indicates that the most likely origin is due to density gradients with a champagne flow. In these numerical models the effect of stellar winds was included. They sweep the H II region into a thin shell, resulting in the limb-brightened appearance of many cometaries. Once again John had considered such a problem many years before when he examined how a stellar wind bubble expands in an H II region with a density gradient (Dyson 1977).

The motions of stars have a role in the structuring of both H II regions and clouds. The fact that a star moves back and forth from high to low densities, as it moves within the gravitational field of the cloud, leads to a wide variety of transient and complex structures being created (see Franco et al. 2006).

6 The Impact of Massive Stars

The birth of a massive star leads to a strong source of energy that excites, stirs and ionizes the parental molecular cloud. The UV photon flux results in the expansion of H II regions, while the strong winds create expanding bubbles that are internal to the main body of the H II region. In addition, the non-ionizing UV radiation is able to dissociate the molecular gas creating a layer of atomic hydrogen that envelops the H II region. Thus, several distinct regions form around the young massive star, creating a nested structure with a wind-driven bubble, a photoionized H II region, and a photo-dissociation region (PDR), which is also used to indicate photon dominated regions. The details, extent, and structure of these regions depend on the mass of the young star, the ambient density structure, and the age of the star. Regardless of these details, however, the combined action of radiation and winds evaporates the gas surrounding the new stars and finally destroys the molecular cloud, effectively shutting-off the star formation process. The evaporation of the cloud occurs via peripheral blisters, or champagne flows.

As discussed by Franco et al. (1994) and Diaz-Miller et al. (1998), then, the star-forming capacity of molecular clouds is limited by cloud destruction from massive stars. The limit on the number of massive stars is set by the overlap of internal H II and PDR regions. Thus, one of the main roles played by the energy injection from massive stars is to control star formation by regulating the number of stars formed within any star-forming cloud. This, in turn, implies that such a process also regulates the actual star-forming cycle of galactic systems.

After the parental cloud is destroyed, the expanding shells continue their evolution in the general interstellar medium, creating supershells. Most of the supernovae resulting from the newly formed stellar group actually explode inside these cavities. Thus, at the late stages of massive star lives, the impact is shaping and stirring the general ISM.

References

Acord, J.M., Churchwell, E., Wood, D.O.S. 1998 ApJ **495**, L107
Alvarez, C., Hoare, M.G., Glindemann, A., Richichi, A. 2004 A&A **427**, 505
Alvarez, C., Hoare, M.G. 2005 A&A **440**, 569
Arthur, S.J., Kurtz, S., Franco, J., Albarran, M. 2004 ApJ **608**, 282
Arthur, S.J., Hoare, M.G. 2006 ApJ in press
Banerjee, R., Pudritz, R.E. 2006 ApJ **641**, 949
Behrend, A., Maeder, A. 2001 A&A **373**, 190
Bernasconi, P.A., Maeder, A. 1996 A&A **307**, 829
Beuther, H., Schilke, P., Sridharan, T.K., Menten, K.M., Walmsley, C.M., Wyrowski, F. 2002a ApJ **383**, 892
Beuther, H., Schilke, P., Gueth, F., McCaughrean, M., Andersen, M., Sridharan, T.K., Menten, K.M. 2002b ApJ **387**, 931
Bieging, J.H., Abbott, D.C., Churchwell, E.B. 1989 ApJ **340**, 518
Bonnell, I.A., Bate, M.R., Zinnecker, H. 1998 MNRAS **298**, 93
Bunn, J.C., Hoare, M.G., Drew, J.E. 1995 MNRAS **272**, 346
Campbell, B., Persson, S.E., McGregor, P.J. 1986 ApJ **305**, 336
Churchwell, E.: The Origin of Stars and Planetary Systems, in NATO Science Series vol **540**, ed. by Lada, C. & N Kylafis, Dordrecht, The Netherlands 1999), pp515–552
Curiel, S. et al 2006 ApJ, **638**, 878
Davis, C.J., Moriarty-Schieven, G., Eislöffel, J., Hoare, M.G., Ray, T.P. 1998 AJ **115**, 1118
Davis, C.J., Varricatt, W.P., Todd, S.P., Ramsay Howat, S. K. 2004 A&A **425**, 981
Pree, C. De, Goss, W., Palmer, P., Rubin, R. 1994 ApJ **428**, 670

Pree, C. De, Rodríguez, L.F., Dickel, H.R., Goss, W. 1995 ApJ **447**, 220
Pree, C.G. De et al 2004 ApJ **600**, 286
Diaz-Miller, R., Franco, J., Shore, S.N. 1998 ApJ **501**, 192
Drew, J.E., Bunn, J.C., Hoare, M.G. 1993 MNRAS **265**, 12
Drew, J.E., Proga, D., Stone, J.M. 1998 MNRAS **296**, L6
Dyson, J.E. 1977 A&A **59**, 161
Dyson, J.E. 1983 A&A **124**, 77
Dyson, J.E., Williams, R.J.R., Redman, M.P. 1995 MNRAS **277**, 700
Dyson, J.E., Franco, J.: Encyclopedia of Astronomy and Astrophysics. (MacMillan, London 2001)
Felli, M., Massi, M., Churchwell, E. 1984 A&A **136**, 53
Franco, J., Carramiñana, A.: Interstellar Turbulence, (Cambridge Univ. Press, Cambridge 1999)
Franco, J., Tenorio-Tagle, G., Bodenheimer, P. 1989 RMexAA **18**, 65
Franco, J., Tenorio-Tagle, G., Bodenheimer, P. 1990 ApJ **349**, 126
Franco, J., Kurtz, S., Hofner, P., Testi, L., García-Segura, G., Martos, M. 2000 ApJ **542**, L143
Franco-Hernández, R., Rodríguez, L.F. 2004 ApJ **604**, L105
Garcia-Segura, G., Franco, J. 1996 ApJ **496**, 171
Gibb, A.G., Hoare, M.G., Little, L.T., Wright, M.C.H. 2003 MNRAS **339**, 1011
Hatchell, J., van der Tak F.F.S. 2003 A&A **409**, 589
Hoare, M.G., Drew, J.E., Muxlow, T.B., Davis, R.J. 1994 ApJ **421**, L51
Hoare, M.G., Muxlow, T.B.: Radio Emission from the Stars and the Sun. In ASP Conf. Ser., Vol. **93**, ed by Taylor, A.R., Paredes, J.M. (ASP, San Francisco 1996) pp 47–49
Hoare, M.G.: Hot Star Workshop III: The Earliest Stages of Massive Star Birth. In ASP Conf. Ser. Vol. **267**, ed by in Crowther, P.A. (ASP, San Francisco 2002) pp137–144
Hoare, M.G., Kurtz, S.E., Lizano, S., Keto, E., Hofner, P.: Protostars and Planets V, ed. by Reipurth, B., Jewitt, D., and Keil, K., (Tuscon: University of Arizona Press) in press
Höflich, P., Wehrse, R. 1987 A&A **185**, 107
Hollenbach, D., Johnstone, D., Lizano, S., Shu, F. 1994 ApJ **428**, 654
Indebetouw, R., et al 2005 ApJ **619**, 931
Jaffe, D., Martín-Pintado, J. 1999 ApJ **520**, 162
Jijina, J., Adams, F.C. 1996 ApJ **462**, 874
Kahn, F.D. 1974 A&A **37**, 149
Kippenhahn, R., Meyer-Hofmeister, E. 1977 A&A **54**, 539
Krumholz, M.R., Klein, R.I., McKee, C.F.: Massive Star Birth: A Crossroads of Astrophysics. In IAU Symp. **227**, ed by Cesaroni, R., Felli, M., Churchwell, E., Walmsley, M. (Cambridge University Press, Cambridge 2005a), pp 231–236
Krumholz, M.R., McKee, C.F., Klein, R.I. 2005b ApJ **618**, L33
Kim, K.-T., Koo, B.-C. 1996 JKAS **29**, 177
Kim, K.-T., Koo, B.-C. 2001 ApJ **549**, 979
Kraus, S. et al 2006 astro-ph/0604328
Kurtz, S.E., Franco, J. 2002 RMexAA **12**, 16
Kurtz, S.E., Cesaroni, R., Churchwell, E., Hofner, P., Walmsley, M.: Protostars and Planets IV, ed by Mannings, V. et al. (Univ of Arizona Press, Tucson 2000), pp 299–326

Kurtz, S.E., Watson, A.M., Hofner, P., Otte, B. 1999 ApJ **514**, 232
Kurtz, S.E.: Massive Star Birth: A Crossroads of Astrophysics. In IAU Symp. **227**, ed by Cesaroni, R., Felli, M., Churchwell, E., Walmsley, M. (Cambridge University Press, Cambridge 2005), pp 111–119
Kurtz, S., Churchwell, E., Wood, D.O.S. 1994 ApJS **91**, 659
Lada, C.J. 1985 ARA&A **23**, 267
Larson, R.B. & Starrfield, S. 1971 A&A **13**, 190
Larson, R.B. 1981 MNRAS **194**, 809
McKee, C.F., Tan, J.C. 2003 ApJ **585**, 850
Martí, J., Rodríguez, L.F., Reipurth, B. 1993 ApJ **374**, 169
Martí, J., Rodríguez, L.F., Reipurth, B. 1998 ApJ **502**, 337
Martí, J., Rodríguez, L.F., Torrelles, J.M. 1999 A&A **345**, L5
Martín-Pintado, J., Bachiller, R., Thum, C., Walmsley, M. 1989 A&A **215**, L13
Meyer, J.M., Nordsieck, K.H., Hoffman, J.L. 2002 AJ **123**, 1639
Mundy, L.G., Adelman, G.A. 1988 ApJ **329**, 907
Nakano, T. 1989 ApJ **345**, 464
Norberg, P., Maeder, A. 2000 A&A **359**, 1025
Palla, F., Stahler, S.W. 1992 ApJ **392**, 667
Prinja, R.K., Barlow, M.J., Howarth, I.D. 1990 ApJ **361**, 607
Noriega-Crespo, A. et al. 2004 ApJS **154**, 352
Oasa, Y. et al. 2006 AJ **131**, 1608
Patel, N.A. et al. 2005 Nature **437**, 109
Plume, R., Jaffe, D.T., Evans, N.J., Martín-Pintado, J., Gomez-Gonzalez, J. 1997 ApJ **476**, 730
Redman, M.P., Williams, R.J.R., Dyson, J.E. 1996 MNRAS **280**, 661
Redman, M.P., Williams, R.J.R., Dyson, J.E. 1998 MNRAS **298**, 33
Ridge, N., Moore, T.J.T. 2001 A&A **378**, 495
Rodríguez, L.F., Garay, G., Brooks, K.J., Mardones, D. 2005 ApJ **626**, 953
Schertl, D., Balega, Y., Hannemann, T., Hofmann, K.-H., Preibisch, Th., Weigelt, G. 2000 A&A **361**, L29
Schwartz, P.R. 1989 ApJ, **338**, L25
Shepherd, D.S., Kurtz, S.E. 1999 ApJ **523**, 690
Shu, F.H. 1977 ApJ **214**, 488
Sim, S.A., Drew, J.E., Long, K.S. 2005 MNRAS **363**, 615
Simon, M., Cassar, L. 1984 ApJ, **283**, 179
Solf, J., Carsenty, U. 1982 A&A **113**, 142
Tan, J.C., McKee, C.F.: Star Formation at High Angular Resolution. In IAU Symp. **221**, ed by Burton, M.G., Jayawardhana, R., Bourke, T. (http://www.phys.unsw.edu.au/iau221 2003)
Tenorio-Tagle, G. 1979 A&A **71**, 59
Terebey, S., Shu, F., Cassen, P. 1984 ApJ **286**, 529
Tofani, G., Felli, M., Taylor, G.B., Hunter, T.R. 1995 A&AS **112**, 299
Tomisaka, K. 2002 ApJ **575**, 306
Trinidad, M.A. et al 2003 ApJ **589**, 386
Turner, J.L., Welch, W.J. 1984 ApJ **287**, L81
Van Buren, D., McCray, R. 1988 ApJ **329**, L93
Van Buren, D., MacLow, M.-M. 1992 ApJ **394**, 534
Walmsley, C.M. 1995 Rev. Mex. Ast. Ap. Conf. Ser. **1**, 137
Williams, R.J.R., Dyson, J.E., Redman, M.P. 1996 MNRAS **280**, 667

Wood, D.O.S., Churchwell, E. 1989 ApJS **69**, 831
Wolfire, M.G., Cassinelli, J.P. 1987 ApJ **319**, 850
Wright, A.E., Barlow, M.J. 1975 MNRAS **170**, 41
Yorke, H.W., Sonnhalter, C. 2002 ApJ **569**, 846

Spectropolarimetry and the Study of Circumstellar Disks

R.D. Oudmaijer[1]

School of Physics and Astronomy, University of Leeds, LS2 9JT Leeds, U.K.
roud@ast.leeds.ac.uk

1 Introduction

The disk accretion paradigm has proven to be extremely successful as an explanation of the formation of low mass stars. Due to angular momentum considerations, it is now commonly accepted that during gravitational collapse a disk forms around a newly born star. The disk material accretes onto the star by moving along magnetic field lines and free-falling onto the stellar photosphere, giving rise to shock induced emission. This picture of magnetically controlled disk accretion has been confirmed for the low mass T Tauri stars, where disks have been found in high resolution images (Dutrey et al. 1998), while evidence for magnetic fields and magnetically controlled accretion via such disks has been often found (Bertout 1989, Johns-Krull et al. 1999). This has been reinforced by model simulations of the data (e.g. Muzerolle et al. 2001).

For the "intermediate-mass" (ranging from 2-8 solar masses) objects, the optically visible Herbig Ae/Be stars, the situation is less clear. Although a small number of Herbig Ae stars has now been found to exhibit a (weak) magnetic field (e.g. Wade et al. 2005), these more massive stars have radiative photospheres, and consequently weaker, if any, magnetic fields. Therefore, the main component for the T Tauri paradigm, magnetically controlled disk accretion, is less well established and other scenarios that can form higher mass stars need to be considered. In addition, observationally, the presence of accretion, and disk accretion in particular is much less settled for these objects than for their low mass counterparts. Spectroscopic evidence for at least some transient accretion has now been found (e.g. Mora et al. 2004), mostly towards the A-type objects. Evidence for the presence of extended, flattened disks has emerged recently (e.g. Mannings & Sargent 2000 in mm-CO observations; Fukagawa et al. 2003, Grady et al. 2001 in coronagraphic images). These large $\sim$ arcsec scale structures do not necessarily indicate the presence of accretion disks, which, inevitably, are to be found at much smaller radii.

The picture is even more obscure for the most massive stars, which we loosely define as having masses greater than about 8-10 $M_\odot$. Even a theoretical consensus has not emerged for these objects. The radiation pressure from a newly born, yet still accreting star, may blow away the infalling material, halting further growth (e.g. Yorke & Kruegel 1977; Adams 1993). The maximum possible mass that can be accumulated in simple accretion models is as low as $10M_\odot$ (Wolfire & Cassinelli 1987). Disks around massive stars may help the continued growth of these cores; material accreting via a disk captures less light than in the spherical case, while disks can withstand the radiation pressure much better due to their higher density (Norberg & Maeder 2000; Yorke & Sonnhalter 2002; see section 2 of the chapter by Melvin Hoare and Pepe Franco). Accretion disks may thus be able to supply the necessary material to allow stars to reach higher masses. As an alternative, stellar coalescence due to collisions of lower mass objects has been proposed as a way to form a massive star (Bonnell, Bate & Zinnecker 1998; Bally & Zinnecker 2005).

In addition to the theoretical difficulties, observations of massive young stellar objects (MYSO) are challenging. The steepness of the Initial Mass Function combined with their short lifetimes make them scarce. They are, therefore, found at distances that are on average much larger than those to well-studied lower mass stars, and this hampers studies aimed at probing the circumstellar material close to the objects. The situation is made worse as most young massive stars are deeply embedded in their natal clouds, in general preventing optical studies. Thus, while the presence of disks would go a long way towards explaining the formation of massive stars, data on disks of any size are sparse.

Indirect evidence for the presence of, large-scale, disks comes from high resolution radio observations of the ionized circumstellar material. Hoare et al. (1994, 2002) have seen evidence both for flattened disk-like winds and highly collimated jets, suggesting the presence of inner disks feeding both processes (see section 3 of the chapter by Melvin and Pepe). Among the highest resolution studies of MYSOs we find the exceptional data for the 8-$10M_\odot$ object G192.16-3.82. Shepherd et al. (2001) obtained 30-40 milli-arcsec resolution data at 7 mm with the VLA for their single target. Based on the similarities of the disk found at 100 AU with those around lower mass objects, they assume the disk to be an accretion disk. Only recently have examples at similar scales been published (also see section 2 of the chapter by Melvin and Pepe). Using the sub-millimetre array, Patel et al. (2005) observed Cepheus A HW2, a $15M_\odot$ pre-main sequence star and found a disk or a flattened geometry with a size of order 330 AU. Jiang et al. (2005) observed the Becklin-Neugebauer object, which has an estimated mass of $7M_\odot$. They revealed a disk of a few hundred AU with their adaptive optics assisted polarization images. These studies are representative of the problems faced in the study of the formation of young massive stars. Firstly, the state-of-the-art-studies are single-object studies and secondly, although disks with sizes comparable to the Solar system are observed, it is extremely difficult to convincingly demonstrate the presence of accretion. The latter can only be studied when we can trace structures much closer to the star.

One such technique that is capable of probing the disks and immediate circumstellar environment on very small scales, of order stellar radii, is spectropolarimetry. Indeed, the only evidence for disks around intermediate-mass pre-main sequence stars comes from spectropolarimetry across the optical Hα emission (Oudmaijer & Drew 1999; Vink et al. 2002, 2005a). In the case of HD 87643, Hα could be kinematically resolved and by comparison with basic models (Wood et al. 1993) it was possible to show that the scattering electrons are located in an expanding rotating disk (Oudmaijer et al. 1998). But before discussing such results, let us first consider the basics of spectropolarimetry below.

2 Spectropolarimetry as a Technique

The principle of the method is rather simple: it uses the fact that free electrons in an extended ionized region scatter the continuum radiation from the central star and polarize it. If the projected distribution of the electrons on the sky is circular, for example as when a disk is seen face-on, or when the material is distributed spherically symmetrically, all polarization vectors cancel out and no polarization is observed. If the geometry is not circular as in the case of an inclined disk for example, a net polarization of the stellar continuum will be observed. The polarization due to electron scattering originates from the region closest the star where the electron densities, and thus optical depths, are the largest. It is typically sensitive to scales of a few stellar radii and results in polarizations of order 1-2% (Cassinelli et al. 1987). By measuring the polarization due to ionized material, the method provides a means to detect disks that otherwise cannot be found. An example of the broadband polarization behaviour around a star that is known to have an inclined disk is shown in Fig. 1. The upper panel shows the flux emanating from this system. The near-infrared excess due to bound-free and free-free emission can be clearly seen. The bottom panel shows the low resolution spectropolarimetry. The model fits (solid lines - computed up to $\sim 1 \mu m$) reproduce the data very well. As electron scattering is grey, it is not intuitive that the observed polarization is not constant, but closely follows the Balmer and Paschen jumps instead. This is due to the increased hydrogen opacity shortward of the respective jumps, here the electron-scattered photons will mostly be absorbed, reducing the observed polarization, whereas the lower hydrogen opacity on the red end allows more electron-scattered photons to escape (see e.g. Wood, Bjorkman & Bjorkman 1997). The magnitude of the jump depends on many parameters, as for example the inclination of the disk.

Observing polarization on its own towards an object does not necessarily imply the presence of an aspherical small scale structure however. Aligned and elongated interstellar dust grains can selectively absorb background light resulting in a net polarization (Serkowski, 1962), while circumstellar dust in non-circular geometries at larger distances from the star can also produce net polarization, in much the same way as the free scattering electrons do. As the observed polarization can be a combination of electron-scattering, circumstellar dust scattering and interstellar polarization, the nature of the polarization can be hard to disentangle. However, the latter mechanisms have a different wavelength dependence of the polarization than the fairly flat polarization due to electron scattering, and very broad wavelength coverage observations may help in assessing the polarizing agent (e.g. Quirrenbach et al. 1997). Yet, the interstellar polarization and circumstellar dust scattering can contribute the majority to the observed polarization of an object. Unless the star is closeby and does not have much circumstellar dust, as is the case for the bright Be stars, the broad band behaviour is less conclusive than we might hope.

Here is where line spectropolarimetry comes in. It exploits the fact that the hydrogen recombination line emission arising from within the ionized material will be polarized much less efficiently than the photospheric continuum. There are two reasons for this. Firstly, the line emission will encounter less electrons as it travels a shorter distance through the disk. Secondly, the lines can form over a larger volume than the electron scattering region, which is very much confined to the inner, denser, regions. This implies that emission lines will be less polarized than the continuum or even unpolarized altogether. Such a "line-effect" will be absent for the interstellar polarization as dust grains have a broad wavelength dependence, while circumstellar dust typically resides at much larger distances than the hydrogen recombination line forming region and consequently both star light and line emission will be scattered equally.

Therefore, a change in the polarization spectrum across an emission line such as Hα immediately indicates the presence of a flattened structure close to the star. This technique can, thus, reveal the presence of aspherical structures at scales that can not even be imaged with the most sensitive telescopes. Indeed, it is distance independent, so potentially stars in other galaxies can be studied in much the same way as closeby Be stars.

2.1 An Example : the Be Star ζ Tau

Spectropolarimetry was already explored in the seventies by Poeckert & Marlborough (1976). They used narrow-band polarimetry to prove that Be stars are surrounded by ionized disks. It was only many years later that this result could be directly confirmed with high spatial resolution radio and optical interferometry (Dougherty & Taylor 1992; Quirrenbach et al. 1994, respectively). With the advent of CCD detectors and the installation of stable and efficient polarization optics, it has become possible to make routine spectropolarimetric observations at medium to high resolution.

Data for the Hα line of a typical Be star, ζ Tau, taken in January 1996 are presented in Fig. 2, where the Stokes I (direct intensity) vector is plotted in the bottom panel, and the polarisation percentage and polarisation angle

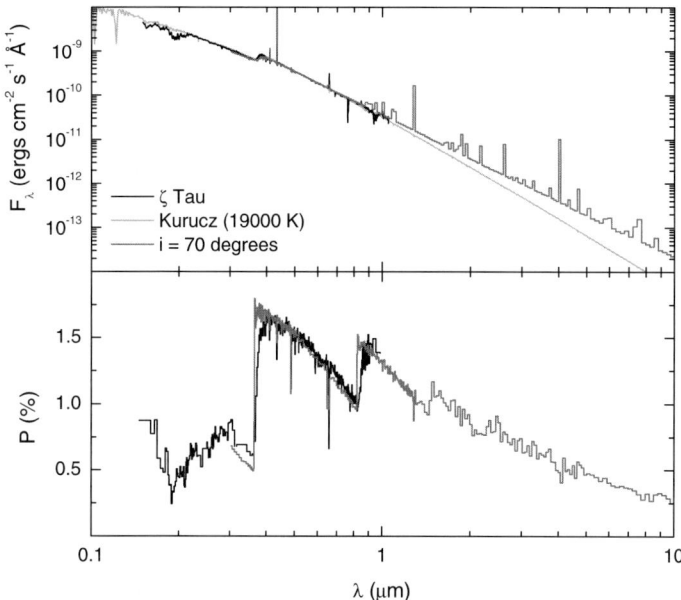

Fig. 1. The broadband polarization due to electron scattering. Data of the Be star ζ Tauri are shown. The upper panel shows the spectral energy distribution and the lower panel the polarization. The dotted lines represent the observations, while the solid line (up to ∼ 1μm) is a model fit to the data. Figure kindly provided by J. Bjorkman.

(PA) are displayed in the middle and upper panels respectively. These data are adaptively binned such that individual bins correspond to 0.1% in polarization. The line is asymetric. If the line is doubly peaked, the blue component is stronger than the red component. The continuum polarisation of 1.3% and the PA of ∼30° follow the trend of increasing polarization observed over the past years as noted by McDavid (1999). The polarisation across the Hα line shows a marked drop with respect to the continuum and is a clear example of the "line-effect" revealing the presence of an aspherical structure, i.e. the disk.

The right hand panel shows the QU polarization vectors plotted against each other, and the line excursion is clearly present as well. The cluster of points in the upper right hand corner of the graph represent the continuum polarization ($p^2 = Q^2 + U^2$) and the excursion to the bottom left shows the polarization when the wavelength moves to and from the line center. The amplitude of the excursion, about 1%, is a measure of the polarization due to electron scattering and in good agreement with theoretical expectations. The individual datapoints are distributed around the linear excursion, and its pattern is not well understood. However, the scatter appears to be of the

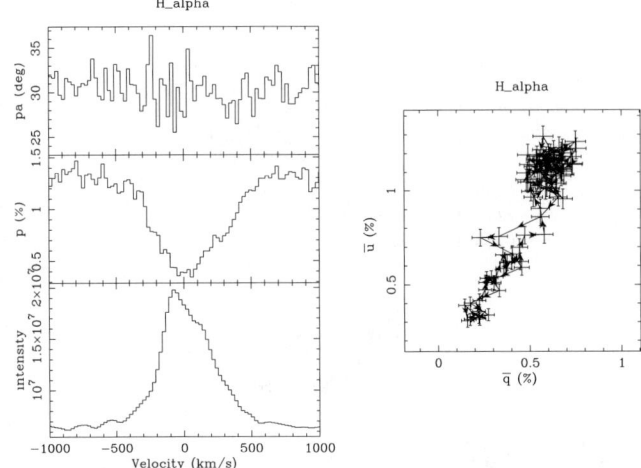

Fig. 2. Hα spectropolarimetry of ζ Tau. On the left hand side we show the so-called "triplot". This presents the intensity spectrum (bottom), and the polarization percentage (middle) and polarization angle (top) rebinned to a given error. The right hand graphs display the polarization in QU space at the same binning. The plots are from Mottram, Vink, Patel, & Oudmaijer (in preparation).

order of the 0.1% error binning applied, and care should be taken when trying to interpret such changes.

The data also allow us to directly determine the *intrinsic* polarization angle (PA) from the excursion across the line profile observed in the QU diagram. The PA can be written as $\Theta = \frac{1}{2} \times \arctan(\Delta U/\Delta Q)$. This results in 32° with an estimated experimental error of around 4°. In the case of optically thin scattering, the situation encountered most often, the polarization angle will be perpendicular to the disk structure itself, and the intrinsic polarization angle implies a disk orientation of -58±4° on the sky.

For comparison, we show Quirrenbach et al.'s (1994) high resolution, reconstructed, image in the Hα line of ζ Tau in Fig. 3. With a V band magnitude of 3 and a distance of 128 pc (as derived from the Hipparcos parallax of 7.8 milli-arcsec) this is one of the brightest and nearest objects in the sky. Yet, the extent of the disk is only a few milli-arcseconds. The second contour, counted from the centre, traces 50% of the peak light, and implies a full-width-at-half-maximum (FWHM) of 5 milli-arcsec. Quirrenbach et al. (1994, 1997) measured the disk's position angle to $-58 \pm 4°$. This value is in very good agreement with the position angle derived from the spectropolarimetry and presents a compelling validation of spectropolarimetry as a means to reveal small scale asphericities.

Let us now extend the wavelength coverage by a factor of a few. In Fig. 4 three other hydrogen recombination lines in the spectrum of ζ Tau are shown, from top to bottom ordered in wavelength, the optical Hγ, Hβ lines and the

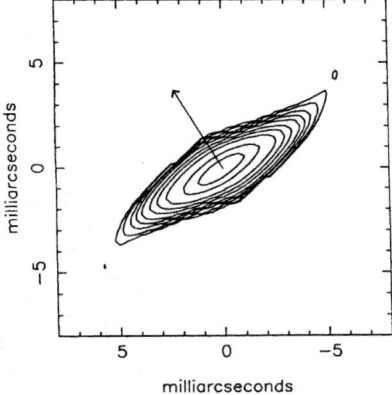

Fig. 3. A maximum entropy reconstruction of ζ Tau in the Hα emission line. The arrow indicates the position angle of the linear polarization. From Quirrenbach et al. (1994).

near-infrared 1.28μm Paβ line. The optical lines were observed 8 years after the Hα line shown in Fig. 2 (September 2004, Mottram, Vink, Patel, & Oudmaijer 2006, in preparation). The Paβ line data were obtained in September 1999 (taken from Oudmaijer, Drew and Vink 2005). As expected from their lower transition probabilities, the lines are much weaker than Hα. The optical lines show stronger blue peaks, like Hα. Paβ data were taken at a different epoch and show both peaks at similar strength. Kaye & Gies (1997) observed the same for Hα and clearly, the profiles are variable. Such blue peak to red peak variability of the line-profile has most often been interpreted as being due to the presence of one-armed density waves within the circumstellar disk (see e.g. Porter & Rivinius 2003). Note also that the optical, simultaneous, continuum polarization is fairly flat, as expected from electron-scattering (see Fig. 1)

The intrinsic angle derived from the QU graphs is 29°, 29°, and 32° for the lines respectively. Both Hγ and Hβ have consistent values, but the Hα and Paβ data, taken a few years earlier, are different by a few degrees. This difference could be explained by the generous errorbars, but may also be a confirmation of the one-sided arm hypothesis of the disks. A axi-symmetric disk can display polarization changes, due for example to changes in its density, but it will normally be expected to show the same orientation. On the other hand, orbiting density enhancements could induce variations in the orientation if they are not always located in the same plane.

The weakest line in the graph deserves particular notice. The Hγ emission does not even exceed the continuum level. Yet, a line-effect is clearly visible in the data. In addition, the position angle derived from the excursion in the QU plot is consistent with the other, stronger, lines. It is because we

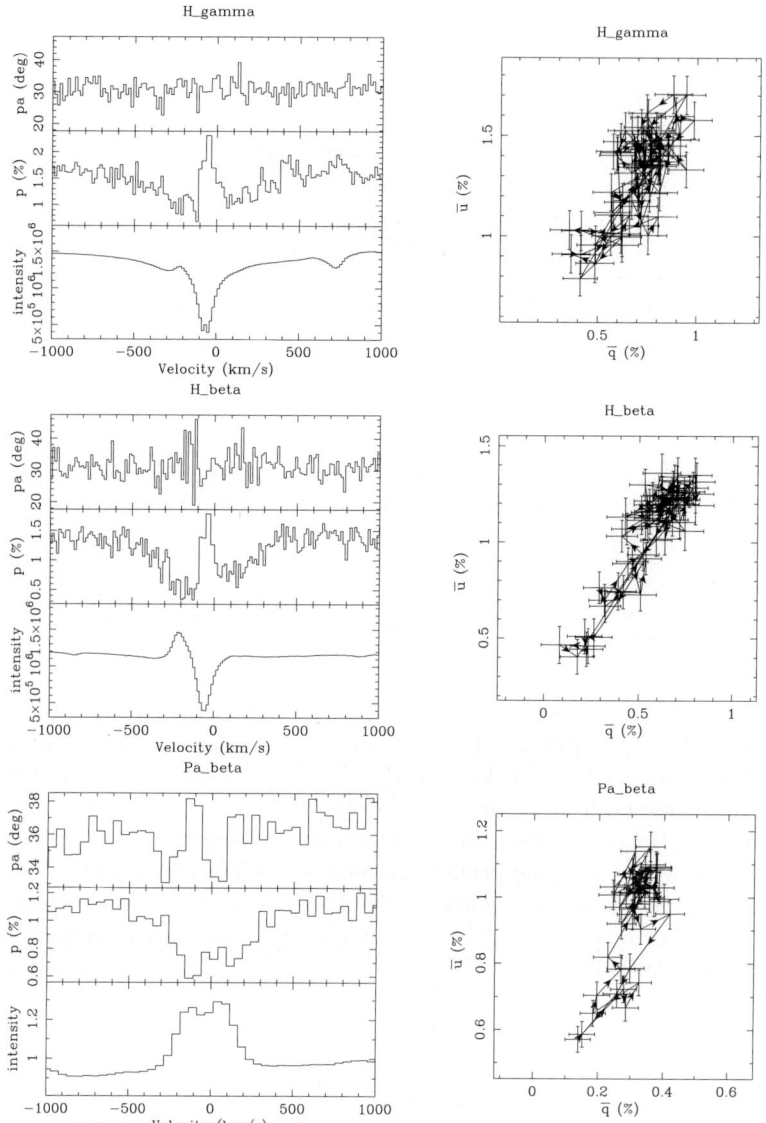

Fig. 4. Medium resolution spectropolarimetry of the Be star ζ Tau. For each of the hydrogen recombination lines, Hγ, Hβ, and Paβ, two figures are shown as in the previous figure. Never before have data such as these, covering so many lines ranging from 4340Å to the near-infrared (1.28 μm) been published, as no comparable data set is in existence. The optical data are from Mottram, Vink, Patel, Oudmaijer & Davies (in preparation), while the near-infrared data are from Oudmaijer et al. (2005).

have knowledge of the other lines in the spectrum of ζ Tau that we can be confident that a true line-effect is present. The, certainly at first sight puzzling, fact that we can see this effect at all is because the disk emission is still substantial. The photospheric hydrogen lines of a B2IV star, such as ζ Tau, show strongly in absorption. Any emission has to fill in the strong absorption line before exceeding the continuum. This emission is unpolarized and significant compared to the underlying remaining photosheric radiation, and gives rise to a line-effect. It is a testament to the spectropolarimetric method that such a weak line still allows a disk to be revealed. This is the first published instance of a "line-effect" in spectropolarimetry, without the presence of an emission line!

In summary, line spectropolarimetry is a powerful method to reveal the presence of circumstellar disks at very small scales. In the following, we give a brief overview of the most recent developments in the field of spectropolarimetry. Inevitably, the topics will be slightly biased towards the interests of the author, and will focus mainly on pre-main sequence stars, and a brief excursion to clumpier material will be presented later.

3 Application to Pre-Main Sequence Stars

We now move towards the pre-main sequence stars. As discussed earlier, one of the major issues in the field of star formation is whether these objects accrete material via a disk or not. Establishing the presence of small scale disks will substantially contribute to settling this issue; to have disk accretion, we first need disks reaching into, or close to, the stellar photosphere. On the longer term, follow-on models have the potential obtain physical parameters for such disks.

Linear spectropolarimetry requires comparatively high spectral resolution to properly sample the emission line profiles and very high signal-to-noise ratios. The binning in the previous figures was done to an accuracy of 0.1% in polarization, which corresponds to measuring the spectrum to a precision of one-thousandth of the total collected light. This means that signal-to-noise ratios (SNR) of order 1000 are needed. If we then also add the requirement of moderately high spectral resolution, it may be clear that we are restricted to observing optically bright objects. In the 4m telescope era (AAT, WHT), objects brighter than about $V = 10$ are routinely observed, objects in the 11-12 magnitude range have been observed, but these become very challenging.

Such bright limits prevent us from observing the most massive, heavily embedded, optically obscured pre-main sequence stars. Having said that, an excellent starting sample to address the issue are the Herbig Ae/Be stars. These objects are intermediate between the lower mass T Tauri stars and the high mass stars. They therefore provide a continuous coverage of the mass spectrum and will allow us to mark the switch from magnetically controlled accretion to other mechanisms. Last but not least, the most massive objects among the Herbig Be stars (>8-10$M_\odot$) are already direct examples of objects which some believe should not have formed via spherical accretion alone.

3.1 On the Presence of the Effect

Although line spectropolarimetry had been performed for a selected number of evolved stars such as AG Car (Schulte-Ladbeck et al. 1994 and references therein), in the early nineties, data for young stars were sparse. The best data are arguably those of Schulte-Ladbeck et al. (1992) of the Herbig Ae/Be star HD 45677. Their emphasis was on the broadband behaviour of the spectropolarimetric data, however, and the spectral resolution was not sufficient to detect a line-effect in the Hα emission line. The first medium resolution (~ 100 km s^{-1}) data were presented by Oudmaijer & Drew (1999) and Vink et al. (2002) who observed a significant sub-sample from the Thé et al. (1994) catalogue of Herbig Ae/Be objects. Some examples of line-effects are shown in Fig. 5, where a T Tauri star from Vink et al. (2005a) is also plotted for comparison. Whenever data on larger scale disks are available, often obtained after the spectropolarimetry, the intrinsic polarization angles are perpendicular to the position angles of the imaged disks, as expected.

To within the sensitivity, more than half of the two dozen objects surveyed show a line effect (16 out of 23 objects). As the systems are oriented randomly in the line of sight it is inevitable that some objects are face-on or close to face-on. A non-detection does not, therefore, necessarily imply the absence of a disk, because a face-on disk would appear circular on the sky, and not produce a line-effect. In the slightly inclined case, only very small effects are present and these may be hard to pick up at the 0.1% sensitivity level. Taking this into account, the high detection rate of a line-effect (70±17%) strongly suggests that all systems are surrounded by disks. This survey therefore provides evidence that the disk accretion scenario is a strong contender to explain the formation of massive stars.

It could be argued that the mere presence of a disk alone cannot be regarded as the smoking gun for disk accretion scenarios. This objection is equally valid for the disks noted in the spectropolarimetry as it is for the high resolution imaging studies by (Shepherd et al. 2001). Recent studies in binary formation may help in this respect. Studies of multiplicity amongst Herbig Ae/Be stars have retrieved large binary fractions up to 70% (Baines et al. 2006, Leinert et al. 1997). As reviewed by Clarke (2001), the formation of a binary system is due to the break-up of the pre-natal cloud into two or more fragments, or due to the capture of a lower mass star. These two competing models predict a different alignment of the binary systems and the disks around the stars. Capture models result in randomly oriented disks with respect to the binary position angle (e.g. Bally & Zinnecker 2005), while fragmentation models predict co-planar disks around the stars (see also Wolf et al. 2001 who find this for T Tauri stars). Baines et al. (2006) compare the orientations of the binary Herbig Ae/Be at their disposal with the intrinsic polarization angle if present in the literature. Five of the six objects that have data for both, have intrinsic position angles perpendicular to within 25° from the binary position angle. From a statistical analysis, they reject the possibility that the sample is drawn from a population of randomly aligned disk-binary systems at the 98.2% level. Instead, as the intrinsic polarization angle is perpendicular to the disk orientation, this result indicates that the circumprimary disks and the much larger binaries are well aligned.

This provides strong evidence in favour of the fragmentation scenario for the formation of binary systems as it predicts aligned disks, and argues against the stellar capture scenario or the stellar merger theory for the more massive stars.

3.2 Resolved Emission Lines

As discussed above, depolarization occuring across the emission line is usually attributed to the electrons scattering the stellar photospheric emission, and is not normally considered a result of photospheric processes.

However, for a large fraction of the stars, more than just a simple depolarization signature is seen, and even instances of enhanced polarization are observed. This occurs more often in the cooler Herbig Ae stars than in the Herbig Be stars. Intrigued by this, Vink et al. (2003, 2005a) observed a sample of T Tauri stars to investigate whether the line polarization properties depend on stellar mass, and if so, at what spectral type there may be a change in properties, perhaps sign posting a different formation mechanism. Fig. 5 shows a representative sequence with data from a Herbig Be star, a Herbig Ae star and a T Tauri star. The earliest type star shows a clear depolarization. The T Tauri star has an additional polarization across the line, and despite many attempts to correct for intervening dust polarization, the enhanced polarization across the line remains. The same applies to the Herbig Ae object. Vink et al. (2005a) quantified these different line profiles using a measure of the width of the polarization feature across the line. There appears to be a trend in that the "shape" of the polarization across the line in the Herbig Be stars is broader than that of the T Tauri stars, with the Herbig Ae stars intermediate between the two, as is also visible in the Figure. The QU graphs also show the marked difference, whereas the Herbig Be object displays a linear excursion over Hα, the later type objects can best be described with a "loop" (Vink et al. 2005c).

The lack of a true line depolarization in T Tauri stars can be explained by the fact that their disks are magnetically truncated and have an inner hole. As most electron-scattering occurs in the inner regions of the disk, the polarization of the photosphere of a T Tauri star due to electron-scattering could be very low. As a result, no true line effect would be expected in in the spectropolarimetry of T Tauri stars. The enhanced polarization can be explained as being due to a compact source of emission that scatters off a rotating disk-type structure. The best candidates for the compact sources

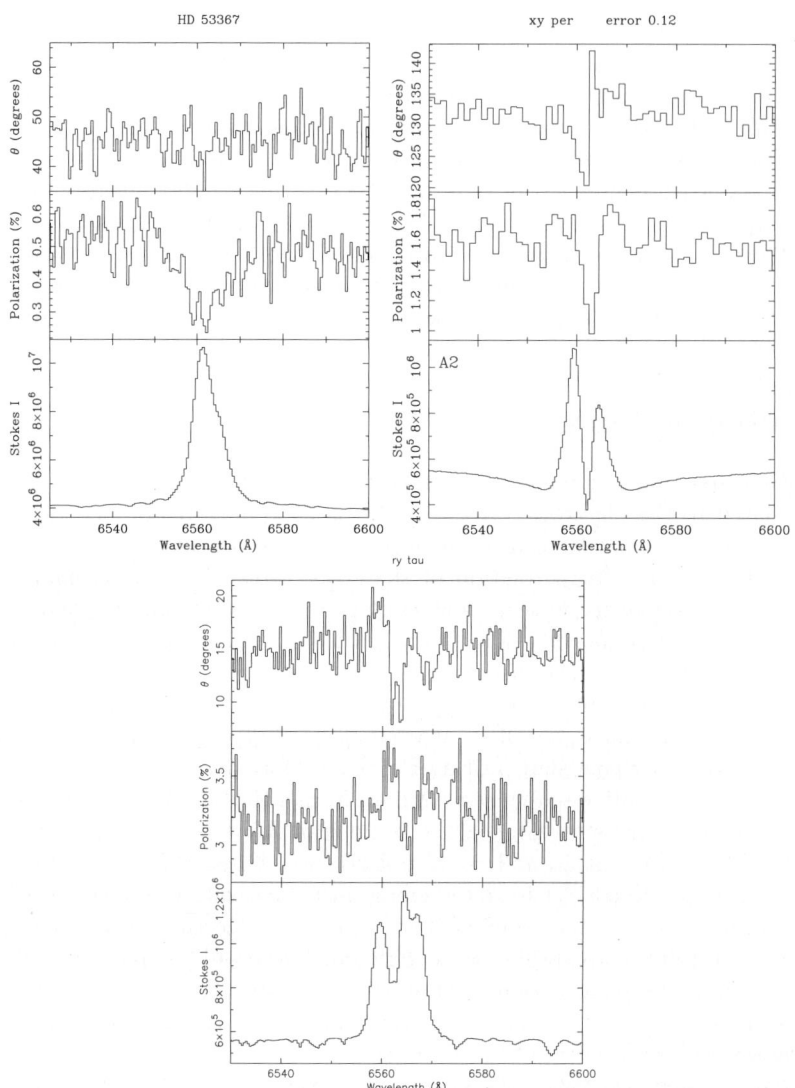

Fig. 5. Spectropolarimetry of a Herbig Be object (top left), a Herbig Ae star (top right) and a T Tauri star (bottom). As discussed in the text, there is a large resemblance in the spectropolarimetric properties between the Herbig Ae and T Tauri stars. Adapted from Vink et al. (2005b).

will be the accretion hot spots where the accretion flow free falls onto the stellar surface.

The similarity in the spectropolarimetric behaviour of the convective T Tauri stars and their hotter Herbig Ae counterparts led Vink et al. (2003) to suggest that magnetically controlled accretion plays a role in stars more

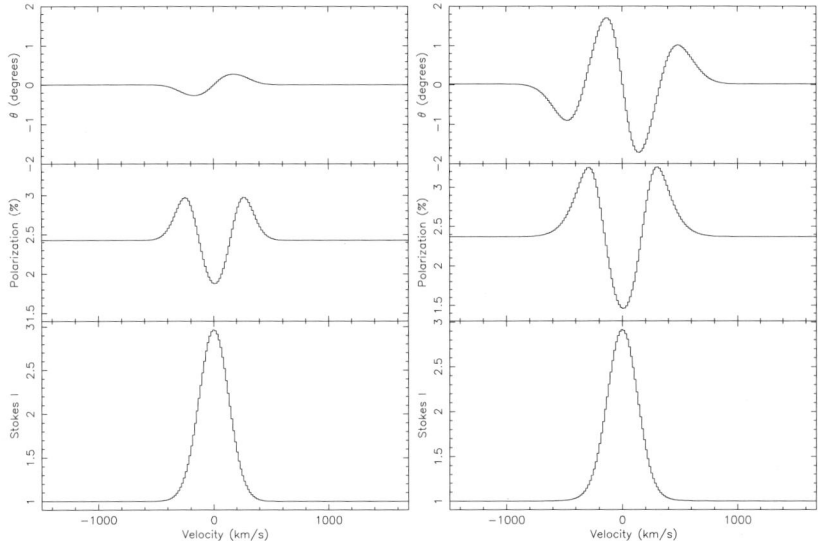

Fig. 6. Model computations of the line polarization in the case of a finite star and a disk with an inner hole (left) and without an inner hole (right). Note the double flip in rotation revealing that the disks reaches into the stellar photosphere. From Vink et al. (2005b).

massive than T Tauri stars. In fact, this was the first indication that the change from magnetically controlled accretion to other processes occurs at higher temperatures than previously thought. Not much later, Hubrig et al. (2004, 2006) and Wade et al. (2005) detected, for the first time, magnetic fields in some Herbig Ae stars, supporting this idea.

3.3 Modelling: Towards Physical Parameters of the Disks

So far, we have only discussed the qualitative, observational, result that disks are present around the target objects. In parallel with the improved instrumentation, theoretical models have matured as well. The ultimate aim of such models is to constrain the astrophysical parameters of the disks giving feedback to the models of the disks' formation. In order to reproduce the observed polarization, line emission and the spectropolarimetric line profiles, one needs full-blown radiative transfer models of such disks that take the dynamics of the system into account as well. At present, this is not readily possible, and currently the problem is approached from several directions. Broadly speaking there are two families of models being developed to interpret spectropolarimetric data.

The first line of attack is based on the line-profiles themselves. The resulting polarization as a function of wavelength gives valuable kinematical information on the velocity field of the scattering electrons. Vink et al. (2005c), using the Monte Carlo code of Harries (2000) and building on the analytical work of Wood et al. (1993), considered the scattering of radiation from a central source by rotating disks. By being scattered off the electrons within the disks the line emission will not only be polarized, but will also carry the Doppler information with them. By making a simple comparison with the Wood et al. (1993) models, Oudmaijer et al. (1998) were able to infer from their observed line profiles that the scattering material is located in an expanding, rotating disk. A remarkable new result from the Vink et al. (2005c) paper is that it proves possible to check whether the disk reaches into the (finite) star or whether an inner hole is present, as expected for the magnetically truncated disks in T Tauri stars. An example of their model results is shown in Fig. 6. The left hand panel shows the polarization profiles for a disk with a hole, a rotation is present in the polarization angle. The right hand panel displays the case for the disk reaching into the star. The polarization itself has qualitatively the same shape, but due to light being eclipsed by the stellar disk, there occurs a double rotation in the polarization angle. This appears to be a robust result, and will allow us to pinpoint the nature of the disks, well before imaging data can resolve them. In fact, when the inclination of the (larger scale) disk is known, it proves possible to estimate the size of the inner hole. Vink et al. (2005a) apply this new technique to a sample of T Tau stars and find that the inner hole of the disk around SU Aur is larger than 3 stellar radii. This is consistent with the interferometric findings of Akeson et al. (2005) who fit ring and disk models to their visibility data of the object, and for the ring model derive an inner radius of 0.18 AU.

This approach is particularly powerful for the interpretation of line profiles from compact emission sources, such as, for example, the accretion hot spots from T Tauri stars. The natural next step is to incorporate radiative transfer in the models to simulate not only the scattering but the line emission itself as well.

This method is perfectly complemented by the radiative transfer models of circumstellar, dense, ionized disks. These models are becoming increasingly increasingly realistic (e.g. Carciofi & Bjorkman 2006). Such 3-D Monte Carlo models not only solve the radiative transfer but, by following the energy packets as they are created and scattered, also compute the observed polarization (see Fig. 1, cf. Wood et al. 1997: note that the line effect is readily visible). Such models, applied to the broad wavelength spectropolarimetry have already allowed the determination of the disks' opening angles (2.5° in the case of ζ Tau), quantities which are impossible to measure by any other observational means. The natural next step is to implement the calculation of line profiles in codes, and it will be only a matter of time that the exciting prospect of full models of line spectropolarimetry is reality.

4 Variability studies

4.1 UXOR Variability

Variability in the observed polarization is an often overlooked property that reveals an object is intrinsically polarized and therefore surrounded by aspherical material. Because interstellar polarization is not expected or observed to change on short timescales, any variability immediately points to either circumstellar dust or electrons being the polarization agents.

A sub-sample of pre-main sequence stars displays a special type of photo-polarimetric variability, commonly referred to as the 'UX Ori' phenomenon (e.g. Grinin et al. 1994; Oudmaijer et al. 2001). From long term monitoring of a small number of young stars, Grinin et al. (1994) identified a group of objects that are photo-polarimetrically variable. This group of stars shows increased polarization when the optical light of the stars is faint. Crucially, the objects become redder with decreasing magnitude until a visual minimum is reached. In extremely deep minima there is a colour reversal, with the observed colours becoming bluer again. Named after their proto-type, UX Orionis, these stars are commonly referred to as UXORs. UX Ori itself is a well-known Herbig Ae/Be star, and indeed many UXORs fall in the intermediate-mass Herbig Ae/Be category.

The main explanation of this phenomenon concerns the existence of dust clumps located in a disk-like configuration rotating around a star. When the dusty clumps are not in our line of sight, the star will be observed at maximum light, with only a slight contribution of radiation scattered off the dusty disk. If the dust intersects the line of sight, light from the star will be absorbed, and the relative contribution of the scattered light to the total light increases, increasing the observed polarization. The fact that the reddening of the star coincides with the faintening, leaves little doubt that dust absorption indeed plays the main role in the process. In cases of extremely deep minima, the light from the star is blocked almost entirely, resulting in a 'blueing' of the energy distribution, as now mostly scattered light dominates the observed light. Depending on the distribution of dust-clouds, the light can be more or less absorbed during a period of photo-polarimetric monitoring. A direct observational consequence is that any variations in QU space predominately occur along a straight line with a slope perpendicular to the orientation of the disk, and is observed.

A more refined explanation of the UXOR behaviour is provided by Dullemond et al. (2003). They suggested that the dusty disks around UXORs are thin and that the inner rims of the disks shadow the outer parts. The inner rim of the disks is heated by the stellar radiation field and is "puffed up". Dullemond et al. discussed and modelled the case where only slight hydrodynamical perturbations can result in an uneven inner rim, and such fluctuations in the density can occasionally obscure the stellar light from view, in exactly the same manner as the qualitative clumps discussed previously in the literature.

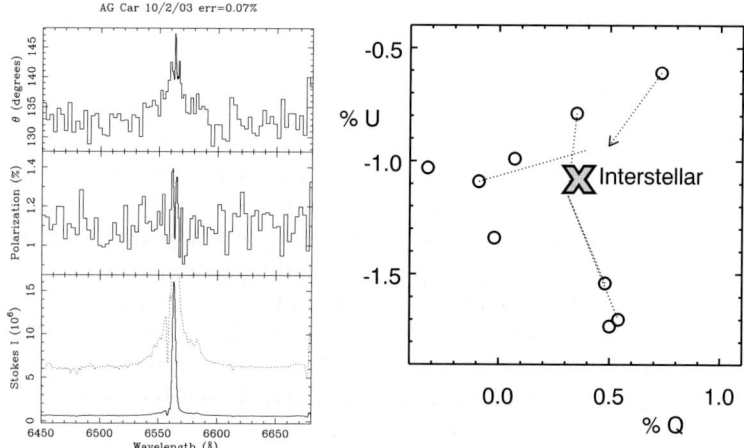

Fig. 7. The polarization spectrum across Hα of the Luminous Blue Variable AG Car. The data are represented as in the previous figures. The right most panel shows the QU graph where existing data are plotted. The big cross in the center denotes the polarization measured in the line centre, which is fairly constant over time, while the other points are the continuum polarization measurements. It is clear that the object is variable. Figure kindly provided by B. Davies.

4.2 Going Clumpy : Luminous Blue Variables

Adding complications to simple observational pictures has always been a habit of theoreticians. Introducing clumps in stellar winds previously assumed to be smooth is but an example. I first met John Dyson at a workshop on evolved stars in Chile in January 1992. In those days, the progenitors to Planetary Nebulae, post-Asymptotic Giant Branch stars, were not very numerous, mostly single and surrounded by spherically symmetric material, or so we thought. At that time, annoying complications such as circumstellar disks and binary central stars were creeping up already (e.g. Waters, Waelkens & Trans 1993), but John Dyson - who gave one of the most hilarious after dinner speeches I've ever witnessed - managed to complicate the issue even further with a flamboyant presentation on clumpy winds (Dyson 1993). I actually proceeded to approach the problem by considering disks around evolved stars. At least the picture represented a step up in complexity from the spherical case. If we fast forward more than a decade, then we arrive at the more massive evolved stars, Luminous Blue Variables (LBVs). These stars are thought to be the link between the main-sequence O stars and the final products of massive-star evolution, such as Wolf-Rayet stars and supernovae (see Lamers et al. 2001). Their large scale structures have bipolar morphologies (Nota et al. 1995). However, it remains unclear whether these bipolar nebulae are due to spherically symmetric winds interacting with a pre-existing density contrast or whether the star is undergoing enhanced mass loss in the equatorial plane, perhaps due to rotation (Dwarkadas & Balick 1998; Dwarkadas & Owocki 2002). The presence of an equatorial density contrast, or a disk, can potentially answer questions related to the nature of the formation of the bipolar flows. To detect these disks, Davies, Oudmaijer & Vink (2005) undertook spectropolarimetric observations of a large sample of both Galactic and Magellanic Cloud Luminous Blue Variables. Around half of the objects show a line effect. As an example, results for AG Car are shown in Fig. 7. Schulte-Ladbeck et al. (1994) observed this line effect and also discovered polarization variability in AG Car. As the points in QU space followed a straight line in the data at their disposal, the presence of a disk was readily inferred. New data are added to this dataset and presented in Fig. 7 as well. The big cross in the middle is the polarization at the line centre. It is not variable and most likely represents the interstellar polarization. This means that the intrinsic polarization angle varies randomly with time, opposed to what would be expected from a circumstellar disk and illustrating that the addition of more data here complicated the issue.

Davies et al. (2005) discussed several mechanisms that may explain the variability, and arrived at the random ejection of clumps in the wind as the most plausible scenario. In fact, this was introduced earlier in the case of P Cygni by Nordsieck et al. (2001), who used the results of Taylor et al. (1991). The variability, at smaller magnitude, was also found to be random. There are not enough variability data available to extrapolate this particular conclusion to the entire class. However, the main result that roughly half of the LBVs show a line effect and, thus, the evidence for either disks or clumpiness is robust. For comparison, Harries et al. (1998) found that only 10% of Wolf-Rayet stars and 25% of O supergiants show the effect (Harries et al. 2002), while the majority of evolved B[e] supergiants do (Schulte-Ladbeck et al. 1993; Oudmaijer & Drew 1999). Pending further analysis, the data may hint at different evolutionary sequences for WR stars and LBVs and further study is warranted. As clumps seem to dominate at least the spectropolarimetry of massive evolved stars, it would appear that John Dyson had indeed been right to warn the unsuspecting community of more complexities!

5 Outlook

The previous sections discussed Hα data and the application of the method to optically visible objects. The earliest type Herbig Be stars sample the highest mass stars already, but even with the advent of the 8m class telescopes, the deeply embedded massive Young Stellar Objects remain elusive. To be able to determine whether (accretion) disks around such objects are present we need study them at longer wavelengths where the objects are brighter. Their infrared hydrogen recombination lines are fairly strong (Bunn et al. 1995) and would be excellent target lines for spectropolarimetry. Little medium resolution spectropolarimetry has been performed in the near-infrared, so far however. Here we present some of the very first data across atomic lines of

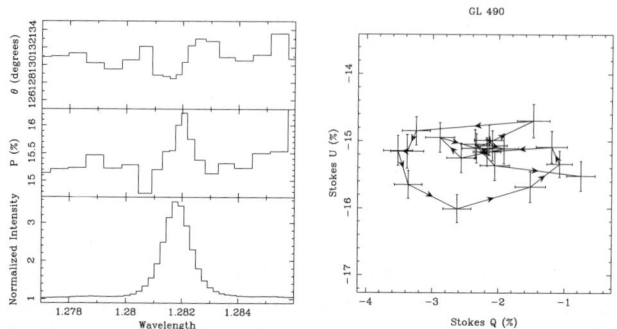

Fig. 8. Paβ polarization data of AFGL 490. The left hand graph shows the polarization data, as in Figure 2, now as a function of wavelength in μm. The bottom panel shows the (normalized) intensity spectrum, the data are rebinned to a corresponding accuracy in the polarization of 0.25% The right hand plot shows the Stokes QU vectors with the same binning applied. From Oudmaijer et al. (2005).

a stellar object. Oudmaijer et al. (2005) undertook proof-of-concept observations at UKIRT and observed, amongst others, the massive YSO AFGL 490. The results around the Paβ line are shown in Figure 8. The data show a clear line-effect, and are consistent with the results of Schreyer et al. (2006) who, in parallel, observed a larger scale rotating disk around the object.

Going to the other end of the evolution of massive stars, much progress has been made over the last years in the study of the geometry of supernova ejecta using spectropolarimetry. For example, studies by Wang et al. (2003) and Leonard et al. (2005 - see references therein), building on the original suggestion by McCall (1984), showed large intrinsic polarizations, and thus asymmetries, towards core-collapse supernovae. Ultimately data such as these, combined with those on their progenitors, such as the LBVs mentioned earlier, will provide strong observational constraints on not only the supernova mechanism itself, but also on the possible Gamma ray bursters resulting from them.

In this review, I have discussed the enormous progress that has been made, both observationally and theoretically, over the past decade in the field of linear spectropolarimetry and its application to star formation and stellar evolution. With the coming of age of spectropolarimetry on 8-m class telescopes and improved modelling, the future holds even more exciting prospects in store.

Acknowledgements

It is a pleasure to thank John Dyson for his inspiration, not only as an astronomer, but also as a friend, and I am happy to have been able to work in Leeds with him. My sincere thanks to my long-term collaborators in spectropolarimetry, Janet Drew, Jorick Vink and Tim Harries. Joe Mottram and

Ben Davies are thanked for their comments on an earlier draft of the manuscript. Jon Bjorkman is thanked for providing figure 1.

References

Adams, F.C. 1993 ASP Conf. Ser. **35**, p56
Akeson, R.L., Walker, C.H., Wood, K. et al. 2005 ApJ **622**, 440
Bally, J., Zinnecker, H. 2005 AJ **129**, 2281
Baines, D., Oudmaijer, R.D., Porter, J.M., Pozzo, M. 2006 MNRAS in press, astro-ph/0512534
Bertout, C. 1989 ARA&A **27**, 351
Bonnell, I.A., Bate, M.R., Zinnecker, H. 1998 MNRAS **298**, 93
Bunn, J., Hoare, M.G., Drew, J.E. 1995 MNRAS **272**, 346
Carciofi, A.C., Bjorkman, J.E. 2006 ApJ in press, astro-ph/0511228
Cassinelli, J.P., Nordsieck, K.H., Murison, M.A. 1987 ApJ **317**, 290
Clarke, C.J. 2001 IAUS **200**, 346
Davies, B., Oudmaijer, R.D., Vink, J.S. 2005 A&A **439**, 1107
Dougherty, S.M., Taylor, R. 1992 Nature **359**, 808
Dullemond, C.P., van den Ancker, M.E., Acke, B., van Boekel, R. 2003 ApJ **594**, L47
Dutrey, A., Guilloteau, S., Prato, L., Simon, M., Duvert, G., Schuster, K., Menard, F. 1998 A&A **338**, L63
Dwarkadas, V.V., Balick, B. 1998 AJ **116**, 829
Dwarkadas, V.V., Owocki, S. 2002 ApJ **581**, 1337
Dyson, J.E. 1993 in "Mass loss on the AGB and beyond", ed. H.E. Schwarz, ESO Conference and Workshop Proceedings **46**, p1
Fukagawa, M., Tamura, M., Itoh, Y., Hayashi, S.S., Oasa, Y. 2003 ApJ, **590**, 49
Grady, C.A., Polomski, E.F., Henning, Th. et al. 2001 AJ, **122**, 3396
Grinin, V.P., Thé, P.S., de Winter, D., Giampapa, A.N., Tambovtseva, L.V., van den Ancker, M.E. 1994 A&A **292**, 165
Harries, T.J., Hillier, D.J., Howarth, I.D. MNRAS **296**, 1072
Harries, T.J. 2000 MNRAS **315**, 722
Harries, T.J., Howarth, I.D., Evans, C.J. 2002 MNRAS **337**, 341
Hoare, M.G., Drew, J.E., Muxlow, T.B., Davis, R.J. 1994 ApJ **421**, L51
Hoare, M.G. 2002 ASP Conf Proc, Vol. **267**, p 137
Hubrig, S., Schöller, M., Yudin, R.V. 2004 A&A **428**, 1
Hubrig, S., Schöller, M., Yudin, R.V., Pogodin, M.A. 2006 A&A **446**, 1089
Jiang, Z., Tamura, M., Fukagawa, M., Hough, J., Lucas, P., Suto, H., Ishii, M., Yang, J. 2005 Nature **437**, 112
Johns-Krull, C.M., Valenti, J.A., Koresko, C. 1999 ApJ **516**, 900
Kaye, A.B., Gies, D.R. 1997 ApJ **482**, 1028
Lamers, H.J.G.L.M., Nota, A., Panagia, N., Smith, L.J., Langer, N. 2001 ApJ **551**, 764
Leinert, C., Richichi, A., Haas, M. 1997 A&A **318**, 472
Leonard, D.C., Weidong, L., Filippenko, A.V., Foley, R.J., Chornock, R. 2005 ApJ **632**, 450
Mannings, V., Sargent, A.I. 2000 ApJ, **529**, 391
McCall, M.L. 1984 MNRAS **210**, 829

McDavid, D. 1999 PASP **111**, 494
Mora, A., Eiroa, C., Natta, A. et al. 2004 A&A **419**, 225
Muzerolle, J., Calvet, N., Hartmann, L. 2001 ApJ **550**, 944
Norberg, P., Maeder, A. 2000 A&A **359**, 1035
Nordsieck, K.H. et al. 2001 in "P Cygni 2000: 400 years of progress" ed. M. de Groot & C. Sterken, ASP Conf. Ser. **233**, 261
Nota, A., Livio, M., Clampin, M., Schulte-Ladbeck, R. 1995 ApJ **448**, 788
Oudmaijer, R.D., Drew, J.E. 1999 MNRAS **305**, 166
Oudmaijer, R.D., Proga, D., Drew, J.E., de Winter, D. 1998 MNRAS, **300**, 170
Oudmaijer, R.D., Palacios, J., Eiroa, C. et al. 2001 A&A **379**, 564
Oudmaijer R.D., Drew, J.E., Vink, J.S. 2005 MNRAS **364**, 725
Patel, N.A., Curiel, S., Sridharan, T.K., Zhang, Q., Hunter, T.R., Ho, P.T.P., Torrelles, J.M., Moran, J.M., Gómez, J.F., Anglada, G. 2005 Nature **437**, 109
Poeckert, R., Marlborough, J.M. 1976 ApJ **206** 182
Porter, J.M., Rivinius, T. 2003 PASP **115**, 1153
Quirrenbach, A., Buscher, D.F., Mozurkewich, D., Hummel, C.A., Armstrong, J.T. 1994 A&A **283**, L13
Quirrenbach, A., Bjorkman, K.S., Bjorkman, J.E. et al 1997 ApJ **479**, 477
Schreyer, K., Semenov, D., Henning, Th., Forbrich, J. 2006 ApJ **637**, L129
Schulte-Ladbeck, R.E., Shepherd, D.S., Nordsieck, K.H. et al. 1992 ApJ **401**, L195
Schulte-Ladbeck, R.E., Leitherer, C., Clayton, G.C. et al. 1993 ApJ **407**, 723
Schulte-Ladbeck, R.E., Clayton, G.C., Hillier, D.J., Harries, T.J., Howarth, I.D. 1994 ApJ **429**, 846
Serkowski, K., 1962 in Adv. Astron. Astroph. **1**, 289
Shepherd, D.S., Claussen, M.J., Kurtz, S.E. 2001 Science **292**, 1513
Taylor, M., Nordsieck, K.H., Schulte-Ladbeck, R.E., Bjorkman, K.S. 1991 AJ **102**, 1197
Thé P.S., de Winter, D., Perez, M.R., 1994 A&AS, **104**, 315
Vink, J.S., Drew, J.E., Harries, T.J., Oudmaijer, R.D. 2002 MNRAS **337**, 356
Vink, J.S., Drew, J.E., Harries T.J., Oudmaijer, R.D., Unruh, Y. 2003 A&A **406**, 703
Vink, J.S., Drew J.E., Harries, T.J., Oudmaijer, R.D., Unruh, Y. 2005a, MNRAS, **359**, 1049
Vink, J.S., Drew, J.E., Harries, T.J., Oudmaijer, R.D. 2005b, ASPC, **343**, 232
Vink, J.S., Harries, T.J., Drew, J.E. 2005c, A&A **430**, 215
Wade, G.A., Drouin, D., Bagnulo, S., Landstreet, J.D., Mason, E., Silvester, J., Alecian, E., Böhm, T., Bouret, J.-C., Catala, C., Donati, J.-F. 2005 A&A **442**, L31
Wang, L., Baade, D., Höfflich, P., Wheeler, J.C. 2003 ApJ **592**, 457
Waters, L.B.F.M., Waelkens, C., Trans, N.R. 1993 in "Mass loss on the AGB and beyond", ed. H.E. Schwarz, ESO Conference and Workshop Proceedings **46**, p298
Wolf, S., Stecklum, B., Henning, Th. 2001 IAUS **200**, 295
Wolfire, M.G., Cassinelli, J.P. 1987 ApJ **319**, 850
Wood, K., Brown, J.C., Fox, G.K. 1993 A&A **271**, 492
Wood, K., Bjorkman, K.S., Bjorkman, J.E. 1997 ApJ **477**, 926
Yorke, H.W., Kruegel, E. 1977 A&A **54**, 183
Yorke, H.W., Sonnhalter, C. 2002 ApJ **569**, 846

How to Move Ionized Gas: An Introduction to the Dynamics of H II Regions

W.J. Henney

Centro de Radioastronomía y Astrofísica, UNAM Campus Morelia,
Apartado Postal 3-72, 58090 Morelia, Michoacán, México
w.henney@astrosmo.unam.mx

1 Introduction

H II regions are volumes of gas surrounding high-mass stars that are ionized and heated by the stellar ultraviolet radiation. This ionization and heating significantly raises the thermal pressure of the gas, which is the principal driver of the dynamics of all but the very smallest and very largest H II regions. The simplest model of H II region evolution predicts a slow and decelerating expansion of the ionized gas with very little internal motion, but this is rarely observed. Instead, when examined in detail, such regions are found to be highly structured with complex internal motions.

For reasons of space, this review is rather narrowly focused on the traditional H II regions found in our galaxy and ionized by one or a handful of O stars. The dynamics of photoionized gas has a much broader domain of application than this, covering such objects as planetary nebulae, novae, Wolf-Rayet nebulae, broad and narrow line regions of active galaxies, and the reionization of the Universe by the first generation of stars. The physical priniciples presented here will still apply in these broader contexts, but care must be taken, as some of the shortcuts and approximations commonly used for H II regions may no longer be valid.

The chapter is divided into three sections. The first section introduces the equations governing the dynamics and physical structures of H II regions and discusses the approximations that are commonly employed. The second section presents the broad physical concepts that provide the building blocks for constructing global models of H II regions. The third section provides an overview of such models, as applied to the different evolutionary stages of H II regions, with particular emphasis on recent models of the evolution in a clumpy, turbulent medium.

2 The Equations

The principal equations necessary for calculating H II region dynamics under the standard assumptions used in the field can be divided into three groups: the Euler equations, which describe the motion of the gas; the radiative transfer equation, which describes the emission and absorption of photons; the rate equations, which describe the transitions between different atomic/ionic/molecular states. For each group, I first present the general form of the equations, in which many difficult bits of physics are hidden away in innocuous looking terms, before discussing the sorts of approximations that might be made in different types of models. All attempts to theoretically model H II regions involve at least *some* level of approximation.

2.1 Euler Equations

These describe the motion of a non-viscous, non-relativistic gas of density ρ, pressure P, and vector velocity $\boldsymbol{u}$, all of which are functions of position, $\boldsymbol{r}$, and time, t. See, for example, Shu (1992) for full derivations. The ratio of specific heats is assumed to be $\gamma = 5/3$, as is appropriate for a monatomic gas. The equations are given here in their *conservation form*. Conservation of mass gives

$$\frac{d\rho}{dt} + \boldsymbol{\nabla} \cdot (\rho \boldsymbol{u}) = 0. \tag{1}$$

Conservation of momentum gives

$$\frac{d}{dt}(\rho \boldsymbol{u}) + \boldsymbol{\nabla}\left(P + \rho u^2\right) = \boldsymbol{g}\rho, \tag{2}$$

where $\boldsymbol{g}$ is the acceleration of the gas due to "body forces" caused by gravity, magnetic fields, or radiation pressure. Conservation of energy gives

$$\frac{d}{dt}\left(\frac{3}{2}P + \frac{1}{2}\rho u^2\right) + \boldsymbol{\nabla} \cdot \left[\boldsymbol{u}\left(\frac{5}{2}P + \frac{1}{2}\rho u^2\right)\right] = \boldsymbol{u} \cdot \boldsymbol{g}\rho + H - C, \tag{3}$$

where H and C are the volumetric rates of gas heating and cooling, respectively, due to atomic processes. These equations are supplemented by the ideal-gas equation of state, which defines the gas temperature, T:

$$P = \frac{\rho k_\mathrm{B} T}{\mu m_\mathrm{H}}, \tag{4}$$

where k_B is Boltzmann's constant, m_H is the mass of a hydrogen nucleus, and μ is the dimensionless mean-mass-per-particle (for solar abundances, $\mu \simeq 1.3$ for neutral gas and $\mu \simeq 0.6$–0.7 for ionized gas, depending on the ionization state of helium).

Although the above *single-fluid* treatment is generally adequate for the gas, it is not always such a good approximation when dust grains are considered in detail. In such a case, one can employ a multifluid treatment (Gail

& Sedlmayr 1979a,b, and see also the chapter by Tom Hartquist and Ove Havnes), in which one uses a separate momentum equation for each fluid, including explicit interaction terms for collisions between particles of the different fluids.

2.2 Radiative Transfer

The fundamental equation of radiative transfer describes the behavior of the *specific intensity*, $I_\nu(\hat{n}, r)$ (Mihalas 1978), which is a function of frequency, ν, position, r, and direction, $\hat{n}$:

$$\frac{1}{c}\frac{dI_\nu}{dt} + \hat{n}\cdot\nabla I_\nu = \eta_\nu - \chi_\nu I_\nu, \tag{5}$$

where c is the speed of light, and η_ν, χ_ν are respectively the emissivity and absorption coefficient, which may contain an arbitrary amount of physics. For many purposes, it is sufficient to work with angle-averaged moments of the specific intensity, such as the *mean intensity*,

$$4\pi J_\nu = \int_{4\pi} I_\nu \, d\Omega, \tag{6}$$

and the radiative flux,

$$\boldsymbol{F}_\nu = \int_{4\pi} \hat{n} I_\nu \, d\Omega. \tag{7}$$

2.3 Rate Equations

The general equation for the evolution of the partial density of a particular state, i, can be written as

$$\frac{dn_i}{dt} + \nabla\cdot(n_i \boldsymbol{u}) = G_i + \sum_{j\neq i} R_{j\to i} n_j - n_i\left(S_i + \sum_{j\neq i} R_{i\to j}\right), \tag{8}$$

where $R_{i\to j}$, $R_{j\to i}$ represent the rates of transitions between state i and other states j, whereas G_i, S_i represent respectively sources and sinks of state i due to other processes. This form of the rate equation can apply to internal states of an atom, an ion, a molecule, or a dust grain. Equally, it may apply to the the total abundance of a given ion stage. In general, the rates may be functions of the local radiation field, electron density, temperature, or indeed any other physical variable.

2.4 How to Avoid the Dynamics

The most drastic simplification from the point of view of the dynamics follows from the assumption that there are no dynamics at all. This is the *static* approximation, which consists in setting $\boldsymbol{u} = 0$ and $dX/dt = 0$ for all quantities

X. In this case, equation (1) is trivially satisfied, while equation (2) reduces to hydrostatic equilibrium, or P = constant if there are no external forces. The energy equation (3) reduces to $H = C$ and the rate equations (8) reduce to sources = sinks. This approximation is commonly employed in *photoionization codes*, which treat the microphysics of the energy and rate equations in great detail, and usually the radiative transfer also. As is shown below, this approximation can be an acceptable one for the ionization and thermal balance of the interior of an HII region in the many instances where the ionization/recombination and heating/cooling timescales are much shorter than the dynamic timescale.

Slightly more realistic is the *steady-state* or *stationary* approximation, in which one allows for non-zero gas velocities, but still ignores all derivatives with respect to time. This adds the *advective* terms, of the form $\boldsymbol{\nabla}\cdot(\boldsymbol{u}X)$, to the static form of the equations. These terms represent the material flow of the quantity X from one point to another and are most important in regions of strong gradients in $\boldsymbol{u}$ or X. In many instances, the flow timescale through the part of the HII region of interest is short compared with the timescale for changes in the parameters of the flow, due, for instance, to changes in the incident ionizing flux or the neutral density outside the ionization front. In such cases, the stationary approximation is often a very good one, provided that a suitable reference frame is chosen. In other cases, especially when the long-term, global evolution of the HII region is of interest, or if dynamical instabilities are expected, then it is necessary to include the non-steady dX/dt terms and solve the *fully time-dependent* equations. Even in this case, it is usually sufficient to treat the radiative transfer equation in the stationary limit, except when light-travel times are significant compared with other timescales of interest (Shapiro et al. 2005).

2.5 How to Avoid the Atomic Physics

In other contexts, one may wish to study the dynamics in detail without worrying unduly about the finer points of the microphysics or radiative transfer. To that end, various rules of thumb have evolved, which give satisfactory approximations in many common situations. However, the use of these approximations should be justified on a case-by-case basis. The simplifications presented here are at the extreme end of what one can get away with in a toy model. For more serious applications, one should consult a text such as Osterbrock & Ferland (2006) for extra ingredients to add.

Thermal Balance

The temperature in galactic HII regions is determined principally by the balance between photoelectric heating (ejection of energetic electrons by photoionization) and cooling due to forbidden lines of the ions of heavy elements, which are excited by electron collisions. The resultant equilibrium temperature is $\simeq 9000$ K and rarely varies by more than 50% throughout the region. It is therefore natural to use an *isothermal approximation* for the ionized gas.

Strong dynamic effects can cause this approximation to break down, but this is rare in H_{II} regions. The dynamical terms in equation (3) will not be greatly important unless the dynamic timescale is shorter than the heating/cooling timescale, which is typically 3–10 times shorter than the recombination timescale. This requires very high gas velocities of > 500 km s^{-1} unless the ionization parameter (see below) is much lower than is typical found.

Transfer of Ionizing Photons

Since hydrogen is (usually) the most abundant element, it is natural to consider a hydrogen-only approximation, in which the only photons of interest are those capable of ionizing hydrogen from its ground state and the only opacity source considered for those photons is the photoionization process itself. If the diffuse field is treated in the on-the-spot approximation, which is that any ionizing radiation emitted due to recombination is absobed where it is emitted (see Osterbrock & Ferland 2006, sec. 2.3), then only the direct stellar radiation need be explicitly considered. If only one ionizing star is present, then the radiation field is monodirectional, so that all angular moments of the radiation field are equal in magnitude, and in particular $4\pi J_\nu = F_\nu = |\boldsymbol{F}_\nu|$. In this case, one can get away with considering only the frequency-integrated flux of ionizing photons: $f = \int_{\nu_0}^{\infty} (F_\nu / h\nu)\, d\nu$, where ν_0 is the frequency corresponding to the Lyman limit. The radiative transfer equation then reduces to $\boldsymbol{\nabla} \cdot \boldsymbol{f} = -\sigma n_\mathrm{n} \hat{\boldsymbol{r}}$, where $\hat{\boldsymbol{r}}$ is the radial direction from the star, n_n is the number density of neutral hydrogen atoms and σ is the frequency-averaged photoionization cross-section, weighted by the local ionizing spectrum. In this approximation, σ is a function only of $\tau = \int \sigma n_\mathrm{n}\, dr$, and can be precomputed for a given stellar spectrum. In reality, dust opacity will often make a significant contribution (e.g., Arthur et al. 2004; Ercolano et al. 2005) and should be included in any realistic treatment.

Ionization Balance

Collisional ionization of hydrogen is unimportant for $T < 20,000$ K and three-body recombinations are negligible at typical H_{II} region densities. The hydrogen ionization balance is then simply between photoionization and radiative recombination, which in the approximation discussed in the previous section is

$$\frac{dn_\mathrm{p}}{dt} + \boldsymbol{\nabla} \cdot (n_\mathrm{p}\boldsymbol{u}) = n_\mathrm{n}\sigma f - \alpha n_\mathrm{p} n_\mathrm{e}, \qquad (9)$$

where n_p, n_e ($\simeq n_\mathrm{p}$), and n_n are the number densities of ionized hydrogen, electrons, and neutral hydrogen, respectively, and α is the appropriate recombination coefficient (if the on-the-spot approximation is used the so-called

Case B coefficient should be adopted). If the radiative transfer is treated in more detail, then σf should be replaced by $\int_0^\infty \sigma_\nu (4\pi J_\nu/h\nu)\,d\nu$.

3 Physical Concepts

Before considering models for H II regions as a whole, it is useful to break down the problem into distinct physical ingredients, which can be studied separately.

3.1 Static Photoionization Equilibrium

In order to compare and contrast with the results of dynamical models, it is instructive to first consider the static case. The algebra is simplest if one considers a uniform, plane-parallel slab, of density n, one face of which is illuminated by a perpendicular ionizing flux, f_0, and in which all the extreme simplifying assumptions of the previous sections are assumed to hold. The fractional ionization or degree of ionization is defined as $x = n_{\rm p}/n$ and, for reasonable values of f_0 and n, this fraction is very close to unity at the illuminated face: $1 - x_0 \simeq \alpha n/(\sigma f_0) \simeq 3 \times 10^{-6}\,\Upsilon^{-1}$, where $\Upsilon = f_0/(cn)$ is the *ionization parameter*, or ratio of photon density to particle density, which typically takes values from 10^{-4} to 10^{-2}. As one moves to greater depths, z, in the slab, then integration of the ionization balance and radiative transfer equations shows that the ionizing flux decreases approximately linearly with depth, $f(z) \simeq f_0(1 - z/z_0)$, until one reaches a depth $z_0 \simeq f_0/(\alpha n^2)$, where the ionization fraction swiftly falls from $x \simeq 1$ to $x \simeq 0$ over a short distance, $\delta z \simeq 1/(n\sigma)$. This depth to the ionization front, z_0, is referred to as the *Strömgren distance* and it is apparent that $\delta z/z_0 \simeq (1 - x_0)$, confirming that the front is indeed thin. The ionization parameter, Υ, is the single most important parameter describing the H II region; as well as determining the degree of ionization and the relative thickness of the ionization front, it is also proportional to the column density through the ionized gas: $n z_0 = f_0/(\alpha n) \simeq 10^{23}\,\Upsilon$ cm^{-2}.

This last result allows one to see how the ionized region in approximate photoionization equilibrium must respond to a change in density. If the density in the slab increases, perhaps as the result of compression by a shock, then the ionization parameter decreases and so must the column of ionized gas, meaning that some of the gas must recombine. On the contrary, if the density decreases, perhaps due to expansion,[1] then the ionization parameter and the ionized column will increase, leading to the advance of the ionization front into previously neutral gas.

---

[1] The same holds for expansion in a spherical geometry since n ($\sim r^{-3}$) falls more rapidly than f ($\sim r^{-2}$).

3.2 Ionization Front Propagation

The speed of propagation, U, of a plane, steady ionization front can be determined from a consideration of the jump conditions in the physical variables across it (Kahn 1954). Variables on the neutral side of the front are given a subscript '1', and those on the ionized side a subscript '2'. The velocity in the frame of reference of the ionization front of the neutral gas, v_1, is equal in magnitude to U, but opposite in direction. It is apparent that one has $f_1 \simeq 0$ and $(1 - x_2) \simeq 0$, so that by integrating equation (8) across the front[2] one has $v_1 = f_2/n_1$. When one then considers mass and momentum conservation across the front, assuming isothermal sound speeds of a_1 and a_2 on the two sides, one finds two classes of solution (for example, Shu 1992). Fronts with $v > v_R \simeq 2a_2 \simeq 20$ km s^{-1} move supersonically into the neutral gas and are called R-type (for "rare"). Fronts with $v < v_D \simeq a_1^2/(2a_2) \simeq 0.05$ km s^{-1} move subsonically into the neutral gas and are called D-type (for "dense"). If conditions are such that f_2/n_1 falls between v_D and v_R, then the ionization front will be preceded by a shock, which compresses the neutral gas to a higher density, n_1', so that $f_2/n_1' < v_D$ and the ionization front becomes D-type. Fronts can be further divided into those that contain an internal sonic transition, which are termed strong, and those that do not, which are termed weak. The only stable R-type fronts are weak. A weak R-type front moves supersonically with respect to both the neutral and the ionized gas, and such fronts show a low density contrast between the two sides: $1 \leq n_1/n_2 \leq 2$. The limit of extreme weak D-type fronts as $v_1 \to 0$ corresponds to the static case considered above, in which the gas pressure is constant across the front so that $n_1/n_2 = a_2^2/a_1^2 \simeq 100$. When $v_1 = v_D$, the front is termed D-critical, with an ionized velocity that is exactly sonic with respect to the front ($v_2 = a_2$), and an even higher density contrast, $n_1/n_2 = 2a_2^2/a_1^2 \simeq 200$. Strong D-type fronts can only occur if the sound speed has a maximum $a_m > a_2$ inside the front and give $a_m < v_2 < 2a_m$. They are probably most relevant at very high metallicities, where the equilibrium temperature of fully ionized gas can be much lower than that of partially ionized gas.

When a volume of gas is initially exposed to ionizing radiation, the flux is usually high enough that the ionization-front is R-type and propagates supersonically through the gas. However, as the front progresses to greater distances, an increasing proportion of the flux is used up in balancing recombinations in the ionized gas. After a time of order $1/(\alpha n)$ and a distance of roughly one Strömgren distance, f_2 has become low enough that $U < 2a_2$ and a preceding shock detaches from the ionization front, which becomes D-type. Soon after this point, the speed of the ionization/shock front falls below the sound speed in the ionized gas, so that the subsequent evolution of the front becomes sensitive to the internal conditions in the HII region, which determine the boundary conditions on the ionized side of the front. In particular, if the ionized gas is free to flow away from the front, then the front is liable to

---

[2] In this approximation, recombinations in the front are ignored.

remain approximately D-critical. On the other hand, if the ionized gas is confined, perhaps by a closed geometry (nowhere for the gas to go to) or a high imposed pressure (e.g., from a stellar wind), then the front will be weak D. The latter case corresponds to the classical evolution of a Strömgren sphere, which is described in many textbooks (e.g., Dyson & Williams 1997).

The relative importance within the front itself of dynamic terms in the ionization balance is large, of order $u_2 \sigma / \alpha \simeq 10M$, where $M = u_2/a_2$ is the Mach number of the ionized gas ($M = 1$ for a D-critical front). However, given that the front is generally thin compared with the H II region as a whole (section 3.1), the relative importance, λ_{ad}, of dynamics in the *global* ionization balance is much smaller, $\lambda_{\text{ad}} = u_2/(\Upsilon c) \simeq 0.003$–$0.3$. This is why photoionization models that ignore dynamics are often a reasonable approximation for the interior of an H II region.

3.3 Structure of a D-type Front

When examined in detail, an ionization front is found to have considerable internal structure. An example of an approximately steady D-critical front is shown in Fig. 1, which results from a radiation-hydrodynamical simulation that uses rather simplified atomic physics and an artificially lowered ionization parameter in order to resolve the ionization front on a two-dimensional grid (Henney et al. 2005). However, the principal results compare well with much more detailed one-dimensional calculations of weak-D fronts using a state-of-the-art plasma physics code (Henney et al. 2005b). In order to emphasize the regions of the front where interesting changes occur, the physical variables are plotted as a function of the mean optical depth for ionizing radiation from the star. The velocity is in the direction of lower optical depths and is given relative to the ionization front, which is moving at 0.95 km s^{-1} away from the star.

The net rate at which the gas gains energy from atomic processes (photoelectric heating minus radiative cooling; solid gray line) has two separate peaks in the front. The deeper peak occurs at $\tau \simeq 8$ and results in the heating of the gas from 100 K to 9000 K and its acceleration from 0.1 km s^{-1} to 3 km s^{-1}, but leaves the gas still largely neutral ($x \simeq 0.01$). The shallower, broader peak occurs at $\tau \simeq 3$ and results in the complete ionization of the gas and its acceleration to $\simeq 10$ km s^{-1}, with very little change in temperature.[3] The gas then passes through a sonic point at $\tau \simeq 1$, at which point $x \simeq 0.95$, and continues accelerating gradually up to $\simeq 30$ km s^{-1} as it flows back past the star (not shown). Although the far downstream velocity of the ionized gas is supersonic, the front is still D-critical since there is no internal maximum in the sound speed. Instead, it is the divergence of the gas streamlines,

---

[3] This can be compared with a simple analytic estimate for the optical depth to the ionization front (see Sec. 3.2): $\tau_{\text{if}} = -\ln \lambda_{\text{ad}} \simeq 10.3 + 2.3 \log_{10} \Upsilon$, giving $\tau_{\text{if}} \simeq 1$ to 6 for ionization parameters in the range $\Upsilon = 10^{-4}$–10^{-2}.

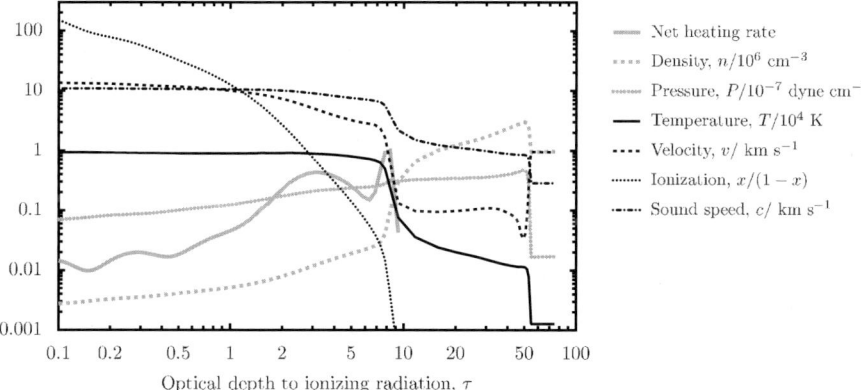

Fig. 1. Physical variables in a typical D-critical ionization front taken from a cut along the axis of a two-dimensional hydrodynamical simulation of a champagne flow. Stellar photons are moving from left to right, but the gas is moving from right to left. Figure adapted from Henney et al. (2005).

as the geometry deviates from plane-parallel, that allow the gas to continue accelerating once it is fully ionized and isothermal.

The shock that precedes the ionization front is also visible, at depth of $\tau \simeq 50$. The shock is propagating into the neutral gas slightly faster than the ionization front velocity, so this part of the model is not exactly steady, but rather the column of shocked neutral material grows with time. This part of the structure is the least well-modeled in the simulation, since the physics of the photodissociation region is not treated adequately, but see Hosokawa & Inutsuka (2005a,b) for dynamical models that do treat this properly.

3.4 Photoablation Flows

If the neutral density is similar in all directions from the ionizing star, then one would expect the radius to the ionization front to be also similar, which would lead to a global geometry for the front that is closed, or concave. Despite this, the opposite case is of great relevance to Hɪɪ region dynamics, when the ionization front is convex; it curves away from the ionizing star. This can arise, for example, if dense condensations exist in the neutral gas, which slow down the propagation of the ionization front, causing it to "wrap around" the obstacle. If this situation persists, then a steady flow of ionized gas that accelerates away from the convex front will establish itself. The density in this ionized *photoablation flow* increases sharply towards its base, where it may greatly exceed the mean density in the Hɪɪ region. Models for such flows were originally developed in the context of the photoevaporation of small, self-gravitating globules (Dyson 1968; Kahn 1969) but have a much wider range of application. If a cylindrical, rather than spherical geometry is adopted, similar models can be applied to cases where the ionization front wraps around a dense filament, as in the Bright Bar in the Orion Nebula, and they can even be extended to cases where the ionization front is flat, or slightly concave, as in the case of champagne flows (see section 4.2 below). In all cases, the structure of the ionized flow is similar, provided only that the gas streamlines diverge and that any pressure imposed at the downstream boundary is low compared with the thermal pressure of the ionized gas at the ionization front. In order for a steady-state approach to be valid, it is also necessary that the timescale for significant changes in the configuration of the neutral gas be long compared with the dynamic timescale of the ionized flow. This condition is easily satisfied in some cases, such as the Orion proplyds, but only approximately satisfied in others, such as the radiation-driven implosion of a neutral globule (Bertoldi 1989).

The simplest models proceed by patching together the solution for a plane-parallel D-critical ionization front (see previous sections) with an isothermal wind solution for the fully ionized flow. This is valid so long as the width of the ionization front is very small compared with its radius of curvature, r_c, which is almost invariably the case. The gas leaves the ionization front at the sound speed but accelerates rapidly (see Fig. 6 of Henney 2003), so that the ionized density falls off more quickly than it would in a constant velocity wind, approximately mimicking an exponential profile. As the Mach number of the flow increases, the acceleration gradually lessens and a terminal velocity is reached when the isothermal approximation breaks down, which occurs when the expansion and heating timescales becoming comparable. For a spherical flow with $\lambda_{ad} = 0.01$, this occurs at velocity of $\simeq 40$ km s^{-1} and a radius of $\simeq 10 r_c$.

Unless the ionization parameter is rather low, λ_{ad} will be small and the flows are *recombination dominated*, which means that the ionized flow has a high optical depth to ionizing radiation and is in approximate static ionization balance.[4] The effective recombination thickness of the flow[5] is $h_{eff} = 0.12 r_c$ for a spherical geometry or $h_{eff} = 0.3 r_c$ for a cylindrical geometry, so long as the curvature is positive (convex) and $r_c \ll r_\star$, where $r_\star$ is the distance of the front from the ionizing star. For the case where $r_c > r_\star$, one instead finds $h_{eff} \simeq 0.2$–$0.7 r_\star$ (see Fig. 7 of Henney et al. 2005). In either case, the entire flow structure is uniquely determined by the incident ionizing flux and the curvature of the front. If $r_c \lesssim 0.2 r_\star$, then the photoevaporation flow is not capable of evacuating the region all the way back to the ionizing star, so the incident flux may be reduced by recombinations in the interior of the H<small>II</small> region.

---

[4] The opposite *advection dominated* case is generally only seen in flows with low surface brightness (Henney 2001), most notably the knots in the Helix planetary nebula (López-Martín et al. 2001; O'Dell et al. 2005), in which a high fraction of incident photons reach the ionization front to ionize new gas.

[5] Defined by $n_0^2 h_{eff} = \int n^2 \, dr$, where n is the ionized density with value n_0 at the ionization front.

3.5 Other Ingredients

Stellar Winds

High mass stars invariably drive fast ($\simeq 1000$ km s^{-1}) stellar winds (Lamers & Cassinelli 1999), Wind mass loss rates are still very uncertain due to the poorly understood effects of clumping on observational diagnostics (Fullerton et al. 2006), but are probably in the range 10^{-9}–$10^{-7} M_\odot$ yr^{-1} for main-sequence O stars. The effect of a stellar wind on the dynamics of an H II region depends crucially on whether or not a hot bubble of shocked stellar wind can persist, which determines whether the shell of shocked H II region gas is *energy-driven* or *momentum-driven* (see the chapter by Jane Arthur). For the energy-driven case, Capriotti & Kozminski (2001) showed that the shell swept up by the stellar wind can dominate the expansion energy of the region at early evolutionary times and when the ambient density is $> 10^4$ cm^{-3} (see also Dyson 1977; Garcia-Segura & Franco 1996). The wind-blown shell is also capable of trapping the ionization front in some instances. In the momentum-driven case, the wind never dominates the global energetics of the region but can nevertheless have an important local influence on the internal dynamics. Examples include the bowshocks found around proplyds (which are protostellar disks) in the inner Orion Nebula (García-Arredondo et al. 2001). Various mechanisms may prevent the formation of the hot shocked bubble that is required for the energy-driven case; they include thermal conduction (e.g., Dorland & Montmerle 1987) or dynamical instabilities (Breitschwerdt & Kahn 1988) at the contact discontinuity, or mass-loading of the wind due to embedded photoablating clumps or proplyds (Dyson et al. 1995; García-Arredondo et al. 2002). All these can cause enhanced cooling in the shocked wind, leading to the loss of its thermal pressure. Specific instances of the importance of winds in the evolution of H II regions are described in section 4 below.

Radiation Pressure

The transfer of momentum between the radiation field and the gas can sometimes affect the dynamics of H II regions. This arises in three different ways: first, through trapped resonance line radiation (mainly Lyman α); second, through the momentum of stellar photons transferred during the photoionization process; and, third, indirectly through collisional coupling with dust grains that are accelerated by the absorption of stellar radiation.

Although the diffusion of resonance line photons can in principle introduce non-local couplings between disparate parts of an H II region, this is severely hampered by the presence of dust absorption (Hummer & Kunasz 1980), which allows a purely local approach to be used if the medium is homogeneous on scales $< 0.1/\kappa_\mathrm{d}$, where κ_d is the dust absorption coefficient. By this means, Henney & Arthur (1998) showed that the resonance line radiation pressure is proportional to the gas pressure, making a contribution of only $\simeq 5\%$ for a standard dust-to-gas ratio.

The radiative acceleration due to absorption of ionizing photons of mean energy $\langle h\nu \rangle$ is $g_{\rm rad} \simeq (\langle h\nu \rangle/c)(f/\rho h_{\rm abs})$, where $h_{\rm abs}$ is the thickness over which the photons are absorbed. The acceleration due to pressure gradients in a photoablation flow (section 3.4) is $g_{\rm flow} \simeq a^2/h_{\rm eff}$. Comparing the two, and putting $h_{\rm abs} = h_{\rm eff}$, one finds $g_{\rm rad}/g_{\rm flow} \simeq \Upsilon \langle h\nu \rangle/(2kT) \sim 10\Upsilon$, so that the pressure of the ionizing radiation is of only secondary importance unless the ionization parameter is higher than is typically found in HII regions. Assuming that the gas and grains are effectively coupled, one finds a similar result for the absorption of radiation by dust, although this can become more important for cooler stars, where the relative luminosity at non-ionizing wavelengths is higher. Radiation pressure on dust can also be important in the inner parts of HII regions due to the increased flux close to the ionizing source,[6] where it may be responsible for central cavities in some cases (Mathews 1967; Inoue 2002).

Magnetic Fields

The magnetic field makes a significant, perhaps dominant, contribution to the pressure both in molecular clouds and in the diffuse ISM. HII regions are generally over-pressured with respect to their undisturbed surroundings, but are within a factor of two of pressure equilibrium with the shocked neutral gas that surrounds them. The dynamical importance of magnetic fields inside the HII region depends sensitively on how the field strength, B, responds to compression in the shock and rarefaction at the ionization front. This can be approximately characterized by an effective adiabatic index, $\gamma_{\rm m}$, such that $B^2 \propto \rho^{\gamma_{\rm m}}$. Possible values of this index are bounded by $\gamma_{\rm m} = 0$ for compression/rarefaction parallel to the field lines of an ordered field and $\gamma_{\rm m} = 2$ for compression/rarefaction perpendicular to the field, whereas $\gamma_{\rm m} \lesssim 1$ seems to be indicated by observations. The order-of-magnitude increase in sound speed between the ionized and neutral gas means that the plasma β-parameter (ratio of gas pressure to magnetic pressure) will be between 5 times (if $\gamma_{\rm m} = 0$) and 2000 times (if $\gamma_{\rm m} = 2$) higher in the HII region than in the undisturbed neutral gas, assuming the HII region to be five times overpressured. Thus, it is plausible that the magnetic field should play a much less important role in the ionized gas than in the neutral gas. Nevertheless, the magnetic fields can still have dramatic effects on HII region dynamics, particularly for the ionization front, and magnetohydrodynamic models of these are presented in detail in the chapter by Robin Williams in this volume.

---

[6] This is not the case for radiation pressure on the gas because of its dependence on the product of the flux and *neutral* density, which is roughly constant throughout the ionized region.

Instabilities

The dynamics of the photoionized gas will also be influenced by many different kinds of instabilities (see Williams 2003 for an overview), which may contribute to the detailed structure observed in HII regions. Different modes of instability at the ionization front are found, depending on whether the shell of shocked neutral gas outside it is thick (Williams 2002) or thin (Giuliani 1979; Garcia-Segura & Franco 1996). Recombinations in the fully ionized gas are found to damp the instabilities in some (Axford 1964; Mizuta et al. 2005) but not all (Williams 2002) cases. The interaction between streams of gas inside the HII region, such as stellar winds or photoablation flows, provides opportunities for further instabilities (e.g., Rayleigh-Taylor, Kelvin-Helmholtz) and these may interact with ionization front instabilities in complicated ways. An example is shown in the simulations of Arthur & Hoare (2005), where the fragmention of a wind-driven shell triggers shadowing instabilities of the type discussed by Williams (1999).

4 HII Region Evolution

While the classical scenario for the expansion of an ionized Strömgren sphere in a constant density medium has clear didactic value, it is not necessarily relevant to the evolution of real HII regions. Massive stars form in dense and highly inhomogeneous molecular clouds and begin to emit ionizing photons while they are still accreting mass, possibly via a disk (see the chapter by Melvin Hoare and Pepe Franco in this volume for an overview of high-mass star formation). Various mechanisms may act to prolong the duration of the early phases of evolution, either by physical confinement or by providing a reservoir of neutral gas. Eventually the HII region may break out from the molecular cloud and become optically visible, but its evolution will still be strongly affected by the strong density gradients in its environment. Since high-mass star formation is strongly clustered, extended HII regions will tend to be excited by multiple OB stars. As the region's age begins to exceed the main-sequence lifetime of the highest mass stars, then the powerful winds from evolved stars and subsequent supernova explosions will have a dramatic effect on the dynamics.

4.1 Early Phases: Hypercompact and Ultracompact Regions

The smallest observed HII regions are of size $\sim 10^{-3}$ parsec, and are classified as hypercompact regions (Carral et al. 1997; Sewilo et al. 2004). At these small sizes, the escape velocity at the ionization front is larger than the ionized sound speed, and so gravitational effects are important. If the central star is still accreting via a spherical Bondi-Hoyle flow, then the accretion velocity just outside the ionization front exceeds the R-critical velocity of $2c_i$, so the gas suffers only a mild deceleration in the front and continues its accretion onto the star (Keto 2002). During this phase, the ionization front does not expand dynamically, but can only increase its radius if the ionizing luminosity of the central star increases (Keto 2003). This phase is therefore rather long-lived, with a duration given by the timescale for the nascent star to grow by accretion, of order 10^5 years.

HII regions with sizes ~ 0.01–0.1 parsec and densities $> 10^5$ cm^{-3} are classified as ultracompact and these are much more numerous and well-studied than the hypercompact regions (e.g., Wood & Churchwell 1989; Kurtz et al. 1994; de Pree et al. 2005). They show a diverse range of morphologies, such as cometary, shell-like, bipolar, or irregular, although roughly half are spherical or unresolved. The high number of observed ultracompact HII regions in the galaxy has been taken to imply a lifetime for this phase of $\simeq 10^6$ years (Wood & Churchwell 1989), although it should be noted that other estimates are as short as 5×10^4 years (Comeron & Torra 1996). A remarkable diversity of dynamical models have been proposed in order to explain this lifetime, which is longer than the time for a classical Strömgren sphere to expand past 0.1 parsec. In some models, the ionized gas is physically confined to ultracompact sizes, either by the ram pressure of the ambient medium as the star moves through the molecular gas (van Buren et al. 1990), or by an accretion flow onto the star (Keto 2002; González-Avilés et al. 2005), or merely by the high static pressure (thermal plus magnetic plus turbulent) of the molecular cloud core (de Pree et al. 1995; Garcia-Segura & Franco 1996). Other models allow the ionized gas to expand freely but provide a reservoir of neutral gas, the photoablation of which sustains the brightness of the ultracompact core: either an accretion disk around the high-mass star (Hollenbach et al. 1994; Yorke & Welz 1996; Richling & Yorke 1997; Lugo et al. 2004), or dense neutral clumps (Dyson et al. 1995; Lizano et al. 1996; Redman et al. 1996). In many instances, ultracompact HII regions are seen to be embedded in larger-scale, more diffuse emission (Kurtz et al. 1999), which would tend to favor the second class of models if the extended emission were physically connected with the ultracompact region. However, it is entirely possible that each of these models is applicable to some subclass of ultracompact regions.

4.2 Later Phases: Compact and Extended Regions

At least some ultracompact HII regions eventually expand to larger sizes, forming the class of compact HII regions (Mezger et al. 1967), with sizes 0.1–0.5 parsec and densities $\simeq 10^4$ cm^{-3}. Most compact regions are embedded inside extended HII regions, which are typically several parsecs in size, with densities $\simeq 100$ cm^{-3}. The distribution of morphological types is similar to that for ultracompact regions (Hunter 1992; Fich 1993), although most modeling effort has gone into explaining the cometary shaped regions. These are similar in appearance to many optically visible HII regions, which are classified as blister-type (Israel 1978), and of which the best-known example is the

Orion Nebula (O'Dell 2001). The low extinction to these optical regions proves that they must be on the near side of any accompanying molecular gas, and they generally show blueshifted velocities of order the ionized sound speed in optical emission lines, indicative of flow away from the molecular cloud (Zuckerman 1973). A broad class of models, commonly referred to as champagne flows, has been proposed to account for these objects. These models share the property that strong density gradients in the neutral/molecular gas allow the ionization front to break out in some directions, leading to a flow of high-pressure ionized gas in the same direction. The original model of Tenorio-Tagle (1979) considered the one-dimensional propagation of an ionization front inside a dense cloud with a sharp edge. Once the ionization front reaches the edge of the cloud, it rapidly propagates through the much rarer intercloud medium and is followed by a strong shock that is driven by the higher pressure ionized cloud material. A rarefaction wave simultaneously travels back into the ionized cloud, initiating the champagne flow that accelerates the ionized gas up to several times the ionized sound speed as it flows away from the cloud. Following studies extended this work to two dimensions (see references in Yorke 1986, section 3.3) and considered the effects of disk-shaped clouds and more gradual cloud boundaries.

Early work on the dynamics of champagne flows considered scenarios that were intrinsically non-steady. However, it is also possible to construct quasi-stationary champagne flow models (Henney et al. 2005) in which the structure of the ionized flow remains constant over several times its dynamical timescale. These models are valid so long as the density of the neutral gas that confines the ionization front on one side is high enough that the ionization front moves slowly and encounters constant upstream conditions during the evolution. It can be shown that in the steady-state limit a divergence of the ionized flow is necessary in order to produce significant acceleration. This is because in a strictly plane-parallel geometry, and in the absence of body forces, the acceleration is proportional to the gradient of the sound speed, which is almost zero in the isothermal ionized gas. Examples of such flows with different degrees of divergence are shown in Fig. 2. The degree of divergence is controlled by the neutral density profile in the lateral direction (perpendicular to the sharp density gradient at the surface of the cloud). When this density profile is constant, as in the upper panels of Fig. 2, the ionization front is concave (negative curvature), leading to weak divergence of the flow and slow acceleration. For a steep lateral density profile, as in the lower panels of Fig. 2, the ionization front becomes convex (positive curvature), giving a strong, almost spherical, divergence to the flow, which now accelerates much more strongly. This configuration is similar to that seen in globule flows. Approximate analytic calculations suggest that the ionization front should be flat when the lateral density profile is proportional to $1/(1 + r^2)$, in which the distance, r, from the symmetry axis has been scaled by the axial offset between the star and the ionization front. Numerical simulations confirm that this is indeed the case, as shown in the central panels of Fig. 2.

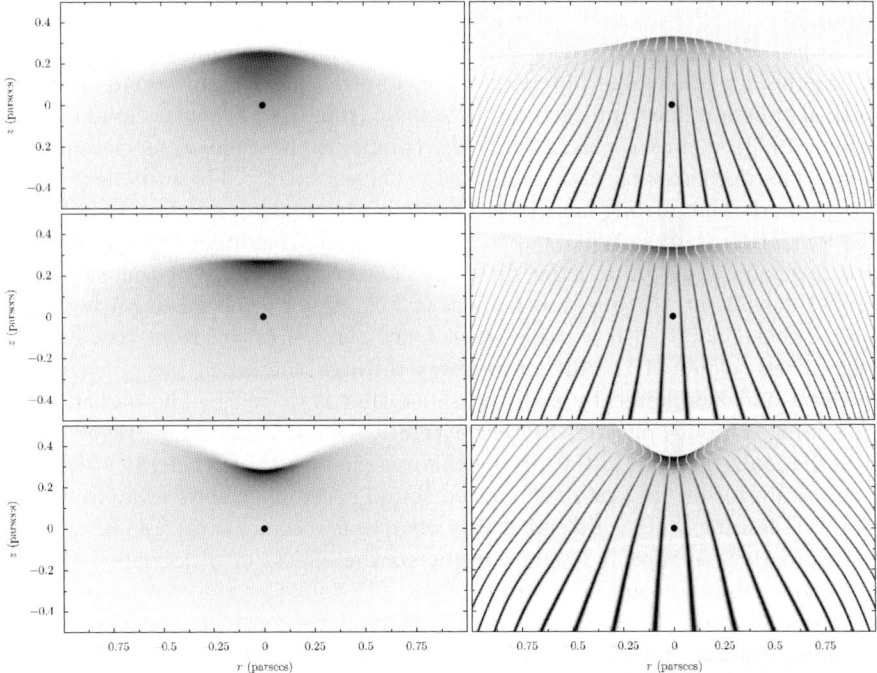

Fig. 2. Simulations of quasi-steady champagne flows, produced by the ionization of the interface of a dense molecular cloud (adapted from Henney et al. 2005). Three different models (top to bottom) are shown with different lateral density profiles in the neutral gas. The left panels shows the ionized density, while the right panels show the total pressure with superimposed streamlines, the darkness of which indicates the gas velocity. The position of the ionizing star is shown by a black circle in each case.

The stellar wind from the ionizing star may also be expected to have an effect on the dynamics of the champagne flow and this was first modelled by Comeron (1997). An alternative explanation for the cometary shapes shown by many regions is the bowshock model of van Buren et al. (1990), previously discussed in the context of ultracompact regions. In this model, the ionizing star moves supersonically through the molecular gas and the ionization front is trapped in the dense shocked shell formed by the interaction between the stellar wind and the ambient gas. An extension of this scenario was studied by Franco et al. (2005). In it the ionizing star moves in and out of a high density molecular cloud core. An exhaustive study of the interplay between champagne flows, stellar winds, and stellar motion was carried out by Arthur & Hoare (2005), who found a great variation in the resultant morphology, depending on the strength of the stellar wind. All the Arthur & Hoare models have concave ionization fronts, as in the top row of Fig. 2. If the stellar wind is weak, as in the first panel of Fig. 3, then the champagne flow is hardly

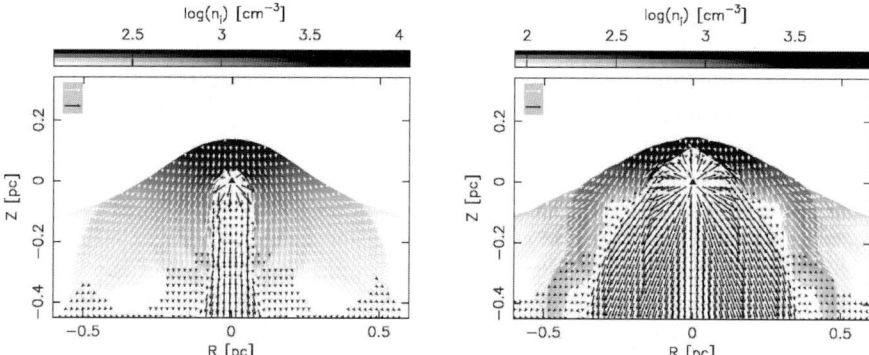

Fig. 3. Simulations of the interaction of a champagne flow with a stellar wind (from Arthur & Hoare 2005). Grayscale shows log of ionized density and arrows show gas velocity on two different scales: the longest white arrow represents 30 km s^{-1} while the longest black arrow represents 2000 km s^{-1}. The left panel shows a model with a weak wind (mass loss rate of $10^{-7} M_\odot$ yr^{-1}) and the right panel shows a model with a strong wind ($10^{-6} M_\odot$ yr^{-1}). Both models have a wind velocity of 2000 km s^{-1} and an ionizing photon luminosity of 2.2×10^{48} s^{-1}.

affected, except in a narrow region around the axis, where it forms a weak bowshock around the stellar wind. A stronger stellar wind, as in the second panel of Fig. 3 has a much more dramatic effect. The point of pressure balance along the axis between the stellar wind and the H II region gas is pushed closer to the ionization front than where the sonic point would have been in the champagne flow. As a result, there is no champagne flow on the axis, although away from the axis a transonic champagne flow does develop, which then shocks against the stellar wind. The large obstacle provided by the wind means that the champagne flow acceleration is predominantly parallel to the ionization front in this case. The inclusion of a stellar velocity in the direction of increasing gradient produces only small changes in the results unless the stellar velocity is higher than the ionized sound speed. Such models always show a greater resemblance to champagne flow models than to classical bowshock models unless the density gradient in the neutral cloud is very shallow. Either cometary or shell-like morphologies can be produced, depending on the steepness of the neutral density distribution.

Another class of models considers the radial expansion of an H II region in a spherical cloud with a density distribution steeper than $r^{-3/2}$ (Franco et al. 1990; Shu et al. 2002; Lizano et al. 2003), in which no static equilibrium solution exists for the position of the ionization front. When the star turns on, an R-type ionization front quickly propagates to large radii, and the ionized gas adopts a configuration in which a roughly uniform density core, bounded by a shock, expands slightly faster than the ionized sound speed. Such a region is doomed to rapidly decline in brightness after a few sound crossing times. Confusingly, these entirely density-bounded models are also referred to as champagne flows, although they are significantly different from the models considered in the previous paragraphs. A classical champagne flow is ionization-bounded on at least one side and will maintain a high brightness indefinitely, as long as sufficient neutral gas exists to confine the ionization front close to the ionizing star on that side.

At even larger scales, any expansion of the HII region will stop when it comes into pressure balance with the local interstellar medium. The mean pressure in spiral arms at the radius of the solar circle is $\simeq 5 \times 10^4$ K cm^{-3}, which is 90% non-thermal (Cox 2005). This would balance that of the ionized gas when its density has fallen to a few cm^{-3}. The equivalent Strömgren radius is $\simeq 30$ parsec for a typical O star, but would be larger for a region ionized by a cluster of stars, in which case vertical density gradients in the disk may become important. However, the time required for an HII region to expand to such sizes is comparable with the evolutionary timescales for high-mass stars, so that the powerful winds from the later stages of stellar evolution and the expansion of supernova remnants will have a profound effect on the dynamics of the region. Similar considerations apply to HII regions seen in external galaxies (Shields 1990), although the largest HII regions (> 500 parsec) show highly supersonic velocities (e.g., Terlevich & Melnick 1981; Gallagher & Hunter 1983; Relaño et al. 2005), which correlate with the luminosities of the regions and may be indicative of virial equilibrium of the ionized gas in the gravitational potential of the star cluster.

4.3 Clumps and Turbulence

One shortcoming of the models mentioned in the previous section is that they consider only smooth density distributions for the neutral gas into which the HII region propagates. In reality, this is unlikely to be the case, since the molecular clouds in which high-mass stars form are known to be highly structured, which is probably the result of supersonic turbulence (Elmegreen & Scalo 2004). Dynamic models that attempt to take into account the clumpy nature of the neutral gas were first proposed by John Dyson and collaborators (Dyson et al. 1995; Williams et al. 1996; Redman et al. 1996, 1998) in order to explain the morphology and lifetimes of ultracompact HII regions. In these models dense neutral clumps are assumed to be distributed throughout the ionized volume and to act as mass-loading sites for a stellar wind from the ionizing star (see the chapter by Julian Pittard for further details on mass-loading). The models are agnostic with respect to the physics of the mass injection process, which is not modeled directly, but which may be ablation by photoionization or hydrodynamic processes. Instead, the flow is treated statistically under the assumption of prompt and complete mixing of the ablated gas with the stellar wind. Similar models have also been presented that treat the ablation process in more detail and explicitly allow for the finite lifetime of the clumps (Lizano et al. 1996; Arthur & Lizano 1997). Both sets of models result in a shell-like morphology for the HII region (if the clumps

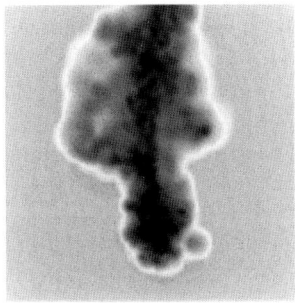 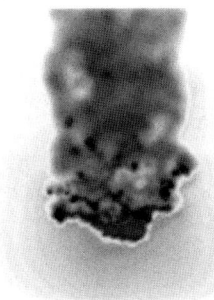 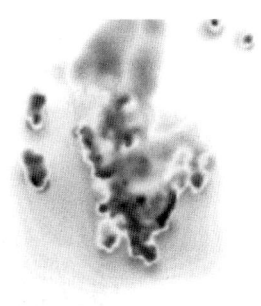

Fig. 4. Three-dimensional simulation of the photoevaporation of a clumpy self-gravitating globule (adapted from Kessel-Deynet & Burkert 2003). *Left:* Phase of maximum compression, 1.6×10^5 years. *Center:* Equilibrium cometary phase, 2.8×10^5 years. *Right:* Fragmentation phase, 4.2×10^5 years. The grayscale measures the absolute value of the difference between the column density of the ionized and neutral gas (white corresponds to the ionization front). Box size is 0.85 parsec in the left panel and 1.8 parsec in the other panels.

are distributed uniformly), bounded by a recombination front through which the ionized wind flows.

A more detailed treatment of the interaction of ionizing photons with a clumpy medium is possible only with three-dimensional numerical simulations. The first such simulation to be carried out studied the response of a static equilibrium self-gravitating spherical globule to an external ionizing flux (Kessel-Deynet & Burkert 2003). The density distribution of the initial globule (of mass 40 $M_\odot$ and radius 1 parsec, on the margin of Jeans instability) is perturbed by Gaussian fluctuations of amplitude $\simeq 50\%$. It is found that the turbulence generated behind the shock driven into the globule by the ionization front is sufficient to prevent gravitational collapse of the globule at its point of maximum compression, despite collapse having occurred at this point in a companion simulation of a globule without density perturbations. After $\simeq 5 \times 10^5$ years, the neutral gas that has survived photoablation has broken up into smaller sub-globules (Fig. 4). Similiar simulations have also been carried out of the photoevaporation of much smaller, non-self-gravitating, clumpy globules (González et al. 2005). In this case, although some fragmentation of the globule occurs, the fragments remain physically attached to one another until the globule is completely evaporated.

Simulations of hydrodynamical and magnetohydrodynamical turbulence in molecular clouds (e.g., Mac Low 1999; Vázquez-Semadeni et al. 2005) have now reached a level of refinement that makes them suitable to use as initial conditions for the evolution of H_{II} regions. Li et al. (2004) carried out a preliminary investigation of this problem by studying only the initial R-type propagation of the ionization front, before the dynamics of the ionized gas becomes important. This approach was extended by Mellema et al. (2005), who carried out a full radiation-hydrodynamic simulation of the evolution of

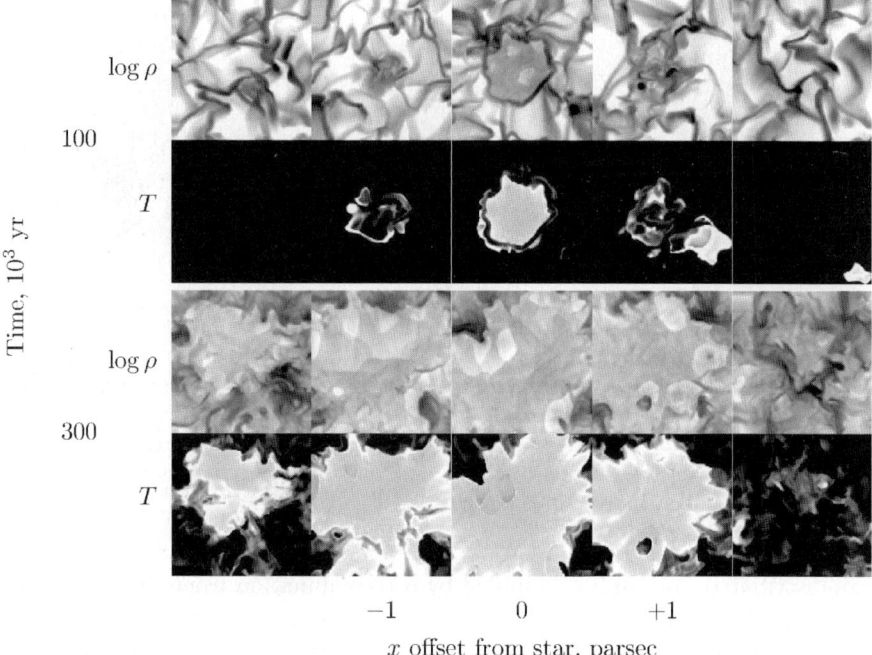

Fig. 5. Dynamical evolution of an H II region in a turbulent molecular cloud, adapted from Mellema et al. (2005). For each of two evolutionary times, 100,000 and 300,000 years, five yz slices through the $(4\ \mathrm{parsec})^3$ computational box are shown, with offsets from the ionizing star as shown on the bottom axis. Density is shown on a negative logarithmic scale between 10 (white) and 10^5 cm^3 (black). Temperature is shown on a positive linear scale between 0 (black) and 10^4 K (white).

an H II region in a turbulent medium, following the birth of an $\simeq 25\ M_\odot$ star inside the densest molecular clump ($\simeq 10^6$ cm^{-3}) that formed in the Vázquez-Semadeni et al. (2005) turbulence simulation.

The results of this simulation are shown in Fig. 5 for two evolutionary times. At the earlier time, the H II region is mainly confined to a compact core of radius $\simeq 1$ parsec, although the ionization front has already broken out to the boundary of the grid through a corridor of low-density neutral gas in one direction, as can be seen in the lower-right corner of the rightmost temperature image. The density variations in the ionized gas are rather mild compared with those in the neutral gas and are of a different nature. Instead of dense sheets and filaments, one finds an almost constant density gas filling roughly half of the ionized region, with the remainder occupied by low-density cavities carved out by transonic photoablation flows, as described in section 3.4 (see also Williams et al. 2001; Henney 2003). This structure of the ionized gas can be better appreciated at the later evolutionary time, where multiple photoablation flows are seen, streaming off dense neutral condensations. The highest ionized densities are found at the ionization fronts of these flows and their mutual collisions produce many weak shocks, which generate density structure in the interior of the H II region. Some of these structures become compressed sufficiently to significantly absorb the ionizing photons, causing the ionized gas beyond them to temporarily recombine. Regions where this has occurred can be seen as intermediate-temperature gas in the bottom row of Fig. 5. The photoablation flows help maintain a high velocity dispersion of the ionized gas, which remains roughly equal to the ionized sound speed during the entire evolution, even though the net radial expansion velocity falls to much lower values (Mellema et al. 2005, Fig. 4). This mechanism is similar to that proposed by Dyson (1968) to explain the velocity dispersion observed in the Orion Nebula.

The predicted appearance of these simulations in optical emission lines is shown in Fig. 6. The [N II] line traces the low-ionization gas close to the ionization front, and these images are dominated by the bright rims of photoablation flows, while the [O III] line traces the higher ionization gas in the interior of the nebula, which shows a more diffuse aspect in the images. The predicted emission line structure is very similar to that observed in real nebulae, such as the Lagoon Nebula shown in Fig. 7. The broad-band filters used in that image include optical emission lines of both low and high ionization, so it should be compared with the superposition of the [N II] and [O III] images

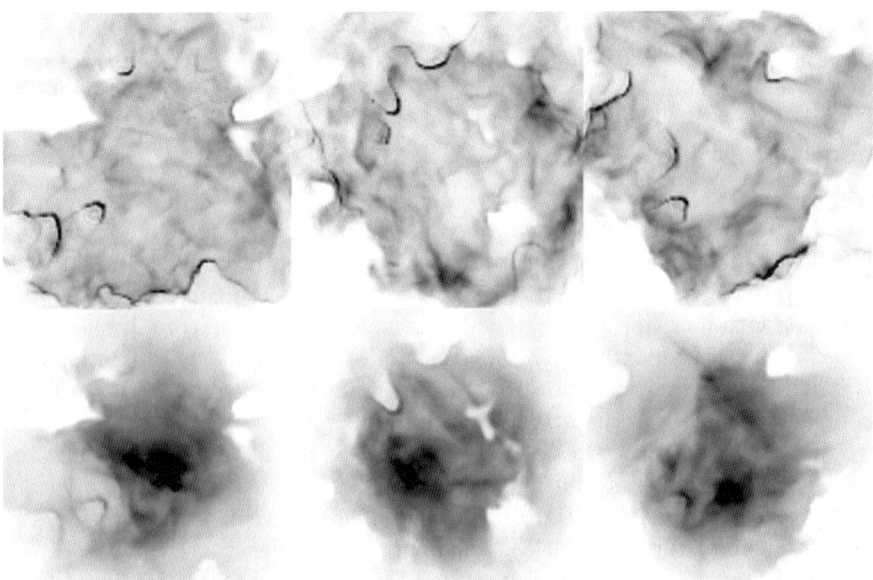

Fig. 6. Fake emission line images from the simulation shown in Fig. 5 at an evolutionary time of 400,000 years. *Top row:* [N II] 6584 Å. *Bottom row:* [O III] 5007 Å. Views along the x, y, and z axes are shown from left to right, respectively.

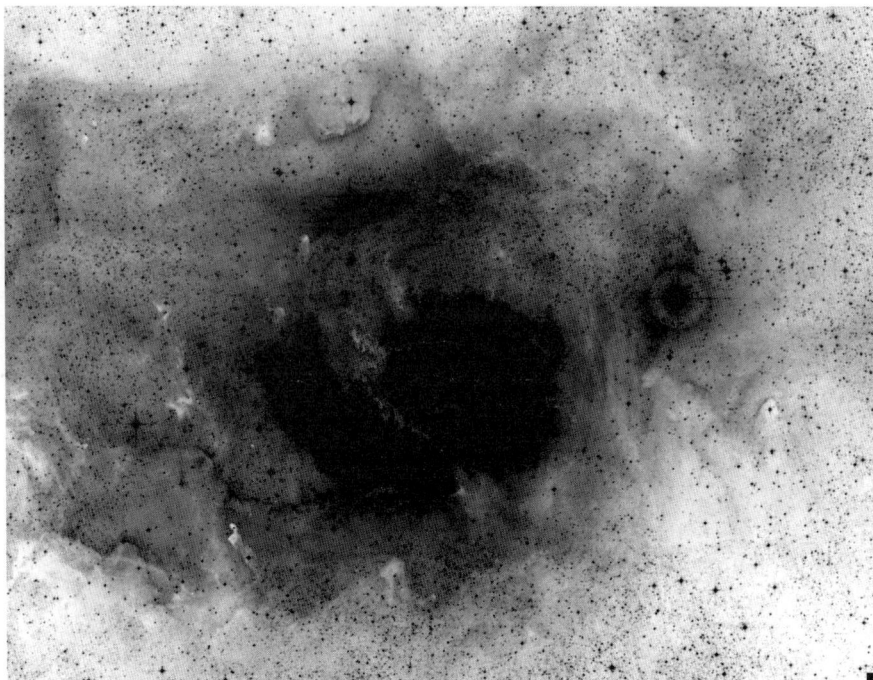

Fig. 7. The Lagoon Nebula (M8). Combination of red and infrared continuum images from the Second Epoch Sky Survey of the UK Schmidt Telescope, operated by the Anglo-Australian Observatory, digitized by the Space Telescope Science Institute. A highly non-linear grayscale mapping has been used to show both bright and faint structures at the same time. The field of view is $\simeq 27 \times 21$ parsec, assuming a distance of 1500 parsec.

(see Fig. 1 of Dufour 1994 for pure emission line images of the inner regions of this nebula). As in the simulations, one sees sharply bounded bright arcs and dark extinction features in the low-ionization periphery of the nebula, together with more diffuse emission in the higher ionization core. It should be noted that the Lagoon Nebula is several times larger than the simulation and is ionized by several massive stars, but a similar morphology is seen in more compact nebulae, such as M16, M20, and M42.

Such sculpted structures at the boundary of the nebula can also be produced via instabilities, even if the ambient neutral gas is initially smoothly distributed (e.g., Garcia-Segura & Franco 1996). A clear example of this is seen in Fig. 3 of Williams (2003), where large-scale density gradients lead to a thin-shell instability and the formation of multiple photoablation flows from the now clumpy shocked neutral layer. Whether ionization front instabilities or pre-exisiting structures in the neutral gas are more important in a given region is still an open question. However, the density contrasts found in molecular clouds on parsec scales are much more extreme than those seen in HII regions, due to the highly supersonic nature of the turbulence and the effects of self-gravity. Thus, it seems likely that the structure of the neutral gas is dominant in shaping H<small>II</small> regions at these scales. At smaller scales, the turbulent velocity dispersion, and hence the density contrast, is much reduced, so that ionization front instabilities may play a more important role, although these tend to saturate at amplitudes of order the recombination length (Williams 2002), which is approximately ten times the ionization front thickness, or $\simeq n^{-1}$ parsec, where n is measured in cm^{-3}.

5 Summary

H<small>II</small> regions, particularly the extended, optically visible variety, have the reputation of being "messy" objects, ill-suited to the austere theoretical prejudice of perfect symmetry. However, as I have tried to illustrate in this review, even their messiest aspects are beginning to fall under the purview of dynamic modelling. We are still some way from a totally satisfactory model of any particular object, let alone H<small>II</small> regions as a class, but a large amount of progress has been made.

References

Arthur, S.J., Hoare, M.G. 2005 arXiv:astro-ph/0511035
Arthur, S.J., Kurtz, S.E., Franco, J., Albarrán, M.Y. 2004 ApJ **608**, 282
Arthur, S.J., Lizano, S. 1997 ApJ **484**, 810
Axford, W.I. 1964 ApJ **140**, 112
Bertoldi, F. 1989 ApJ **346**, 735
Breitschwerdt, D., Kahn, F.D. 1988 MNRAS **235**, 1011
Capriotti, E.R., Kozminski, J.F. 2001 PASP **113**, 677
Carral, P., Kurtz, S.E., Rodriguez, L.F., de Pree, C., Hofner, P. 1997 ApJL **486**, L103+
Comeron, F. 1997 A&A **326**, 1195
Comeron, F., Torra, J. 1996 A&A **314**, 776
Cox, D.P. 2005 ARA&A **43**, 337
de Pree, C.G., Rodriguez, L.F., Goss, W.M. 1995 Revista Mexicana de Astronomia y Astrofisica **31**, 39
de Pree, C.G., Wilner, D.J., Deblasio, J., Mercer, A.J., Davis, L.E. 2005 ApJL **624**, L101
Dorland, H., Montmerle, T. 1987 A&A **177**, 243
Dufour, R.J. 1994 Revista Mexicana de Astronomia y Astrofisica **29**, 88
Dyson, J.E. 1968 Ap&SS **1**, 388
—. 1977 A&A **59**, 161
Dyson, J.E., Williams, D.A. 1997 The physics of the interstellar medium (Bristol: Institute of Physics Publishing)
Dyson, J.E., Williams, R.J.R., Redman, M.P. 1995 MNRAS **277**, 700
Elmegreen, B.G., Scalo, J. 2004 ARA&A **42**, 211

Ercolano, B., Barlow, M.J., Storey, P.J. 2005 MNRAS **362**, 1038
Fich, M. 1993 ApJS **86**, 475
Franco, J., Garcia-Segura, G., Kurtz, S. 2005 arXiv:astro-ph/0508467
Franco, J., Tenorio-Tagle, G., Bodenheimer, P. 1990 ApJ **349**, 126
Fullerton, A.W., Massa, D.L., Prinja, R.K. 2006 ApJ **637**, 1025
Gail, H.-P., Sedlmayr, E. 1979a, A&A **77**, 165
—. 1979b, A&A **76**, 158
Gallagher, J.S., Hunter, D.A. 1983 ApJ **274**, 141
García-Arredondo, F., Arthur, S.J., Henney, W.J. 2002 Revista Mexicana de Astronomia y Astrofisica **38**, 51
García-Arredondo, F., Henney, W.J., Arthur, S.J. 2001 ApJ, **561**, 830
Garcia-Segura, G., Franco, J. 1996 ApJ **469**, 171
Giuliani, J.L. 1979 ApJ **233**, 280
González, R.F., Raga, A.C., Steffen, W. 2005 Revista Mexicana de Astronomia y Astrofisica **41**, 443
González-Avilés, M., Lizano, S., Raga, A.C. 2005 ApJ **621**, 359
Henney, W.J. 2001 in The Seventh Texas-Mexico Conference on Astrophysics: Flows, Blows and Glows (Eds. William H. Lee and Silvia Torres-Peimbert) Revista Mexicana de Astronomía y Astrofísica (Serie de Conferencias) **10**, 57–60
Henney, W.J. 2003 in Winds, bubbles, and explosions: a conference to honor John Dyson (Eds. S.J. Arthur and W.J. Henney) Revista Mexicana de Astronomia y Astrofisica Conference Series **15**, 175–180
Henney, W.J., Arthur, S.J. 1998 AJ **116**, 322
Henney, W.J., Arthur, S.J., García-Díaz, M.T. 2005a ApJ **627**, 813
Henney, W.J., Arthur, S.J., Williams, R.J.R., Ferland, G.J. 2005b, ApJ **621**, 328
Hollenbach, D., Johnstone, D., Lizano, S., Shu, F. 1994 ApJ **428**, 654
Hosokawa, T., Inutsuka, S.-I. 2005a arXiv:astro-ph/0511165
—. 2005b, ApJ **623**, 917
Hummer, D.G., Kunasz, P.B. 1980 ApJ **236**, 609
Hunter, D.A. 1992 ApJS **79**, 469
Inoue, A.K. 2002 ApJ **570**, 688
Israel, F.P. 1978 A&A **70**, 769
Kahn, F.D. 1954 Bull. Astron. Inst. Netherlands **12**, 187
Kahn, F.D. 1969 Physica **41**, 172
Kessel-Deynet, O., Burkert, A. 2003 MNRAS **338**, 545
Keto, E. 2002 ApJ **580**, 980
—. 2003 ApJ **599**, 1196
Kurtz, S., Churchwell, E., Wood, D.O.S. 1994 ApJS **91**, 659
Kurtz, S.E., Watson, A.M., Hofner, P., Otte, B. 1999 ApJ **514**, 232
López-Martín, L., Raga, A.C., Mellema, G., Henney, W.J., Cantó, J. 2001 ApJ **548**, 288
Lamers, H.J.G.L.M., Cassinelli, J.P. 1999 Introduction to Stellar Winds (UK: Cambridge University Press)
Li, Y., Mac Low, M.-M., Abel, T. 2004 ApJ **610**, 339
Lizano, S., Canto, J., Garay, G., Hollenbach, D. 1996 ApJ **468**, 739
Lizano, S., Galli, D., Shu, F., Cantó, J. 2003 in Winds, bubbles, and explosions: a conference to honor John Dyson (Eds. S.J. Arthur and W.J. Henney) Revista Mexicana de Astronomia y Astrofisica Conference Series **15**, 166–171
Lugo, J., Lizano, S., Garay, G. 2004 ApJ **614**, 807

Mac Low, M.-M. 1999 ApJ **524**, 169
Mathews, W.G. 1967 ApJ **147**, 965
Mellema, G., Arthur, S.J., Henney, W.J., Iliev, I.T., Shapiro, P.R. 2005 arXiv:astro-ph/0512554
Mezger, P.G., Altenhoff, W., Schraml, J., Burke, B.F., Reifenstein, E.C., Wilson, T.L. 1967 ApJL **150**, L157
Mihalas, D. 1978 Stellar atmospheres (San Francisco: W.H. Freeman and Co.)
Mizuta, A., Kane, J.O., Pound, M.W., Remington, B.A., Ryutov, D.D., Takabe, H. 2005 ApJ **621**, 803
O'Dell, C.R. 2001 ARA&A **39**, 99
O'Dell, C.R., Henney, W.J., Ferland, G.J. 2005 AJ **130**, 172
Osterbrock, D.E., Ferland, G.J. 2006 Astrophysics of gaseous nebulae and active galactic nuclei (Sausalito, CA: University Science Books)
Redman, M.P., Williams, R.J.R., Dyson, J.E. 1996 MNRAS **280**, 661
—. 1998 MNRAS **298**, 33
Relaño, M., Beckman, J.E., Zurita, A., Rozas, M., Giammanco, C. 2005 A&A **431**, 235
Richling, S., Yorke, H.W. 1997 A&A **327**, 317
Sewilo, M., Churchwell, E., Kurtz, S., Goss, W.M., Hofner, P. 2004 ApJ **605**, 285
Shapiro, P.R., Iliev, I.T., Alvarez, M.A., Scannapieco, E. 2005 arXiv:astro-ph/0507677
Shields, G.A. 1990 ARA&A **28**, 525
Shu, F.H. 1992 Physics of Astrophysics, **II** (University Science Books)
Shu, F.H., Lizano, S., Galli, D., Cantó, J., Laughlin, G. 2002 ApJ **580**, 969
Tenorio-Tagle, G. 1979 A&A **71**, 59
Terlevich, R., Melnick, J. 1981 MNRAS, 195 839
van Buren, D., Mac Low, M.-M., Wood, D.O.S., Churchwell, E. 1990 ApJ **353**, 570
Vázquez-Semadeni, E., Kim, J., Shadmehri, M., Ballesteros-Paredes, J. 2005 ApJ **618**, 344
Williams, R.J.R. 1999 MNRAS **310**, 789
—. 2002 MNRAS **331**, 693
Williams, R.J.R. 2003 in Winds, bubbles, and explosions: a conference to honor John Dyson (Eds. S.J. Arthur and W.J. Henney) Revista Mexicana de Astronomia y Astrofisica Conference Series **15**, 184–189
Williams, R.J.R., Dyson, J.E., Redman, M.P. 1996 MNRAS **280**, 667
Williams, R.J.R., Ward-Thompson, D., Whitworth, A.P. 2001 MNRAS, **327**, 788
Wood, D.O.S., Churchwell, E. 1989 ApJS **69**, 831
Yorke, H.W. 1986 ARA&A **24**, 49
Yorke, H.W., Welz, A. 1996 A&A **315**, 555
Zuckerman, B. 1973 ApJ **183**, 863

MHD Ionization Fronts*

R.J.R. Williams

AWE Aldermaston, Reading, RG7 4PR, UK
robin.williams@awe.co.uk

1 Introduction

The structure and evolution of H II regions has been described by Will Henney in this volume. Young massive stars emit copious fluxes of ultraviolet radiation, forming regions of almost entirely ionized gas in their surroundings. The dynamics of an H II region depend on the history of the illumination and the initial structure of the surrounding medium.

In the classical theory of ionization fronts, the medium into which one moves behaves as a pure fluid. However, it is clear from observational evidence that the neutral gas in star-forming regions is threaded by dynamically significant magnetic fields (Crutcher 1999), with consistent mean direction over large scales despite the relatively high turbulent velocities in the clouds (Ward-Thompson et al. 2000). Cosmic ray ionization maintains a small density of ions within the neutral clouds, and they behave as a magneto-hydrodynamic (MHD) fluid over large scales. The loss of magnetic field support from the clouds is a crucial part of the star formation process.

John Dyson suggested that we consider the effect that magnetic fields could have on the properties of the H II regions around young massive stars. This topic has been rather neglected since the pioneering work of Lasker (1966), possibly in part due to the difficulty of obtaining detailed observations of the magnetic fields present in these diffuse media. Nevertheless, we found that the effects predicted by dynamical models and analyses should indeed have observable effects on the evolution of H II regions. In this chapter, I will describe these effects.

---

* © British Crown Copyright 2006/MOD

2 Magnetic Field Observations

Magnetic fields in astronomical systems are difficult to observe. Most of the techniques available test observational sensitivity to the limit, and yield only partial information about the field distribution.

Techniques which have been used include observations of the Zeeman effect in absorption, for species such as OH and CN, and in the OH maser lines emitted by clumps around many H II regions, and polarization measurements of dust emission (Crutcher 2005). These results are complementary: Zeeman measurements provide information about the line-of-sight magnetic field strength, while polarization measurements indicate the direction of the field in the plane of the sky. No direct measurement of the field strength is available from the polarization measurements, although a statistical indication of the field strength can be obtained by comparing the line-of-sight velocity dispersion with the angular scatter in polarization direction (Chandrasekhar & Fermi 1953; Vallee & Fiege 2005). These methods are all sensitive to magnetic field distributions only in the neutral gas which surrounds the ionized nebula rather than within the H II region, although OH Zeeman observations may be localized at the interface between ionized and molecular material (Crutcher 1991).

Nevertheless, a fairly consistent picture has emerged for the magnetic fields in nearby star forming regions (Crutcher 1999). The magnetic fields in the neutral gas are strong, with pressures typically 10 times the thermal pressure of the neutral gas. The energy density in random motions within the clouds are comparable to those in the magnetic fields. Dust polarization observations (Ward-Thompson et al. 2000; Matthews et al. 2005) indicate large-scale magnetic field structures in pre-stellar cores, with directions correlated across star forming regions and little evidence for turbulent tangling except for in the most massive cores.

The hot H II region S106 has a good collection of magnetic field measurements (Roberts, Crutcher & Troland 1995; Vallee & Fiege 2005). In this bipolar region, the ionized gas lobes appear to be driving a shocked shell outwards into the the surrounding molecular cloud. Polarimetric data suggest that the magnetic field is primarily directed parallel to the dense rim of molecular material around the bipolar lobes. Closer to the central star, the field moves towards the equatorial plane, in the direction of a residual disk of molecular material. This much is in accord with a simple picture in which the magnetic field is swept up and cushions the shock surrounding the ionized outflow from the protostar. However, the dense shell produced in this manner will be a plasma with a low value of β (the ratio of the thermal and magnetic pressures), and so subject to both small scale instabilities (Falle & Hartquist 2002) and, as we shall see, large scale loss of magnetic support. The result of these large scale instabilities may be the molecular component with low magnetization but *high* relative velocity found in the Zeeman observations (Williams, Dyson & Hartquist 2000).

3 Ionization Dynamics of Magnetized Material

As is the case for ionization fronts propagating into material with negligible magnetic field, ionization physics suggests that the fronts will be sharp, with widths typically $1/(na) \simeq 10^{14} n_4^{-1}$ cm for a front propagating into material with a density of $10^4 n_4$ cm^{-3} hydrogen atoms.

Magnetized ionization fronts, however, could have significant width due to slip between the ions, coupled to the magnetic field, and neutral material, which dominates the mass, in the weakly ionized molecular gas ahead of the shock. Using the results for a 10 km s^{-1} shock in Figure 2 of Ciolek & Roberge (2002) (see also Stone 1997), the ion-neutral drag length is $L_{\rm drag} \simeq 10^{18} n^{-1/2}$ cm. This scaling is approximate for a fixed value of the shock speed and the upstream Alfvén speed, and a fractional ionization varying as $n^{-1/2}$ (cf. Eq. 62 of the chapter by Tom Hartquist and Ove Havnes) and cases in which grain-neutral friction is negligible and the ion flux is conserved. Comparing $L_{\rm drag}$ to the Strömgren radius, we find

$$\frac{L_{\rm drag}}{R_{\rm S}} \simeq 0.015 \frac{n_{\rm H\,II}^{1/6}}{S_{48}^{1/3}} \left(\frac{n_{\rm H\,II}}{n_{\rm mol}}\right)^{1/2}, \qquad (1)$$

where $n_{\rm H\,II}$ is the number density (in cm^{-3}) of the spherical H II region around a source of $S_{48} 10^{48}$ s^{-1} ionizing photons, and $n_{\rm mol}$ is the density of the molecular gas, which will for a D-type ionization front will typically be 100 to 1000 times greater than that in the H II region. Hence, it is still reasonable to treat the fronts as sharp discontinuities within the global flow.

3.1 Jump Conditions for MHD Ionization Fronts

The jump conditions for magnetized ionization fronts are derived from the flow equations in a similar manner to that used to obtain the Rankine-Hugoniot conditions for shocks. Indeed, they are fundamentally the same as the Rankine-Hugoniot conditions, with the exception that the energy flux increases due to photoionization. The gas downstream of the ionization front is assumed to be held at an equilibrium temperature determined by the balance of heating, by photoionization, and radiative cooling, with the highly temperature-dependent forbidden line emission being responsible for keeping the equilibrium temperature at $\sim 10^4$ K. The imposed energy flux also determines the ionization front velocity, rather than it being a free parameter as in a shock, and so the number of flow characteristics entering and leaving the front are the same (at least for weak fronts).

Various authors (e.g., Abe & Sakashita 1963; Okamoto 1968; Hill 1977; Redman et al. 1998) present the jump conditions for a flow containing a magnetic field in the plane of the ionization front as a prelude to more detailed work. In this case, flux freezing implies that the magnetic field strength scales directly with the material density, so there are only two dynamical equations and hence typically two allowed solutions, essentially the hydrodynamical solutions cushioned by the additional magnetic pressure. The jump conditions in fact have three solutions in this case, one of which corresponds to a nonphysical negative density. This results from the effective equation of state for the magnetized material, which is

$$p_{\text{eff}}(\rho) = \rho c^2 + \frac{1}{8\pi}\left(\frac{B_0}{\rho_0}\right)^2 \rho^2, \qquad (2)$$

where flux freezing gives the magnetic field as $B = (B_0/\rho_0)\rho$ if the gas originated with uniform density ρ_0 and magnetic field B_0. For a given pressure there are two solutions for ρ – but only one is positive. The overall result is that the effective wave speed in the magnetized gas is

$$c_{\text{eff}} = c\left(1 + \frac{1}{4\pi}\frac{B^2}{\rho c^2}\right)^{1/2}, \qquad (3)$$

which is larger than for isothermal flow. As a result, compared to the hydrodynamical case, the limiting speeds of the D- and R-type fronts will be significantly larger, the reduction in density across a D-type front less significant, and the forbidden region narrower.

A similar argument can be applied for randomly oriented magnetic fields. The anisotropic expansion of magnetic field within the front may mean that the field strength does not exactly follow the $B \propto \rho^{2/3}$ expected for quasi-static flows, but similar classes of magnetically cushioned ionization fronts would certainly be expected.

Okamoto (1968) then uses the jump conditions for the flow and its perturbations to analyze the stability of the ionization fronts, finding that, in contrast to the suggestion advanced by Axford (1964), the magnetic fields destabilize the ionization fronts for wavenumbers at some angles to the surface magnetic field.

For oblique upstream magnetic fields, the jump conditions may be derived from the MHD equations (Williams, Dyson & Hartquist 2000). These conditions were presented by Lasker (1966). Orienting axes so that $\hat{z}$ is normal to the front, we take the upstream velocity and magnetic field to be in the (x, z) plane (without loss of generality). The jump conditions are then

$$[\rho v_z] = 0 \qquad (4)$$
$$[\rho v_z^2 + p + B_x^2/8\pi] = 0 \qquad (5)$$
$$[\rho v_z v_x - B_z B_x/4\pi] = 0 \qquad (6)$$
$$[B_z] = 0 \qquad (7)$$
$$[v_x B_z - v_z B_x] = 0. \qquad (8)$$

Brackets indicate the differences between the upstream and downstream values of the bracketed quantity. We use an isothermal equation of state $p = \rho c_s^2$, where the sound speed, c_s, increases across the front but is constant on either side of it. Subscripts 1 and 2 denote upstream and downstream parameters, respectively, with $v_x = u_{1,2}$, $v_z = v_{1,2}$, $c_s = c_{1,2}$ and $B_x = B_{1,2}$. Hence we can rewrite the conditions as

$$\rho_1/\rho_2 = v_2/v_1 \equiv \delta \tag{9}$$
$$\rho_2(v_2^2 + c_2^2) + B_2^2/8\pi = \rho_1(v_1^2 + c_1^2) + B_1^2/8\pi \tag{10}$$
$$\rho_2 v_2 u_2 - B_z B_2/4\pi = \rho_1 v_1 u_1 - B_z B_1/4\pi \tag{11}$$
$$u_2 B_z - v_2 B_2 = u_1 B_z - v_1 B_1 \tag{12}$$

Equations (9), (11) and (12) give

$$B_2 = \frac{m_1^2 - 2\eta_1}{\delta m_1^2 - 2\eta_1} B_1, \tag{13}$$

where $m = v/c_s$ and we define $\eta = B_z^2/8\pi\rho c^2$ and $\xi = B_x^2/8\pi\rho c^2$ (the z and x contributions to the reciprocal of the plasma β). The dependence on the upstream transverse velocity has disappeared, as expected as a consequence of frame-invariance.

In equation (10), we substitute with (13) for B_2 and use equation (9) to eliminate ρ_2 and v_2 to find that, so long as $\delta \neq 0$ and $\delta m_1^2 \neq 2\eta_1$, the dilution factor δ is given by the quartic equation (see also Lasker 1966)

$$m_1^6 \delta^4 - m_1^4(1 + m_1^2 + 4\eta_1 + \xi_1)\delta^3$$
$$+ m_1^2(\alpha m_1^2 + 4\eta_1(1 + m_1^2 + \eta_1 + \xi_1))\delta^2$$
$$+ (\xi_1 m_1^4 - 4\eta_1(\eta_1 + m_1^2(\eta_1 + \xi_1 + \alpha)))\delta$$
$$+ 4\alpha\eta_1^2 = 0, \tag{14}$$

where $\alpha = (c_2/c_1)^2$ (100 is a typical value).

For a given set of initial conditions, this equation has four solutions, and it is not immediately clear which of these solutions are physically possible. Lasker (1966) suggested that the physical solutions could be isolated by perturbing continuously from conditions where the result was known. However, it is in fact possible to find a simple criterion by using characteristic analysis (cf. Falle & Komissarov 1999).

As has been outlined by Will Henney in this volume, a similar situation occurs for the hydrodynamic equations. Then the relevant solutions are those for the weak cases, in which the flow through the front either remains subsonic (weak D-type) or remains supersonic (weak R-type). In addition, strong D-type solutions, in which internal constraints within the front allow a solution to exist which is underdetermined from the external flow, may occur. These strong D-type fronts take the flow from sub- (on the neutral inflow side) to super-sonic (in the ionized outflow), and are found to be particularly important in divergent flows from dense globules.

In the present case, we have four possible exhaust states. However, the physical solutions can again be isolated by requiring the same number of flow characteristics to enter and leave the front. This suggests that four types of weak MHD ionization fronts should occur, characterized by the flow velocity through the front, compared to that of the slow, Alfvén and fast MHD characteristics: slow D, slow R, fast D and fast R. The flow through a slow D front moves more slowly than the slow mode waves in both the neutral and ionized gas; that through a slow R front travels faster than the slow speed but slower than the Alfvén velocity. The Alfvén velocity is a singular case, at which the ionization front jump conditions predict that the flow passes through the front with constant velocity and density. Above the Alfvén velocity, fast D and fast R type fronts are also possible. Strong solutions may also exist, which take the flow upwards through one or more critical speeds subject to internal constraints on the front structure.

The phase space of allowed solutions is, naturally, somewhat more complex than that found for hydrodynamical ionization fronts. In Figure 1, I show some examples, plotted as a function of the hydrodynamical Mach number of the incoming flow, $m_1 = v_1/c_1$, and the strength of the magnetic field in the plane of the front compared to the thermal pressure, ξ_1. The four regions of solutions can be seen, separated by forbidden regions around the slow and fast magnetosonic speeds.

4 The Microstructure of MHD Ionization Fronts

In the previous section, the physical classes of ionization fronts were isolated on the basis of characteristic constraints. However, it is of interest to verify that the solutions found from these conditions do indeed have realizable internal structures. In particular, in the hydrodynamical case, the region where solutions are forbidden is, in fact, somewhat wider than that implied by the jump conditions. This is because the flow within the ionization front heats rather beyond the ionization equilibrium temperature within the front.

Hill (1977) presented detailed models of the internal structures of fronts including a magnetic field in the plane of the ionization front. For the case of oblique magnetic fields, we integrated the steady state flow equations (Williams & Dyson 2001), for a simplified model of the internal physics given by Axford (1961). We discovered that we could indeed find smooth solutions for the fronts allowed by the characteristic conditions, and that the increase of the width of the forbidden region also occurs for these MHD ionization fronts. An example of our results are shown in Figure 2, where the solutions can be seen not to quite reach the edge of the forbidden region for the fast R case.

The discontinuous solution shown at the lower limit of this band includes a steady internal shock. It is possible to interpret the collection of solutions shown in the diagram, with similar upstream conditions but varying front

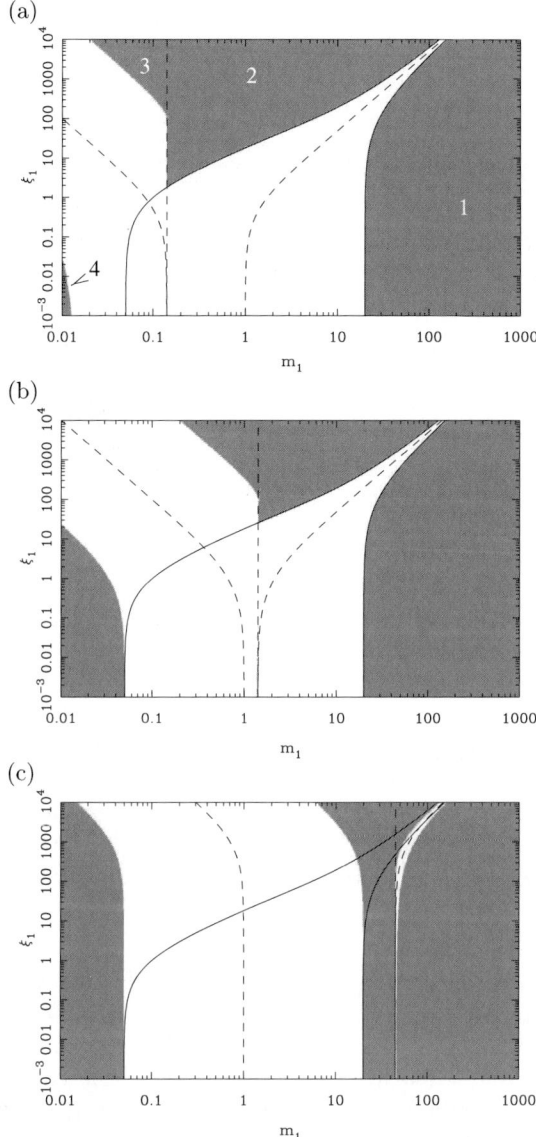

Fig. 1. Allowed obliquely magnetized IFs (shown grey). The solid lines show the edges of the forbidden region for $\eta = 0$, and the dashed lines where the inflowing gas moves at the slow, Alfvén and fast speeds, from left to right. The grey regions correspond to allowed weak solutions for (a) $\eta_1 = 0.01$, (b) $\eta_1 = 1$, and (c) $\eta_1 = 1000$. In (a) the regions are labelled: 1 – the $1 \to 1$, fast-R solutions; 2 – the $2 \to 2$, fast-D solutions; 3 – the $3 \to 3$, slow-R solutions; 4 – the $4 \to 4$, slow-D solutions. The fast-D and slow-R regions are separated by the dashed line at the upstream Alfvén speed. Critical solutions will be expected at the right-most edge of the slow-D type region (a $4 \to 3$ IF) and of the fast-D type region (a $2 \to 1$ IF). From Williams, Dyson & Hartquist (2000).

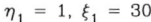
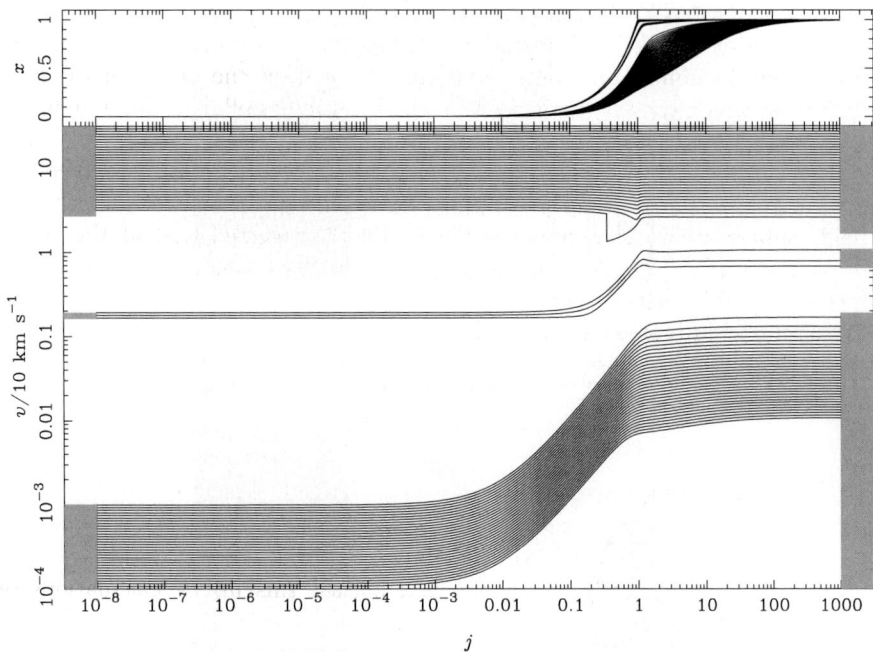

Fig. 2. Internal structure for IFs with $\eta_1 = 1$, $\xi_1 = 30$, for a range of inflow velocities. The panels show fractional ionization, x, and velocity, v as functions of dimensionless radiation intensity, j. The lower ionization curves correspond to the lower velocity curves. The grey regions to the left show the allowed IF for the slow-D, fast-D and fast-R allowed for this value of ξ_1, predicted from Fig. 1(b). The regions to the right show the corresponding range of exit velocities for *weak* IF. From Williams & Dyson (2001).

velocity, as an example of the quasi-static evolution of the front structure as a large scale H II region expands. The static internal shock appears as the first stage of the conversion of the front from one type to another, at the point where the flow within the front first decreases to the characteristic speed. As the front slows further, the shock strengthens and moves further into the neutral gas, eventually escaping the ionization front entirely and moving into the large-scale flow.

The solution shown behind the shock in Figure 2 (i.e. the right-hand part) can also be smoothly extended into the neutral component, reaching a different initial state. This solution takes the flow continuously from a sub-critical to super-critical velocity, i.e. it is of *strong* fast D type. Note that the development of internal shocks and occurrence of strong-D type solutions, where the physical solutions do not reach the edge of the band predicted by the jump conditions, both result from the presence of an internal temperature maximum within the ionization front.

Even when there is no magnetic field in the medium, they can be generated within ionization fronts. This may have been the origin of the seed field from which the magnetic fields in the local interstellar medium grew. At small scales, the differences in the inertia of ion and electron fluids in an ionization front cause an electric field to be produced, and curvature of this field resulting from density inhomogeneities drives the generation of a magnetic field. Subramanian, Narasimha & Citre (1994) suggested that at the epoch of cosmological ionization, surrounding dense clumps within the intergalactic medium would generate fields by a Biermann battery. These results have been corroborated in cosmological simulations including ionization, demonstrating that mean field strengths $B \simeq 10^{-19}$ G can be produced (Gnedin, Ferrara & Zweibel 2000).

5 Global Dynamics

We have argued that only certain solutions of the MHD ionization front jump conditions are physically realizable, and backed this up by calculations of the substructures with suitable physics included. The remaining question is whether these jumps solutions are found in practise when solutions are sought for large-scale problems. In this section, we will present results for two cases, the evolution of plane parallel fronts and that for an ionized globule.

There are some discussions of global dynamics in the literature. For example, Franco, Tenorio-Tagle & Bodenheimer (1989) discussed the development of ionization fronts in magnetized winds from the surface of an accretion disc, assuming that these fields are sufficiently strong to restrain the gas to flow along the initial field lines.

Bertoldi (1989) in his analytical study of the evolution of irradiated globules investigated the cases where the initial globule is threaded by a uniform magnetic field, parallel or perpendicular to the impinging radiation field. He suggested that the leading shock into the globule will compress the field into a direction nearly parallel to the dense shell between shock and ionization front, nearly independent of the direction of the field in the initial configuration. A small amount of flux will escape, nearly parallel to the ionization front surface, but most flux will be compressed into the dense neutral core which ablates in a quasi-steady fashion once the main ionization front has moved far beyond the condensation. Bertoldi provided analytic estimates of some of the properties of these clouds, but suggests that the structures are likely to be sufficiently complex that these may not fully capture the development of these globules.

Ryutov et al. (2005) discussed the effects of magnetic fields on the development of ionized columns, such as those observed in the Eagle Nebula. They suggested that magnetic fields may be able to explain the excess pressure of the ionized flow compared to the thermal pressure in the columns, either as a large-scale uniform field or a random field of 'magnetostatic turbulence'. Numerical studies imply that turbulent motions within the columns should be strongly suppressed by the formation of shocks, although there is some observational evidence for that the random velocities within the columns are large enough to support the clouds (Williams, Ward-Thompson & Whitworth 2001). Uniform upstream fields may suppress the development of the columns, as tension in the swept-back field lines around a condensation will tend to pull it backwards towards the main ionization front. However, they also noted that the compression of the magnetic field lines within the column may also allow reconnection to occur, limiting the magnitude of this restoring force.

5.1 Plane-Parallel Fronts

If we impose an incident radiation field on uniform neutral material in plane parallel geometry, we would expect the ionization front to slow with time. This is, therefore, an excellent way to study the development of plane-parallel ionization fronts.

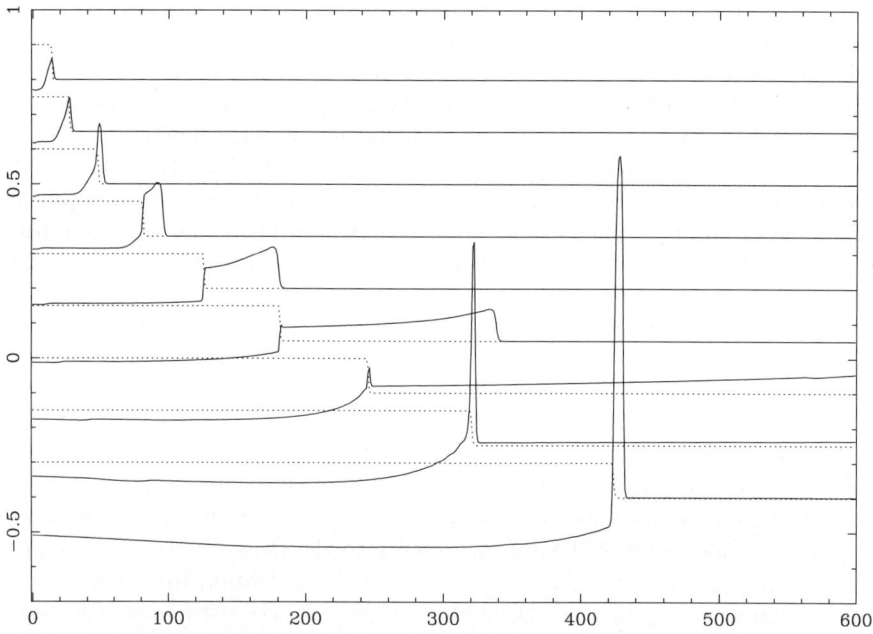

Fig. 3. Time development of the density (solid) and ionization fraction (dotted) in a plane-parallel simulation of the development of an MHD ionization front. The curves are scaled and offset by arbitrary factors, and are presented in time order from top to bottom. From Williams, Dyson & Hartquist (2000).

Figure 3 shows results of time dependent numerical MHD calculations for a plane-parallel flow subject to an increasing incident radiation field (Williams, Dyson & Hartquist 2000). As the ionization front slows, it goes through several types of transitions, as predicted by the regions of allowed solutions found above. Initially, the front is of fast R type, with the density of the flow increasing as it passes through the ionization front, but soon a fast shock escapes the region, leaving the front as a fast D type (with decreasing density). However, once the ionization front slows below the Alfvén speed, the density starts again to increase through the front (now of slow R form) before a slow mode shock is eventually driven out into the surrounding neutral medium, leaving a slow D ionization front. The shell of neutral material behind the outgoing slow shock is of substantially greater density than that behind the fast shock, as the material here does not benefit from cushioning by the swept up magnetic field.

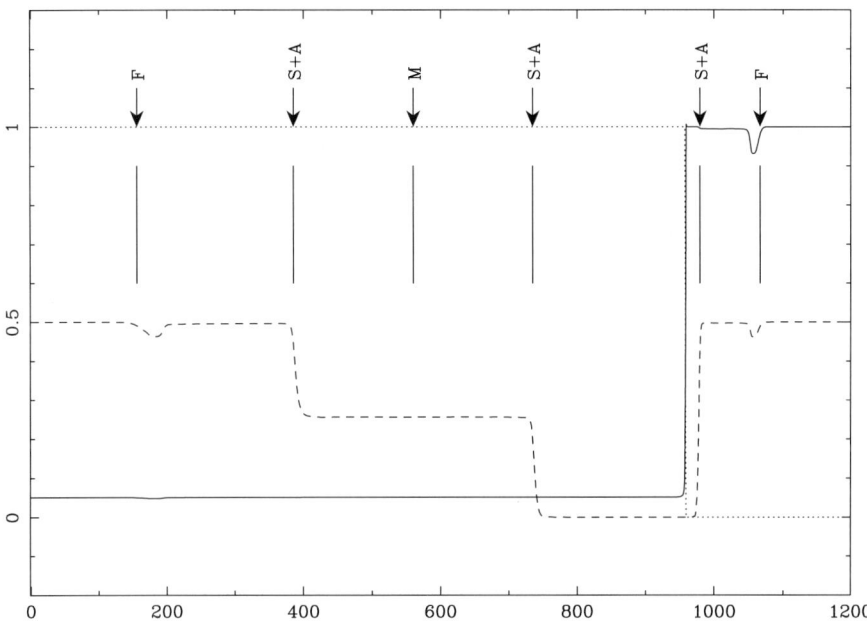

Fig. 4. Relaxation of a front initially close to the hydrodynamic D-critical conditions. From Williams, Dyson & Hartquist (2000).

From Figure 1, it will be noted that the MHD jump conditions often appear to disallow solutions obeying the *hydrodynamic* jump conditions with magnetic fields precisely perpendicular to their surface. It is clear that these fronts will have realizable internal structures, as the magnetic field influence drops out of the structure equations. While these solutions pass through characteristic speeds for MHD waves, this may be possible for singular classes of physical fronts, in a similar manner to hydrodynamic strong D-type fronts. To investigate the existence of these fronts, we calculated the time development of a flow with conditions marginally perturbed from the hydrodynamical critical conditions, see Figure 4. Rather than relaxing to one of the allowed classes of oblique MHD ionization fronts, the IF emitted weak MHD waves to switch off the transverse component of the magnetic field, and returned to hydrodynamic D-critical conditions. However, due to the restricted resolution and geometry of the flow in this simulation, this remains only suggestive of the stability of these 'extra-strong' ionization fronts.

5.2 Globule Photoevaporation

Figure 5 shows the results of a model of the photoevaporation of a magnetized globule in two dimensions, calculated with the MHD code Athena (Gardiner & Stone 2005). In this model, the initial magnetic pressure is approximately 10 times smaller than the thermal pressure. Different orientations of initial magnetic field still have significant effects on the dynamics of the flow, but the overall field topology is remarkably similar for the perpendicular and 45° cases (not illustrated). For a flow with such a complex geometry, it is difficult to unambiguously categorize the class of the ionization front across its surface, but for this relatively weakly magnetized case, the conditions at the front would be expected to be close to those for a hydrodynamic ionization front in any case.

Figure 5(a) shows the logarithm of the flow density, together with contours following the magnetic field direction. The magnetic field has bowed into the column, as expected, although many of the field contours leave the surface in a sheath of intense field at a distance behind the tip of the finger, which surrounds a bubble of material blown by the transonic flow from the clump surface. The magnetic field in the ionized gas close to the ionization front is general nearly perpendicular to the surface, as expected. The density field in the ionized gas is generally fairly smooth, apart from the large-scale decrease expected as a result of the overall divergence.

Figure 5(b) shows the magnitude of the flow velocity, relative to the initial rest frame of the neutral material. The velocity of the flow at the ionization front appears to be fairly constant. Comparing the magnetic field amplitudes shown in Figure 5(c), there is a weak correlation between the highest velocities and the stronger magnetic fields in the weak streamers at the head of the column. For the strongest, and best resolved, magnetic field streamers, the local velocity seems instead to be smaller than elsewhere, except for a thin high-velocity flow at the edge furthest from the ionization source.

These results may be compared to the physical arguments of Bertoldi (1989). In general terms, the picture which emerges is similar to that described by Bertoldi. However, the concentration of magnetic field in the tip of the column is less dramatic than he predicts, probably as a result of reconnection driven by sweeping backwards of the exterior field by the main shock and

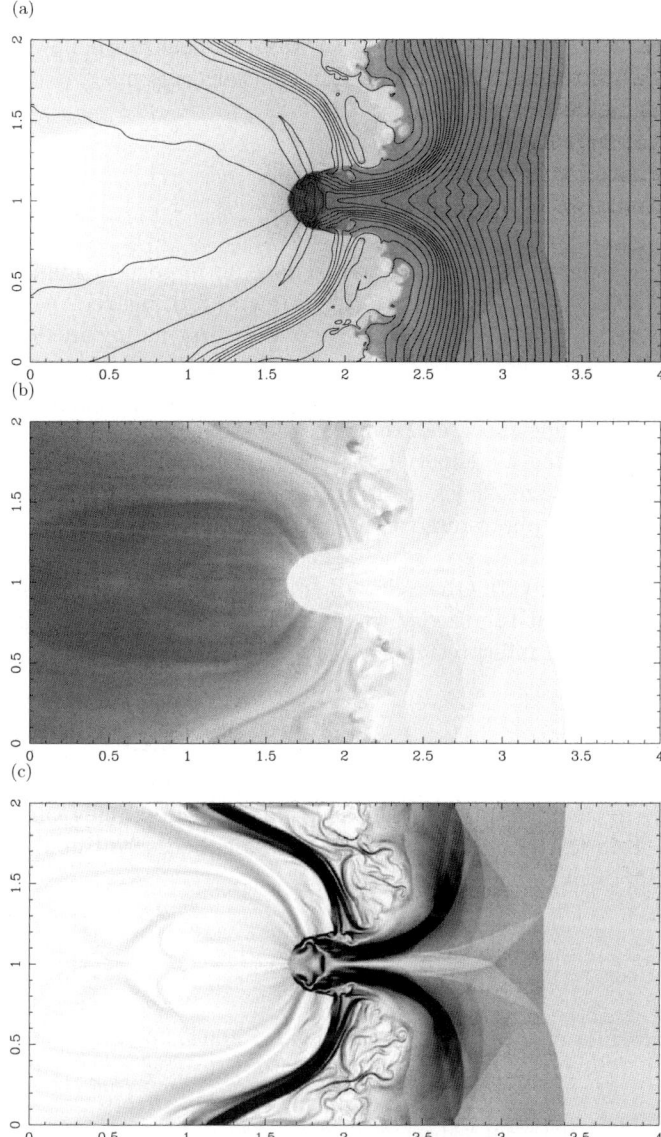

Fig. 5. Photoevaporation of a uniformly magnetized condensation. (a) log density, with magnetic field contours superimposed, (b) flow speed (relative to initial rest frame), (c) magnetic field strength.

ionization front. Where the initial magnetic field is stronger, the initial diffuse medium has a far greater influence on the development of the flow around the globule, as expected since the increase in effective pressure due to ionization is substantially smaller.

6 Conclusions

Ambient magnetic fields are predicted to have an important influence on the structure of astrophysical ionization fronts in star forming regions. The fronts will drive both slow and fast shocks into the surrounding medium. Where the initial magnetic field is strong, a fast-mode shock will propagate into the external medium more rapidly than in a purely hydrodynamical flow, but a succeeding slow-mode shock can remove at least part of magnetic support for the material, allowing its density to rise closer to the value expected for a hydrodynamical flow. This shock history will determine the properties of the gas flowing through the photo-dissociation region. While it is difficult to observe directly, there are indications of these effects in nearby H II regions (Williams, Dyson & Hartquist 2000). Observing the magnetic fields to demonstrate these effects is difficult, but the theory we have developed may help motivate such work and will aid in interpreting the results.

Acknowledgements

I would like to thank John Dyson for inspiring and supporting this work, and Sam Falle, Tom Hartquist and Matt Redman for their scientific contributions to its development.

References

Abe, Y., Sakashita, S., Ôno, Y. 1963 Prog. Theor. Phys. **30**, 816
Axford, W.I. 1961 Phil. Trans. A **253**, 301
Axford, W.I. 1964 ApJ **140**, 112
Bertoldi, F. 1989 ApJ **346**, 735
Chandrasekhar, S., Fermi, E. 1953 ApJ **118**, 113
Ciolek, G.E., Roberge, W.G. 2002 ApJ **567**, 947
Crutcher, R.M. 1991 in Fragmentation of Molecular Clouds and Star Formation: Proceedings of IAU Symp. 147. Edited by E. Falgarone, F. Boulanger, and G. Duvert, Kluwer Academic Publishers, Dordrecht, p61
Crutcher, R.M. 1999 ApJ **520**,706
Crutcher, R.M. 2005 in The Magnetized Plasma in Galaxy Evolution, Proceedings of the conference held in Kraków, Poland, Sept. 27th – Oct. 1st, 2004, Edited by K. Chyży, K. Otmianowska-Mazur, M. Soida, and R.-J. Dettmar, Jagiellonian University, Kraków, p103

Falle, S.A.E.G., Komissarov, S.S. 2001 Journal of Plasma Physics **65**, 29
Falle, S.A.E.G., Hartquist, T.W. 2002 MNRAS **329**, 195
Franco, J., Tenorio-Tagle, G., Bodenheimer, P. 1989 RMAA **18**, 65
Gardiner, T.A., Stone, J.M. 2005 J. Comp. Phys **205**, 509
Gnedin, N.Y., Ferrara, A., Zweibel, E.G. 2000 ApJ **539**, 505
Hill, J.K. 1977 ApJ **212**, 685
Lasker, B.M. 1966 ApJ **146**, 471
Matthews, B.C., Lai, S.-P., Crutcher, R.M., Watson, C.D. 2005 ApJ **626**, 959
Okamoto, I. 1968 PASJ **20**, 122
Redman, M.P., Williams, R.J.R., Dyson, J.E., Hartquist, T.W., Fernandez, B.R. 1998 A&A **331**, 1099
Roberts, D.A., Crutcher, R.M., Troland, T.H. 1995 ApJ **442**, 208
Ryutov, D.D., Kane, J.O., Mizuta, A., Pound, M.W., Remington, B.A. 2005 Ap&SS **298**, 183
Stone, J.M. 1997 ApJ **487**, 271
Subramanian, K., Narasimha, D., Citre, S.M. 1994 MNRAS **271**, L15
Tytarenko, P.V., Williams, R.J.R., Falle, S.A.E.G. 2002 MNRAS **337**, 117
Vallee, J.P., Fiege, J.D. 2005 ApJ **627**, 623
Ward-Thompson, D., Kirk, J.M., Crutcher, R.M., Greaves, J.S., Holland, W.S., André, P. 2000 ApJ **537**, L135
Williams, R.J.R., Dyson, J.E., Hartquist, T.W. 2000 MNRAS **314**, 315
Williams, R.J.R., Dyson, J.E. 2001 MNRAS **325**, 293
Williams, R.J.R. Ward-Thompson, D., Whitworth, A.P. 2001 MNRAS **327**, 788

Herbig-Haro Jets from Young Stars

T.P. Ray

School of Cosmic Physics, Dublin Institute for Advanced Studies, 5 Merrion Square, Dublin 2, Ireland
tr@cp.dias.ie

1 Historical Introduction

It is over half a century since the independent discovery by George Herbig (Herbig 1950, 1951) and Guillermo Haro (Haro 1952, 1953) of the nebulous patches in the sky that bear their names. It was clear, right from the very beginning, that they had something to do with star formation. The precise nature of the link, however, remained a mystery until a few decades ago. It was even suggested at one time that Herbig-Haro (HH) objects were actual sites of star formation harbouring newborn stars. This error was repeated, up until fairly recently, in such esteemed tomes as the *Cambridge Encyclopedia of Astronomy*!

Donald Osterbrock (Osterbrock 1958) in a seminal paper suggested HH objects could be driven by outflows from young stars. This paper, however, was well before its time and received scant recognition, even to this day. It was not until Richard Schwartz (Schwartz 1975) noticed the similarity of HH spectra to those of supernova remnants that their true nature began to unfold. In particular the spectra of HH objects are dominated by forbidden lines (e.g. from species such as [OI] and [SII]) suggesting low electron densities $n_e \leq 10^4 \text{cm}^{-3}$. Moreover, as in the case of supernova remnants, the ratio of various lines showed that they could not arise from photoionization, i.e. a HII region, but instead come from a radiative shock. Modelling of shocks in HH objects parallelled developments in supernova remnant shock theory (Raymond 1979; the chapter by John Raymond in this volume) although it was obvious from very early on that the typical velocities of HH objects (at most a few hundred km s^{-1}) were much smaller than those of most supernova remnants. The lower velocities of HH objects could be deduced not only by observation of their radial velocities (Cohen & Fuller 1985) but also from measuring their tangential velocities through proper motion studies (Herbig & Jones 1981). Moreover lines from high excitation species present in many supernova remnants (e.g. [OIII]λ5007) were faint or absent in HH objects.

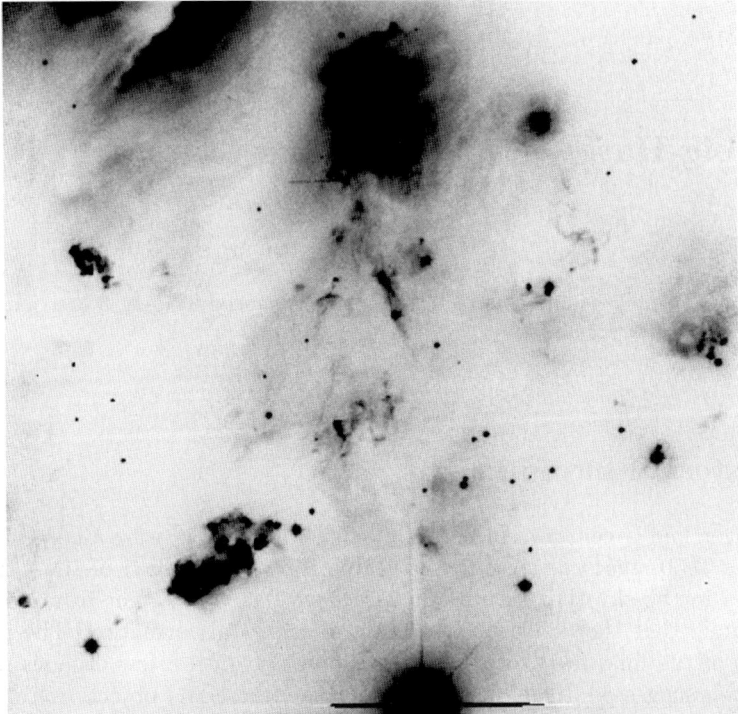

Fig. 1. This [SII]λλ6716,6731 image of the NGC 1333 region shows not only the well-known HH 7–11 outflow (centre and bottom left) but a large number of other outflows in its vicinity. Image courtesy of Jochen Eislöffel.

Such species are not expected to be present in low velocity shocks ($V_{shock} \leq 100$ km s^{-1}).

The realisation that HH objects are due to shocks, and the discovery that young stars produce winds (Kuhi 1964), led to the suggestion that HH objects were caused by clumps, from the parent molecular cloud, getting 'in the way' of a stellar wind (Schwartz 1978). In this case the shock would be expected to be concave outwards from the parent YSO and between the clump and the star. An alternative scenario, suggested by Norman & Silk (1979), is that the star emits high velocity dense supersonic bullets into the ISM. In that case the shock would be concave inwards towards the YSO and lie beyond the bullet.

The recognition that these theories are not correct did not come until the widespread use of the CCD camera. Due to its much higher quantum efficiency, use of such cameras allowed much deeper imaging than possible before with photographic plates. It was then discovered that known HH objects are either parts of jets or apparent terminal bow shocks marking the 'end' of an outflow (see Figure 2 and Mundt, Brugel & Bührke 1987).

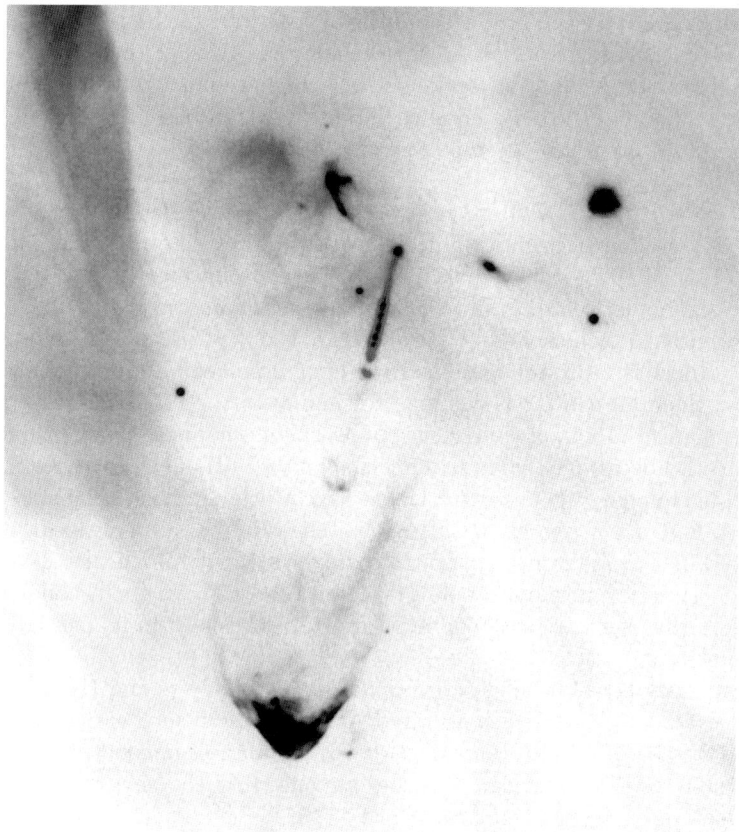

Fig. 2. ESO image of the HH 34S bow shock with the HH 34 blue-shifted jet pointing towards it. A counter bow shock (HH 34N) lies to the north (outside of this image). At one time it was thought that HH 34S represented the 'terminal shock' where the outflow from the young star rammed into ambient material. We now know this idea to be incorrect and that the outflow is much bigger than previously thought.

The first HH jet to be discovered appears to have been from 1548C27, a cometary nebula in the NGC 6820/6823 region associated with an embedded Herbig Ae star (Craine, Boeshaar & Byard 1981)[1]. The presence of a jet however was inferred solely on the basis of morphology and no spectroscopic data was available at that time to confirm its nature. Subsequent imaging and spectroscopic studies (e.g. Mundt & Fried 1983; Reipurth et al. 1986; Ray 1987) show HH jets to be abundant in star forming regions such as Orion and Taurus-Auriga.

---

[1] Herbig Ae/Be stars are the intermediate mass counterparts ($2 \, M_\odot \leq M_\star \leq 10 \, M_\odot$) to the T Tauri stars

In what follows I will first explore the propagation of HH jets and outflows on large-scales, i.e. greater than a few hundred AU, (§2) before looking at models for their generation and how new high resolution observations are helping to distinguish between such models (§3). In the last section (§4), I will speculate on where we might expect advances to be made in the near future.

Before closing this section, it is worth putting YSO jets in context. Currently jets are known from a whole host of astrophysical objects ranging from AGNs (Ferrari 1998), micro-quasars (Mirabel & Rodríguez 1999), GRBs (e.g. Granot & Kumar 2003), PNs (e.g. Vlemmings, Diamond & Imai 2006) and, very recently, brown dwarfs (Whelan et al. 2005). The span in mass of the central object is truly remarkable stretching from tens of Jupiter masses to 10^8 $M_\odot$. Such an enormous range is testimony to the robustness of the outflow mechanism: I am, of course, assuming one mechanism is responsible. The race is on to determine what that mechanism is, although there seems to be general agreement that magnetic fields play a primary role in launching jets (see §3). It seems to me highly likely that the origin of astrophysical jets will be solved first in the context of young stars for a variety of reasons. The major ones concern the enormous amount of detail that we have about conditions in these jets through their line emission and the relatively large angular size of their central engine.

I will start however by describing observations of jets and HH objects on extended (greater than a few hundred AU) scales. On such scales the jet has reached its asymptotic form as dictated by its launching conditions (see §3). Subsequent development can then be influenced by the environment the jet finds itself in.

2 Large-Scale Observations of HH Outflows

In John Godfrey Saxe's poem *"The Blind Men and the Elephant"*, an elephant is inspected by six blind men. Depending on what part of the elephant they touch first, each man has an entirely different opinion of what constitutes an elephant! In the same way, we see different aspects of an outflow depending on the wavelength used. Outflows from young stars were first detected in rotational molecular transitions of CO at mm wavelengths (see Bachiller 1996 for an excellent review of molecular outflows). This is contrary to what was originally expected, i.e. observers thought they would see the spectral signature of accretion i.e. infall. More precisely poorly collimated, blue-shifted and red-shifted outflow lobes were usually found straddling either side of a protostar. Typical velocities of tens of km s^{-1} were observed and the total extent of the outflow varied between 0.1 – 1 parsecs. Although not immediately obvious, it quickly became clear that molecular outflows are due to ambient material being pushed by a wind or jet from the YSO rather than gas that originates close to the star. One reason for this is that the total mass of the molecular outflow is often much more than that of the star. If such massive outflows were made up of material that had initially fallen inwards but were subsequently deflected outwards, star formation would be a very inefficient process indeed! Another, perhaps more compelling, reason is that molecular outflows are not seen near YSOs lacking ambient molecular material. Such YSOs nevertheless, as we shall see, possess atomic/ionized outflows (see below). If we examine higher order CO rotational transitions, a somewhat different picture emerges. Higher order transitions trace hotter gas and such gas is found to be moving much more quickly (at velocities up to 100 km s^{-1}, see, e.g., Palau et al. 2006). Moreover the molecular outflow appears more collimated. Thus there seems to be a general trend from poorly collimated molecular outflows to more highly collimated, higher velocity gas. It is also clear that all of these outflows, slow and fast, are highly supersonic, with Mach numbers of 100 or more.

The origin of molecular outflows is a matter of considerable debate. Several models have been proposed including turbulent entrainment of ambient material along the edges of an atomic jet (see, e.g. Cantó et al. 2003), so-called 'prompt entrainment' in the wings of jet driven bow shocks (Ostriker et al. 2001; Downes & Cabrit 2003) and the sweeping-up of ambient material in a thin shell by a poorly collimated wind (e.g. Matzner & McKee 1999; Lee et al. 2001). The possibility that the underlying HH jet turbulently entrains ambient material along its length to produce a CO outflow seems the least likely scenario. There are a number of reasons for this but perhaps the major one is that such models predict decreasing outflow momentum and velocity with distance from the YSO, contrary to what is observed. The other two scenarios have their advocates and a number of observations in their favour (see, for example, Arce et al. 2006) but there are also problems. Perhaps, as suggested by Arce et al. (2006), a synthesis model combining the jet and wind driven models, is our best hope for explaining the CO data.

As mentioned earlier on, the realisation that HH objects are often parts of jets emerged as a result of deep imaging with CCDs. A classic example that illustrates this phenomenon is HH 34 (see Fig. 2). HH 34 was originally discovered as a bow-shaped object in Orion (Herbig 1974). Early CCD observations (Reipurth et al. 1986; Bührke, Mundt & Ray 1988) showed a jet, consisting of several knots, pointing towards the bow and a counter-bow shock (HH 34N) in the opposite direction to HH 34 (now renamed HH 34S). No counterjet has been detected optically to date. Closer examination of the knots in the HH 34 jet showed them, in effect, to be small-scale HH objects with spectra similar to that of HH 34 itself. The jet points back to an embedded source HH34-IRS where optically nothing is seen apart from a small reddened reflection nebula. This is typical of many striking HH jet sources, i.e. they are often optically obscure, so-called Class-I YSOs. For such protostars most of their energy is absorbed in the optical by the disk and envelope that surrounds them and re-radiated at longer wavelengths.

HH jets/outflows are also known from less obscured, more evolved, young stars, i.e. Class II YSOs, in particular classical T Tauri stars (CTTSs) and their more massive counterparts, the Herbig Ae/Be stars (e.g. Dougados et al. 2000; Corcoran & Ray 1998; Grady et al. 2004). The outflows from these sources tend to be less striking and are sometimes, rather disparagingly, referred to as micro-jets. This name is rather unfortunate, and in fact inappropriate since the total spatial extents of such outflows are at least comparable, and theoretically exceed, those of outflows from Class I YSOs (McGroarty & Ray 2004; McGroarty, Ray & Bally 2004).

In all cases the pattern seems to be the same: a string of knots close to the source, and thereafter knots, or even large HH complexes, at increasing distances. HH 34 is a classical example although there are many others: e.g. HH 111, HH 212, (Hartigan et al. 2001; Zinnecker, McCaughrean & Rayner 1997). Spectral diagnostics has shown us that the velocity of the shocks responsible for the emission of jets is considerably smaller than the actual jet velocity V_{jet}. This has led to the suggestion that the knots are due to 'internal working surfaces' (Raga, Beck & Riera 2004). The concept is easy to understand: YSO jets are highly supersonic. This follows from the measured temperatures (around 10^4 K) implying a sound speed of about 10 km s^{-1} and V_{jet} being 150-300 km s^{-1}. Thus, their Mach numbers are typically 15-30. In such an outflow, it takes only a moderate increase in the flow velocity from the source (say 10%) to induce a pair of internal shocks: one where the slower material upstream is accelerated and another where the faster material downstream is slowed down. Such 'working surfaces' would also explain the low degree of excitation observed in jet knots (e.g. Reipurth et al. 2002).

A striking finding, often neglected by those attempting to model YSO outflows, is the observed asymmetry in the properties of the blue and redshifted jets from the same source. Generally speaking the outflow directed towards us is usually brighter than the one directed away. This, of course, can be readily explained by extinction, as the blueshifted jet is directed out from the parent cloud and the redshifted jet inwards. The asymmetries however that I am referring to cannot be explained by extinction. The bipolar outflow from RW Aur is an excellent example (Hirth et al. 1994). Here the blue and redshifted jets differ in radial velocity by a factor of two. More recently this asymmetry has been seen in transverse velocity measurements (i.e. observed proper motions) and in the same relative ratio as the radial velocities. This incidentally shows that the observed proper motions measure true bulk motion rather than some form of outward moving emissivity wave (López-Martín, Cabrit & Dougados 2004). Other asymmetries exist, for example in electron density and excitation. Combined uncertainties however in the derivation of basic physical parameters (see below) through spectral diagnostics and measurements from imaging, lead to poorly determined estimates of quantities such as mass and momentum flux. Thus, it is hard to be certain that such fluxes are different in both jets.

The first HH jets to be discovered appeared to be at most a few thousand AU in length. One or two notably larger outflows were recorded (e.g. Ray 1987) but, it was only with the development of large-format CCDs, that the full extent of typical HH outflows was realised (e.g. Bally & Devine 1997; Eislöffel & Mundt 1997). Many that stretch up to ten or more parsecs are now known. In retrospect such large sizes should have been predicted. After all the outflow phase lasts about 10^6 yrs; thus, for a typical transverse velocity of 50 km s^{-1}, an average HH outflow will extend for 30 pc! On such large scales, the meandering of the outflow, or influence of its environment, on the outflow is much more evident than for the portion close to the source. For example, many HH outflows have been shown to display C-type morphologies and S-type point symmetries around their source (Bally & Reipurth 2002). C-type distortions can, in principle, be due to the movement of the jet source through the parent cloud reminiscent of the wide-angle tailed (WAT) radio galaxies (e.g. Sakelliou & Merrifield 2000). In comparison, S-type symmetries could be due to precession by the jet source (Reipurth et al. 2002).

Although initially many HH jets were discovered in rather quiescent molecular clouds such as Taurus Auriga, others were found in the hasher environments of Orion for example. More recently a whole new class of 'irradiated' HH jets have been found. Here the jets are fully or partially exposed to UV radiation from nearby OB-stars, i.e. they are photo-ionised (e.g. Andrews et al. 2004). An advantage of such jets is that quantities such as the mass loss rate can be estimated from the strength of the their Hα emission in combination with their known velocities (e.g. Bally et al. 2006).

The large sizes of HH outflows, which are often comparable, if not bigger than their parent cloud, has led to speculation that they generate turbulence in their surroundings (Arce, 2003). As is well known observations of suprathermal CO line widths in molecular clouds, in combination with empirically derived scaling laws (e.g. Mac Low & Klessen; Heyer & Brunt 2004), suggest supersonic turbulence is present. At the same time simulations (e.g. Mac Low et al. 1999), tell us that this turbulence should decay in a few cloud sound crossing times if it is not somehow renewed. HH outflows could provide turbulence as it is clear that they have enough momentum and energy flux (power). The basic problem however is their high degree of collimation which may lead to only a small fraction of the cloud being disturbed. Precessing of an outflow (akin to the dentist drill model for AGN jets) and the presence of a wide angle wind, in addition to the collimated outflow, have been suggested as ways of increasing the dynamical coupling of a HH jet with its environment (Arce et al. 2006). It should be noted however that recent work on molecular cloud lifetimes suggest they may survive for much shorter times than previously thought (e.g. Hartmann 2003) and that a cloud might persist for a period comparable to its free-fall time. If this is the case, then there may be no need to regenerate the turbulence. Sam Falle has addressed issues relevant to these in his chapter.

The rich spectra of HH jets, with lines from S^+, O, N^+, H, C^+, Fe^+, etc., permit detailed analysis of their basic physical parameters. As this line emission comes from the cooling zones of radiative shocks, precise analysis is best performed using line ratio fits to radiative shock models (e.g. Hartigan, Morse & Raymond 1994). Such models are themselves an approximation, since in reality one should allow for complex shock geometries and clumpiness in the shocked gas, not to mention radiative transfer effects (e.g. the presence of a radiative precursor). See the chapter by John Raymond for further discussion of shock models. More recently a number of 'rough-and-ready' techniques have been employed to estimate parameters. For example, Bacciotti & Eislöffel (1999) have shown how the level of O^+, and hence O, should be primarily determined through charge-exchange with hydrogen. Using this assumption, and a number of line ratios, it is possible to derive not only standard numbers such as the electron density but also quantities such as the ionization fraction, total density and temperature. An example of the power of such diagnostic techniques is illustrated in Fig. 3 for the case of the HH 30 jet from Bacciotti, Eislöffel & Ray (1999). Of course, realistically a range of temperatures, densities, etc., are present. Thus, techniques, such as the BE method, only give an approximation valid in the zone where emission from the employed lines peak.

With these caveats, estimates of such quantities as mass and momentum flux rates have been made. Typically for CTTS outflows, values of $\dot{M}_{jet} \approx 10^{-8} - 10^{-7}$ $M_\odot$ yr^{-1} are found. It was clear from very early studies (Cabrit et al. 1990; Hartigan, Edwards & Ghandour 1995) that mass outflow rates scale with circumstellar disk accretion rates. The numbers are subject to a large degree of uncertainty, but typically $\dot{M}_{jet} \approx 1 - 10\%$ $\dot{M}_{acc}$. As the line emission is optically thin, outflow rates can be determined from line strengths, measured velocities and the known distances to the sources. Accretion rates can be derived independently from the amount of 'filling-in' of photospheric absorption lines by the accretion-generated continuum (an effect known as line veiling). Thus, outflow and accretion go hand-in-hand.

So far, apart from a few references to molecular outflows, we have concentrated on the optical appearance of outflows from YSOs. Strictly speaking the term Herbig-Haro jet refers only to those producing optically visible emission. Of course, particularly in the case of more embedded sources, the optical manifestation of shocks may not be seen. As already pointed out, however, YSO outflows can be observed at a variety of wavelengths due to transitions of various ionised, neutral, atomic and molecular species. Thus, for example, there is no shortage of lines in the near-IR, e.g. [CI]λ(9824+9850), [FeII]1.64μm, H_2 ν=1-0 S(1) line at 2.12μm etc., which can be used as probes (e.g. Micono et al. 1998; Podio et al. 2006).

Recently combined optical/NIR spectroscopic studies of a number of HH jets have been made (Nisini et al. 2005; Podio et al. 2006). The advantage of such an approach is that a wide range of line ratios can be used to determine electron density, temperature, extinction, etc. Note, however, that

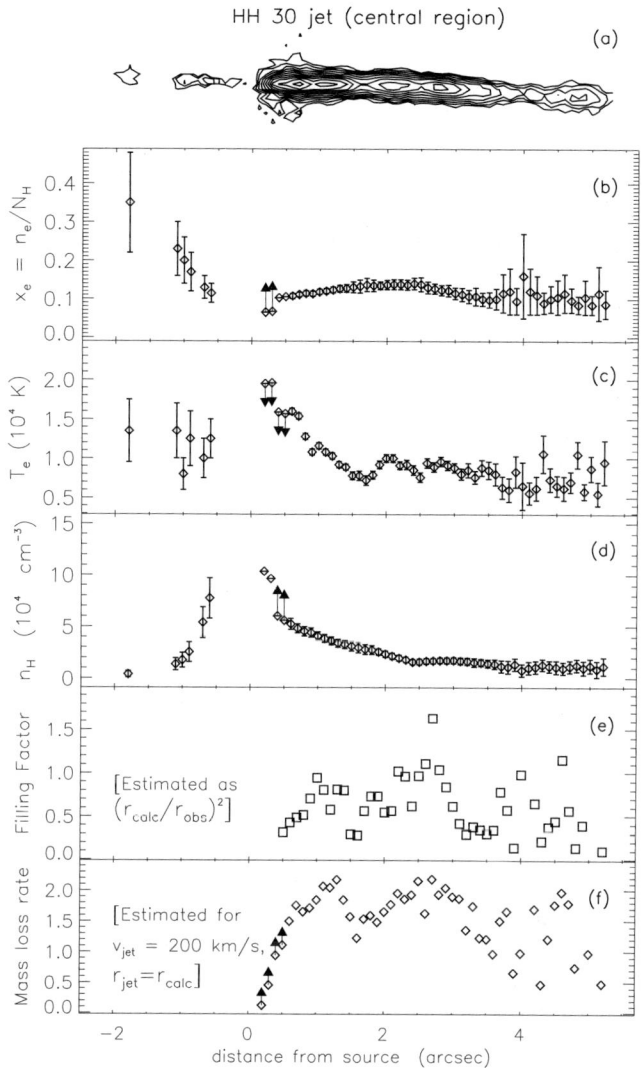

Fig. 3. Parameters along the HH 30 jet derived using the Bacciotti & Eislöffel (1998) technique (see text) from HST Wide Field Planetary Camera 2 (WFPC2) narrow-band images. From top to bottom: [SII] emission, ionization fraction, electron temperature, total density, filling factor and mass flux in units of 10^{-9} $M_\odot$ yr^{-1} along the jet. Adapted from Bacciotti, Eislöffel & Ray (1999).

different line ratios for physical quantities, e.g. electron density, can give very different results (e.g. Podio et al. 2006). As stated above the reason for this is that we are not dealing with a homogeneous structure: individual lines give results typical of the region in which they peak. For example, use of the [FeII]λ7155/[FeII]λ8617 ratio consistently gives higher electron densities than those from the standardly used [SII]λ6731/[SII]λ6716 ratio. This is not too surprising when one considers that according to shock models the extent of the [FeII] emitting region in the post-shock cooling zone is much larger than that giving rise to the red [SII] doublet. Thus, on average, the density derived from [FeII] will be larger as it samples the higher density regions of the radiative shock. Even higher densities are traced by [CaII] lines.

A very interesting result from combined optical/IR analysis is that the dust seems to at least partially survive the process that gives rise to jets. Fig. 4 shows a comparison of the [CaII]$\lambda\lambda$7290,7324/[SII]$\lambda\lambda$6716,6731 and [CI]$\lambda\lambda$9824,9850/[SII]$\lambda\lambda$6716,6731 line ratios for the HH 111 and HH 34 jets with predicted values assuming solar abundances and no depletion. While it is clear that most of the carbon has been unlocked from the dust grains, this is not true of the calcium. We also see that the level of Ca depletion reduces as the outflow goes through further shocks. This is in line with shock theory (see, e.g. Draine 2002) which predicts grains will only be totally destroyed in shocks with velocities $\geq$ 100 km s^{-1}, i.e. only in major bow shocks like HH 34S.

3 Testing Models for the Generation of HH Jets

Current models for the the generation of YSO jets are based on the assumption that the jet is launched centrifugally along magnetic field lines and focused through magnetic hoop stresses. Although there is general agreement that this is the most likely mechanism, the details are not well understood. For example, is the jet launch region in the innermost 0.1-1 AU region of the circumstellar disk as predicted by disk (or D-)wind models (Pudritz & Banerjee 2005) or is it produced even closer in where the stellar magnetospheres interacts with the disk at the co-rotation radius as suggested by the so-called X-wind model (Shu et al. 2000)?

To start to answer such questions, high spatial resolutions observations are needed using Adaptive Optics (AO) with ground-based telescopes, HST and, in the future, interferometry. Although such studies are in their infancy, they are already beginning to produce some very interesting results. For example AO and HST narrow-band images have been used to measure the variation in jet width versus distance from the source (see Fig. 5). Measurement of this parameter, particularly close to the source, is an important discriminant between models. As has already been mentioned, HH jets are highly collimated with average opening angles, over tens of thousands of AU, of only a few degrees. In the case of the nearest star forming regions (at around 150 parsecs),

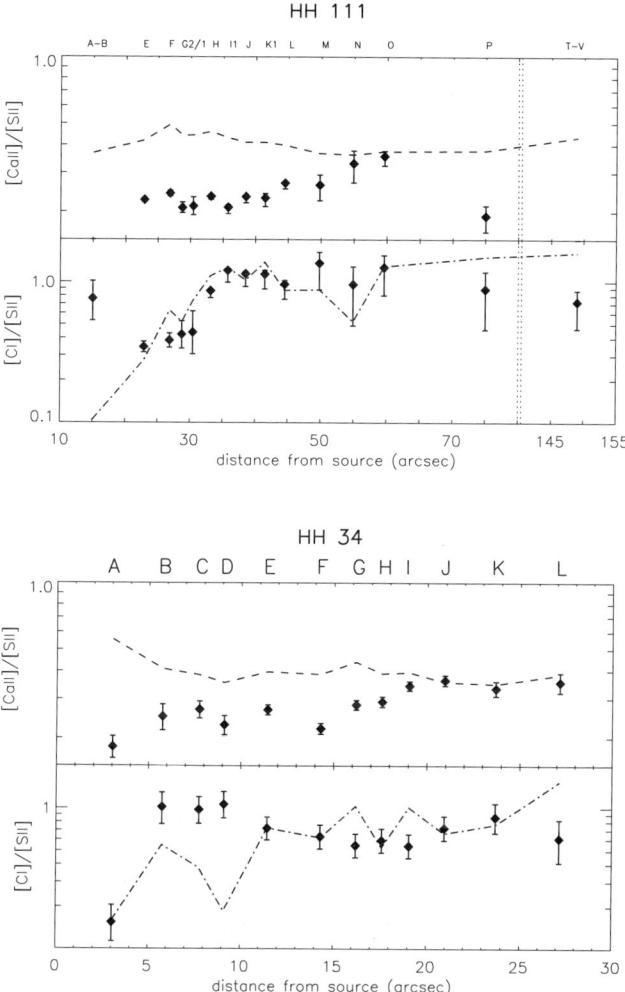

Fig. 4. Evidence that dust is not totally destroyed in HH jet shocks. Here a comparison is made between predicted (dashed and dashed-dot lines) and observed (diamonds) [CaII]$\lambda\lambda$7290,7324/[SII]$\lambda\lambda$6716,6731 and [CI]$\lambda\lambda$9824,9850/[SII]$\lambda\lambda$6716,6731 line ratios for the HH 111 and HH 34 jets. Solar abundances for Ca and C have been assumed. The observed line ratios suggest that Ca is depleted in both jets although the amount of depletion apparently decreases with increasing distance from the source, i.e. when it has gone through more shocks. Note the vertical broken lines in the HH 111 plot indicate a gap in the distance axis. From Podio et al. (2006)

this translates into typical average jet widths of at most a few arcseconds. Within a few hundred AU of the source, however, opening angles tend to increase and it is the shape of this increase that can help us differentiate between models. For example, according to D-wind theory, cold wind solutions can fit the observed variations (see Fig. 5). However, such solutions predict higher poloidal velocities than observed. Moreover, they are of very low efficiency, i.e. the ratio of accreted to outflowing mass is very high ($\approx 100 - 1000$). Again this contradicts what is seen (see §2). In contrast, warm solutions, i.e. ones in which the wind is launched with a significant thermal as opposed to kinetic energy, seem more appropriate (e.g. Dougados et al. 2004). Such a solution is valid, for example, if there is significant coronal heating of the disk.

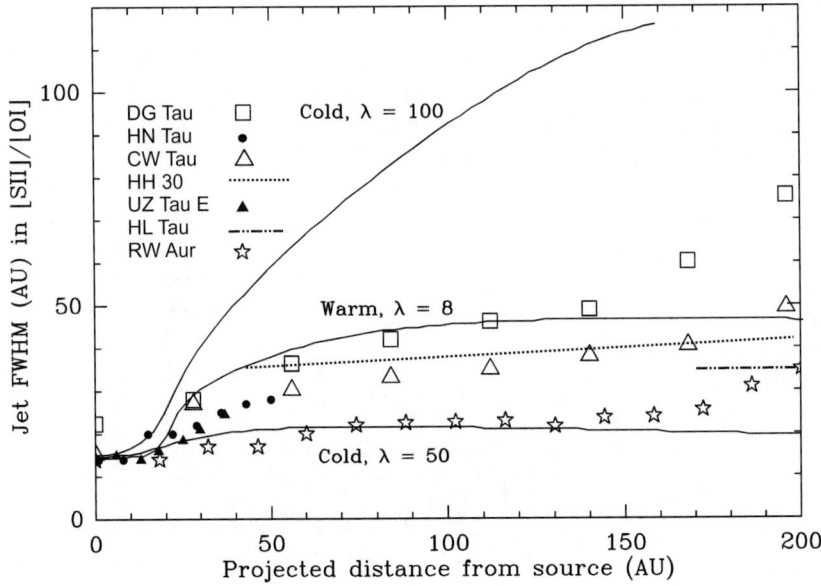

Fig. 5. Jet diameter (FWHM), derived from [OI] and [SII] images, as a function of distance from the parent YSO. Data is from ground based AO and HST/STIS observations. Overlaid (solid lines) are predicted variations based on two cold disk wind models with low efficiency (high λ) and a warm disk solution for comparison from Dougados et al. (2004). Note that moderate to high efficiency is favoured, i.e. warm solutions. Here efficiency is measured in terms of the ratio of mass outflow to mass accretion. Models are convolved with a 14 AU (FWHM) gaussian beam. For full details see Ray et al. (2006), Garcia et al. (2001) and Dougados et al. (2004). From Ray et al. (2006).

Moreover, it is not just high spatial resolution studies of morphology that are yielding valuable information on jet launching. Spectroscopy is as well. For example Bacciotti et al. (2000) used the Space Telescope Imaging Spectrograph (STIS), on board the HST, to examine the kinematics of the DG Tau outflow. The method they employed was to place a 0.″1 wide slit in a series of overlapping positions *parallel* to the outflow. Although time consuming, this allowed them to build up 'images' of the outflow in various emission lines, e.g. in the [OI]$\lambda\lambda$6300,6363 and [SII]$\lambda\lambda$6716,6731 doublets, Hα, and in different velocity intervals. A number of significant findings were deduced. First of all it was found that the spatial width of the outflow varied with velocity in the sense that the width decreased with increasing velocity. In other words, the jet contained a high velocity spine with lower velocity gas confined to the peripheries. Moreover, the lower velocity gas was also less extended in the outflow direction.

Both the X and D-wind models suggest that outflows should rotate since an outflow carries away angular momentum from the underlying disk. The first suggestion that outflows may be rotating was found by Davis et al. (2000) who discovered velocity differences across a large bow-shock in HH 212. These regions, however, are far from the source (thousands of AU) and, thus, may be significantly affected by interactions with their environments. It is, thus, more convincing if rotation is found closer to the YSO. Returning to the DG Tau jet, Bacciotti et al. (2002) also noticed that there was a net velocity difference between opposite sides of the jet in the direction *transverse* to the outflow (see Fig. 6). This was tentatively interpreted in terms of rotation. Further studies followed: in some cases with the slits parallel to the outflow (Woitas et al. 2005) while in others they were transverse (Coffey et al. 2004). Although the sample is relatively small, in almost all cases a velocity difference between opposite sides of the jet is seen. Moreover, in the case of DG Tau, the sense of rotation of the jet and its circumstellar disk were found to match as expected (Testi et al. 2002). The disks associated with other outflows, where possible rotation has been observed, are now being examined to see if they rotate in the same way. One case, RW Aur, in which the disk might be rotating in the opposite sense (Cabrit et al. 2005) has been found. This, however, is a complex hierarchical system, the dynamics of which could affect the apparent sense of rotation. Moreover, a number of authors have put forward alternative explanations for the observed velocity asymmetries, including jet precession (Cerqueira et al. 2006) and disk warping (Soker et al. 2005). Although in principle such models are capable of explaining the observations in any one case, they seem a little contrived when one looks at the whole sample. Further study, as the cliche says, is required.

If, however, we assume that it is rotation that we are observing, then in combination with the known densities, jet velocities and widths, we can deduce the angular momentum flux through the jet. It has to be emphasised that there are great uncertainties in doing such a calculation as their are significant errors associated with the basic physical quantities. Nevertheless, it is interesting to note that the angular momentum flux derived (Bacciotti et al. 2002) is at least comparable to what has to be removed from disk material to fuel the existing rate of accretion. If this is confirmed, it demonstrates that disk angular momentum, at least during the CTTS phase, is removed

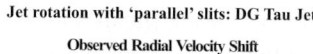
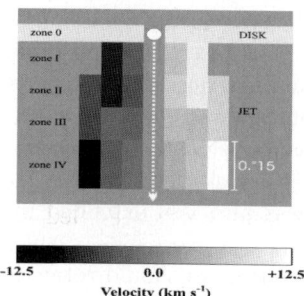
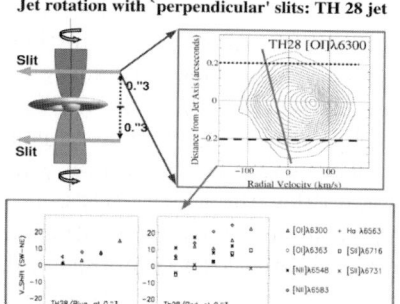

Fig. 6. This figure shows the transverse velocity shifts in emission lines as found by the Space Telescope Imaging Spectrograph (STIS) on board HST across the jets from DG Tau (Left) and Th28 (Right). These shifts were measured at about 50 - 60 AU from the source and 20 - 30 AU from the outflow axis. Use of both gaussian fitting and cross-correlation to the line profiles from diametrically opposite positions centred on the jet axis show velocity shifts of 5–25 km s^{-1}. Such values suggest the jet might be rotating at speeds of 10-20 km s^{-1} at its boundaries. From Ray et al. (2006).

via MHD outflows rather than, for example, through turbulence-enhanced viscosity (e.g. Papaloizou & Lin 1995). As pointed out earlier, and as should be stressed here, the accretion rate, and thus the angular momentum loss rate, are determined entirely independently from the outflow through line veiling (Hartigan, Edwards & Ghandour 1995). It should also be noted that the observed amount of angular momentum is consistent with D-wind models rather than X-wind models, as the latter predicts a lower average specific angular momentum than observed.

4 Future Prospects

We have made significant advances in understanding the propagation of HH jets on large scales, i.e. thousands of AU, not only through observations such as those described here but also through generic jet simulations (e.g. Downes & Cabrit 2003) and modelling of individual outflows (e.g. Coffey, Downes & Ray 2004). More recently laboratory simulations of highly supersonic jets are beginning to contribute to our knowledge (e.g. Ampleford et al. 2005; Frank et al. 2005). Although such jets are limited in physical extent (typically 1 cm) what is important, of course, is not their overall dimensions but that their basic numbers, e.g. Mach number, the ratio of the cooling length to overall length, etc., match those seen in YSO jets.

To understand the launching of jets however and, to distinguish which, if any, of the currently available models are correct, it is necessary to probe the so-called 'central engine'. In this regard, optical and near-infrared interferometry will help us by routinely studying kinematics and morphology of jets on scales of ≈ 1 AU, i.e. well within the jet launching zone. Both Michelson systems, e.g. ESO's VLT Interferometer (VLTI), as well as Fizeau systems, e.g. the Large Binocular Telescope on Mount Graham, should play a leading role. Moderate spectral resolution will be required and in this regard instruments such as AMBER on the VLT are already starting to give us the first glimpse of how interferometry can transform this field (e.g. Benistry et al. 2005). One should strike a note of caution however in that initial (u, v) plane coverage, particularly in the case of the Michelson systems, will be poor so that modelling will play a very important role (Bacciotti et al. 2003).

Of course interferometry has been used extensively in the past to study YSO jets at radio wavelengths (e.g. Ray et al. 1997; Girart et al. 2002) with angular resolution comparable to what can be achieved currently with the HST. In most cases the emission is thermal although there are a number of examples where non-thermal spectra are observed or suspected (e.g. Reid et al. 1995; Ray et al. 1997). A major difficulty however in carrying out radio continuum observations of YSO jets is their typically low flux. Moreover, even in those cases where the YSO is detected, often nothing more is seen than an extension in the direction of the known HH jet/molecular outflow. This situation will change in the next few years with the roll-out of e-MERLIN (Garrington et al. 2004) in the UK and EVLA (Rupen 2003) in the US. As these systems are more broadband than before, using fiber optics to replace microwave links and state-of-the-art correlators, they have much higher sensitivities than their current counterparts. This should lead not only to more YSO jets being detected but, more importantly, deeper 'imaging' at radio wavelengths. The future is indeed very bright for this subject.

Acknowledgments

TR would like to thank John Dyson for all his help and encouragement: he is a true friend and mentor. He also wishes to acknowledge support for this research through the Marie Curie Research Training Network JETSET (Jet Simulations, Experiments and Theory) under contract MRTN-CT-2004-005592 and from Science Foundation Ireland under contract 04/BRG/P02741.

References

Ampleford, D.J., et al. 2005 Astrophysics and Space Science, **298**, 241
Andrews, S.M., et al. 2004 Astrophysical Journal, **606**, 353

Arce, H.G. 2003 Revista Mexicana de Astronomia y Astrofisica Conference Series, **15**, 123
Arce, H. et al. 2006 In *Protostars and Planets V* (ed by B. Reipurth, D. Jewitt, K. Keil) University of Arizona Press, Tucson, in press
Bacciotti, F., Eislölffel, J. 1999 Astronomy and Astrophysics, **342**, 717
Bacciotti, F., Eislöffel, J., Ray, T.P. 1999 Astronomy and Astrophysics, **350**, 917
Bacciotti, F., et al. 2000 Astrophysical Journal, **537**, L49
Bacciotti, F., et al. 2002 Astrophysical Journal, **576**, 222
Bacciotti, F., et al. 2003 Astrophysics and Space Science, **286**, 157
Bachiller, R. 1996 Annual Review of Astronomy and Astrophysics, **34**, 111
Bally, J., Devine, D. 1997 In *IAU Symp. 182: Herbig-Haro Flows and the Birth of Stars*, (ed by Bo Reipurth and Claude Bertout) Kluwer Academic Publishers, p29
Bally, J., Reipurth, B. 2002 Revista Mexicana de Astronomia y Astrofisica Conference Series, **13**, 1
Bally, J., et al. 2006 Astronomical Journal, **131**, 473
Benisty, M., et al. 2005 In *Protostars and Planets V*, LPI Contribution No. 1286., Poster 8395,
Cabrit, S., et al. 2005 In *Protostars and Planets V*, LPI Contribution No. 1286., Poster 8103
Bührke, T., Mundt, R., Ray, T.P. 1988 Astronomy and Astrophysics, **200**, 99
Cabrit, S., et al. 1990 Astrophysical Journal, **354**, 687
Cabrit, S., et al. 2005 In *Protostars and Planets V*, LPI Contribution No. 1286., Poster 8103
Cantó, J., Raga, A.C., Riera, A. 2003 Revista Mexicana de Astronomia y Astrofisica, **39**, 207
Cerqueira, A.H., et al. 2006 Astronomy and Astrophysics, **448**, 231
Cohen, M., Fuller, G.A. 1985 Astrophysical Journal, **296**, 620
Corcoran, M., Ray, T.P. 1998 Astronomy and Astrophysics, **336**, 535
Coffey, D., et al. 2004 Astrophysical Journal, **604**, 758
Coffey, D., Downes, T.P., Ray, T.P. 2004 Astronomy and Astrophysics, **419**, 593
Craine, E.R., Byard, P.L., Boeshaar, G.O. 1981 Astronomical Journal, **86**, 751
Draine, B.T. 2002 In *The Cold Universe* (ed by Daniel Pfenniger, Yves Revaz) Saas-Fee Advanced Course **32**, Swiss Astronomical Society, p213
Dougados, C., et al. 2000 Astronomy and Astrophysics, **357**, L61
Dougados, C., et al. 2004 Astrophysics and Space Science, **292**, 643
Downes, T.P., Cabrit, S. 2003 Astronomy and Astrophysics, **403**, 135
Eisloffel, J., Mundt, R. 1997 Astronomical Journal, **114**, 280
Frank, A., et al. 2005 Astrophysics and Space Science, **298**, 107
Garcia, P.J.V., et al. 2001 Astronomy and Astrophysics, **377**, 609
Garrington, S.T., et al. 2004 Proceedings of the SPIE, **5489**, 332
Girart, J.M., et al. 2002 Revista Mexicana de Astronomia y Astrofisica, **38**, 169
Grady, C.A., et al. 2004 Astrophysical Journal, **608**, 809
Granot, J., Kumar, P. 2003 Astrophysical Journal, **591**, 1086
Hartigan, P., Edwards, S., Ghandour, L. 1995 Astrophysical Journal, **452**, 736
Hartigan, P., Morse, J.A., Raymond, J. 1994 Astrophysical Journal, **436**, 125
Hartmann, L. 2003 Astrophysical Journal, **585**, 398
Haro G. 1952 Astrophys. J. **115**, 572
Haro G. 1953 Astrophys. J. **117**, 73

Herbig G.H. 1950 Astrophys. J. **111**, 11
Herbig G.H. 1951 Astrophys. J. **113**, 697
Herbig, G.H. 1974 Lick Observatory Bulletin, **658**, 1
Herbig, G.H., Jones, B.F. 1981 Astronomical Journal, **86**, 1232
Hirth, G.A., et al. 1994 Astrophysical Journal, **427**, L99
Kuhi, L.V. 1964 Astrophysical Journal, **140**, 1409
Lee, C.-F., et al. 2001 Astrophysical Journal, **557**, 429
López-Martín, L., Cabrit, S., and Dougados, C. 2004 Astrophysics and Space Science, **292**, 531
Mac Low, M.-M., et al. 1999 In *Interstellar Turbulence, Proceedings of the 2nd Guillermo Haro Conference* (ed by Jose Franco and Alberto Carraminana), Cambridge University Press, p.256
Matzner, C.D., McKee, C.F. 1999 Astrophysical Journal, **526**, L109
McGroarty, F., Ray, T.P. 2004 Astronomy and Astrophysics, **420**, 975
McGroarty, F., Ray, T.P., Bally, J. 2004 Astronomy and Astrophysics, **415**, 189
Micono, M., et al. 1998 Astrophysical Journal, **494**, L227
Mundt, R., Brugel, E.W., Bührke, T. 1987 Astrophysical Journal, **319**, 275
Mundt, R., Fried, J.W. 1983 Astrophysical Journal, **274**, L83
Nisini, B., et al. 2005 Astronomy and Astrophysics, **441**, 159
Norman, C., Silk, J. 1979 Astrophysical Journal, **228**, 197
Osterbrock, D.E. 1959 Publ. Astron. Soc. Pac. **70**, 399
Ostriker, E.C., et al. 2001 Astrophysical Journal, **557**, 443
Palau, A., et al. 2006 Astrophysical Journal, **636**, L137
Papaloizou, J.C.B., Lin, D.N.C. 1995 Annual Review of Astronomy and Astrophysics, **33**, 505
Podio, L. et al. 2006 Astronomy and Astrophysics, in press
Pudritz, R.E., Banerjee, R. 2005 In *Massive Star Birth: A Crossroads of Astrophysics, Proceedings of IAU Symposium 227* (ed. by R. Cesaroni, M. Felli, E. Churchwell, M. Walmsley) Cambridge University Press, p163
Raga, A.C., Beck, T., Riera, A. 2004 Astrophysics and Space Science, **293**, 27
Ray, T.P. 1987 Astronomy and Astrophysics, **171**, 145
Ray, T.P. et al. 2006 In *Protostars and Planets V* (ed by B. Reipurth, D. Jewitt, K. Keil) University of Arizona Press, Tucson, in press
Ray, T.P. et al. 1997 Nature, **385**, 415
Raymond, J.C. 1979 Astrophys.J. Suppl. **39**, 1
Reid, M.J., et al. 1995 Astrophysical Journal, **443**, 238
Reipurth, B., et al. 2002 Astronomical Journal, **123**, 362
Reipurth, B., et al. 1986 Astronomy and Astrophysics, **164**, 51
Rupen, M. 2003 In *Future Directions in High Resolution Astronomy: A Celebration of the 10th Anniversary of the VLBA* (ed. by J.D. Romney, M.J. Reid) National Radio Astronomy Observatory, Socorro, p67
Sakelliou, I., Merrifield, M.R. 2000 Monthly Notices of the Royal Astronomical Society, **311**, 649
Schwartz, R.D. 1975 Astrophys. J. **195**, 631
Schwartz, R.D. 1978 Astrophysical Journal, **223**, 884
Shu, F.H. et al. 2000 In *Protostars and Planets IV* (ed. by V. Mannings, A.P. Boss, and S.S. Russell) University of Arizona Press, Tucson, p789
Soker, N. 2005 Astronomy and Astrophysics, **435**, 125
Testi, L., et al. 2002 Astronomy and Astrophysics, **394**, L31

Vlemmings, W.H.T., Diamond, P.J., Imai, H. 2006 Nature, **440**, 58
Whelan, E.T., et al. 2005 Nature, **435**, 652
Woitas, J., et al. 2005 Astronomy and Astrophysics, **432**, 149
Zinnecker, H., McCaughrean, M., Rayner, J. 1997 In *IAU Symp. 182: Herbig-Haro Flows and the Birth of Low Mass Stars*, (ed by F. Malbet and A. Castets) Grenoble Observatory, p198

Hypersonic Molecular Shocks in Star Forming Regions

P.W.J.L. Brand[1]

Institute for Astronomy, University of Edinburgh
pwb@roe.ac.uk

1 A Little History

Shocks are everywhere in the Universe, and with hindsight it is easy to see why.

In the John Dyson era one might be tempted to ask if indeed there is anything else? Such a coherent view of so much of the goings-on between the stars has been imparted by John and his colleagues that most interesting phenomena now have a Dyson – or Dyson-style – similarity solution to describe them.

But the discovery of shocks in the cold clouds where stars are born caused widespread surprise.

The earliest clear indications came in a rush. As noted in the chapter by Tom Ray, it was pointed out by Schwartz (1975) that the enigmatic Herbig-Haro (HH) objects were shock-excited; and in 1976, nearly simultaneously, the new techniques of millimetre and infrared astronomy led to important discoveries: 2.3mm CO line profiles with a width of 150 km s^{-1} were observed by Kwan & Scoville (1976) and by Zuckerman, Kuiper & Kuiper (1976) around the position of the Orion IR cluster, and in a similar location infrared spectra obtained by Gautier et al. (1976) showed H_2 quadrupole lines, revealing molecular gas with an apparent temperature of 2000 K (Beckwith et al. 1978).

Very soon a detailed map of shock excited H_2 by Beckwith, Persson & Neugebauer (1979) appeared, showing two lobes centred around the Becklin Neugebauer/Kleinmann Low infra-red cluster; and the important measurement by Nadeau & Geballe (1979) of extremely large H_2 line-width in the Orion outflow set the scene.

Almost immediately several important models were published (Hollenbach & Shull 1977; London, McCray & Chu 1977; Kwan 1977). Kwan (1977) demonstrated that a hydrodynamic shock travelling faster than 24 km s^{-1} through molecular gas would completely dissociate the H_2. This created a major difficulty in explaining the wide observed profiles.

The two-fluid magnetic shock ('C type') model introduced by Draine (1980) and Draine, Roberge & Dalgarno (1983) was far less destructive to H_2 and provided reasonable fits to the line intensities measured at that time (Cherneff, Hollenbach & McKee 1982; Draine & Roberge 1982).

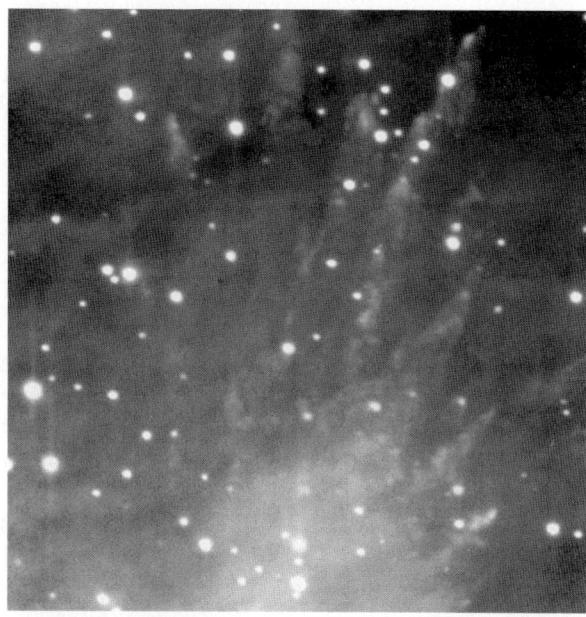

Fig. 1. The northern part of the Orion molecular outflow showing the H_2 emission in a series of bow shocks. The bright tips show up in the light of [FeII], due to strong J shocks. Image courtesy Michael Burton.

Meanwhile the hunt for hot H_2 raced on. Many HH objects were observed: Mundt (1987) listed fourteen such sources; and the Orion outflow, the brightest source, had been studied in detail.

A plot of column densities from the near IR lines of H_2 measured in the Orion outflow (Brand et al. 1988) demonstrated that a single C type shock could not explain the excitation there, and maps of line ratios (Brand et al. 1989) in the outflow, showing little change in excitation with position, further clouded the issue.

As well as these star-forming regions, supernova remnants were mapped in shock-excited H_2. The old remnant IC443 shows shocked H_2 emission round a large part of the shell, and the line excitation appears strikingly similar to that found in Orion (Graham, Wright & Longmore 1987; Moorhouse et al. 1991). More surprisingly, H_2 was found in the heart of the Cygnus loop (Graham et al. 1991).

In his review Mundt (1987) suggested one possible explanation for the H_2 emission often found near, but not exactly at the optical knots in HH objects (there was initially some uncertainty because of the low spatial resolution of the early IR measurements). Amongst other possibilities he proposed that the H_2 excitation might come from the flanks of a bow shock, the head of which produced the optical emission.

In this he was following on from the original Schwartz (1978) suggestion that a bow shock model might be used to explain the range of different excitations seen in the optical spectra of HH objects. Subsequently, Hartmann & Raymond (1984) proposed that a bow shock model could replicate the observations of HH1 and HH2; several papers demonstrated that bow shock model position/velocity diagrams fitted the observations (Choe, Böhm & Solf 1985; Raga & Böhm 1985; Raga 1986); and the important paper by Hartigan, Raymond & Hartmann (1987) explored in detail the excitation and line profile predicted for bow shock observations.

This idea was applied to the observations of the Orion outflow (Smith & Brand 1990a,b; Smith, Brand & Moorhouse 1991b), demonstrating that single plane C-shocks could not explain the observations, but that a C shock propagating into material with a very high magnetic field could do so. The magnetic field required was extremely high, implying an unusual pre-shock magnetic field.

The review by Draine & McKee (1993) summarised the theoretical and observational state at the time.

Before going further, we shall peruse the properties of these shocks.

2 Shock Basics

2.1 Hydrodynamic Jumps, Dissociation, Isothermal Shocks, Magnetic Fields

As a prelude, we examine the simplest possible case of a steady strong plane shock in an ideal gas (e.g., Draine & McKee 1993).

Consider the frame in which the shock front is static and the upstream gas flows perpendicularly towards it, at a speed V_{shock} much greater than the random thermal motions – hence hypersonically. This is a parallel flow of independent particles which encounters a wall of more slowly moving post-shock material. Each incoming particle collides elastically (within one or two mean free paths) with a post-shock particle, sharing its energy and randomizing its direction of travel. Thus, downstream of the shock the gas is hot ($\frac{3}{2}kT \simeq \frac{1}{2}mV_{shock}^2$) and, because of the redirection of velocity, slower.

It is evident that the shock is a 'sudden' transition in the hydrodynamic sense, since hydrodynamics is applicable only to scales much greater than a mean free path for elastic collisions; and it remains sudden because to smooth itself it would have to propagate pressure waves upstream at the speed of sound through gas arriving at a speed much greater than that.

At a downstream distance considerably larger than the elastic mean free path (in fact of the order of the *in*elastic mean free path) there has been time for collisions to excite the internal states of the gas (which could for example lead to ionization and/or dissociation if the shock is strong enough) and to change the effective γ from $5/3$ to a smaller number $N+5/N+3$ where N is the effective number of internal degrees of freedom of the gas particles. The internal states then de-excite by radiating photons and the gas cools. The size of this cooling zone depends on the rate of this process, and is large compared with the mean free path but usually small compared with the global structure of the shocked region.

More formally, in this frame the mass, momentum and energy are conserved through the front. We denote upstream gas by suffix '0', downstream by '1'; ρ is density, p is pressure and v is velocity in this frame, and we assume that the gas is ideal with constant adiabatic index $\gamma = 5/3$. This value corresponds to velocity thermalization by elastic collisions but no internal excitation. Hence, the Rankine-Hugoniot shock jump conditions are:

$$\rho_1 v_1 = \rho_0 V_{\text{shock}}$$
$$p_1 + \rho_1 v_1^2 = p_0 + \rho_0 V_{\text{shock}}^2$$
$$\frac{\gamma}{\gamma-1}\frac{p_1}{\rho_1} + \frac{1}{2}v_1^2 = \frac{\gamma}{\gamma-1}\frac{p_0}{\rho_0} + \frac{1}{2}V_{\text{shock}}^2$$

Further behind the shock when cooling sets in at a net rate $\mathcal{L}$ W kg^{-1}, the first two equations, expressing conservation of mass and momentum, continue to apply while the third becomes the initial condition for

$$v\frac{d}{dx}\left(\frac{\gamma}{\gamma-1}\frac{p_1}{\rho_1} + \frac{1}{2}v_1^2\right) \equiv v\frac{dw}{dx} = -\mathcal{L}$$

in the cooling gas, and γ may be regarded as a variable to take account of internal excitation.

The various processes giving rise to cooling behind shocks has been treated in detail in two important papers by Hollenbach & McKee (1979) and McKee & Hollenbach (1980). Very roughly, in order of decreasing temperature, the major coolants are dissociation, collisional excitation of H Lyman levels, [OI] line emission, H_2, H_2O (to which any free oxygen is rapidly converted in the hot gas) and CO line emission.

Both processes, shocking and cooling, increase the entropy of the gas (the first adiabatically) from which we infer the obvious: a shock cannot run in the opposite sense. Furthermore, as we saw, a shock travels supersonically with respect to the gas ahead. Downstream the gas has to be subsonic with respect to the shock so that the high pressure which drives it continues to reach the front (the sound speed has been increased to a fraction of V_{shock} and the exit speed has been reduced to a fraction of that speed).

The quantity being differentiated in the last equation, w, is the *stagnation enthalpy*. If the net cooling function $\mathcal{L}$ is expressible as a function of ρ and T then the equation is integrable by straightforward quadrature.

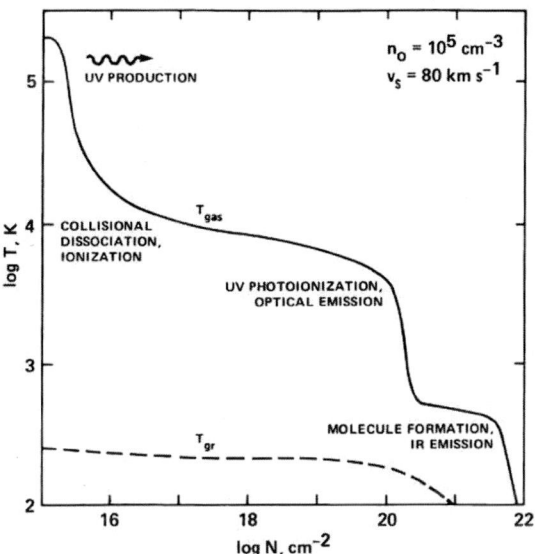

Fig. 2. A fast (80 km s^{-1}) fully dissociative J shock showing T as a function of column density N through the shock. The plateau at 10^4 K is where the atomic gas is heated by UV radiation from the front of the shock; the plateau at 400 K extending to $N = 10^{22}$ cm^{-2} is due to H$_2$ reformation (from Hollenbach & McKee 1989).

If the shock is *strong*, i.e. the pressure ratio p_1/p_0 or equivalently the initial Mach number $M = V_{\text{shock}}/(\text{upstream sound speed})$ is large, then the R-H conditions give (with $\gamma = 1.4$, for diatomic molecules without other internal states)

$$\rho_1/\rho_0 = V_{\text{shock}}/v_1 = 6, \quad kT_1/\overline{m} = (5/36)V_{\text{shock}}^2, \quad p_1 = (5/6)\rho_0 V_{\text{shock}}^2.$$

As a simple application of these ideas we can show how such shocks have a characteristic velocity for the destruction of H$_2$. At the highest temperatures the cooling is dominated by dissociation until the temperature drops to a value T_{line} at which line cooling dominates and dissociation is much reduced. If energy kT_{diss} is released per dissociation at a rate R kg^{-1} s^{-1} then the cooling rate is $\mathcal{L} = kT_{\text{diss}}R$. Each dissociation produces two H atoms with number density $n(H)$, so $d(n(H)/\rho)/dt = 2R$, whence

$$v\frac{d}{dx}(w + n(H)T_{\text{diss}}/2\rho) = 0$$

For a strong and initially molecular shock, we neglect the bulk energy in the enthalpy, while the pressure is the sum of the partial pressures of atomic hydrogen, molecular hydrogen and helium (number fraction $y = 0.1$), each a product of number density and temperature. Define $f = n(H)/n_H$ where n_H is the number density of H nuclei and is proportional to ρ; use $\gamma = 5/3$ for atoms and $\gamma = 7/5$ for molecules; and note that at the start of the cooling zone $T = T_{\max}$ and $f = 0$. Then it is a matter of algebra to show that through the cooling flow

$$f = \frac{(10y + 7)(T_{\max} - T)}{2T_{\text{diss}} + 3T}.$$

If we take values of 52,000 K for T_{diss} and $T_{\text{line}} = 3500$ K, then f becomes unity – H_2 is completely dissociated – at $T = T_{\text{line}}$ when $V_{\text{shock}} = 24$ km s^{-1}, the Kwan (1977) result.

In the case where we can treat the cooling zone as small compared to other length scales in the region we can lump it together with the shock transition. The jump conditions across this 'isothermal shock' (for simplicity it is assumed that initial and final temperatures - and internal states - of the gas are the same: hence the name) are

$$\rho_1/\rho_0 = M_{\text{iso}}^2, \quad p_1 = \rho_0 V_{\text{shock}}^2.$$

Cooling allows the density to reach very high values set by the 'isothermal Mach number' $M_{\text{iso}} = V_{\text{shock}}/\sqrt{kT_0/m}$. In the cooling zone the pressure has increased by only 17%.

If a magnetic field is flux-frozen to the gas (*i.e.* a fully ionized plasma) and the field is directed at right angles to the shock velocity, then pressure is the sum of that of plasma and magnetic field. The magnetic pressure is $B^2/2\mu_0$. The effective mass density of the magnetic field for non-relativistic shocks is negligible, and the shock is assumed to be strong. The momentum equation is then

$$p + B^2/2\mu_0 + \rho v^2 = B_0^2/2\mu_0 + \rho_0 V_{\text{shock}}^2$$

while mass conservation becomes $\rho v = \rho_0 V_{\text{shock}}$. Since flux freezing is assumed, $B \propto \rho \propto v^{-1}$. Denote the Alfvén speed by V_A ($V_A^2 = B^2/2\mu_0\rho$), and the Alfvénic Mach number by $M_A = V_{\text{shock}}/V_{A,0}$.

Writing $\bar{v} = v/V_{\text{shock}}$ and dividing through by $\rho_0 V_{\text{shock}}^2$ the momentum equation becomes

$$(p/\rho_0 V_{\text{shock}}^2) + (2M_A^2 \bar{v}^2)^{-1} + \bar{v} = (2M_A^2)^{-1} + 1$$

and if downstream the gas is cool enough that magnetic pressure dominates (for more general conditions see Roberge & Draine 1990) then the downstream value of v/V_{shock} is

$$\bar{v} = (1 + \sqrt{1 + 8M_A^2})/4M_A^2.$$

For magnetic pressure domination downstream (*i.e.* the final Alfvén speed greater than the sound speed, assumed equal to the initial sound speed for simplicity), assuming as we have that the Mach number M is large,

$$M_A < \sqrt{2}M^2.$$

In such shocks the cooling zone is supported by the magnetic field stress at a density not hugely different from the initial density (thus slowing cooling processes in comparison with that in the high density achieved in the absence of magnetic field).

The detailed structure of J-shocks is modified by the possible changes of state of the material, and its effect on the surroundings. If the shock is strong enough, radiation generated in this region will propagate ahead of the shock and may alter its state depending on shock velocity. And since hydrogen molecule re-formation is a slow process, dissociated molecules will not reform in the close vicinity of the shock, and a separate re-formation zone may occur (Hollenbach & McKee 1989).

2.2 C Shocks

A quarter of a century ago, following an idea by Mullan (1971), Draine (1980) pointed out that, since molecular clouds are slightly ionized and are permeated by magnetic fields, a different kind of shock is possible and might be prevalent. This he named the C shock, using the name J shock for the hydrodynamic discontinuity just described.

The C shock can be envisaged in a two fluid continuum. One of the fluids is the molecular neutral gas representing most of the mass, while the other fluid is a conducting magnetized plasma consisting of the tiny fraction (typically $10^{-7} - 10^{-5}$) of particles ionized by cosmic rays plus the magnetic field inferred from observation, in which the plasma is presumed to be flux-frozen. If a smooth pressure gradient is generated perpendicular to the magnetic field it will drive a compressive magnetohydrodynamic wave through the plasma. This wave, if unencumbered by interaction with neutral particles, would rapidly steepen into a shock travelling faster than the magnetosonic speed (in these cases roughly equal to the ionic Alfvén speed, much greater than either the thermal sound speed in the neutral gas or the Alfvén speed M_A of the whole medium treated as a single magnetized fluid). But by virtue of gas viscosity the magnetic pressure is transferred slowly to acceleration of the neutral material until far downstream the two components have the same velocity. The viscous drag can ensure that the plasma pressure wave will not accelerate to a shock state. For a wide range of conditions (magnetic field strength, state of ionization) the neutral gas is accelerated *slowly*, and remains supersonic with respect to the structure throughout, without passing through a hydrodynamic shock. The whole process reaches a stationary state for suitable initial conditions far upstream and boundary conditions downstream. The relatively wide

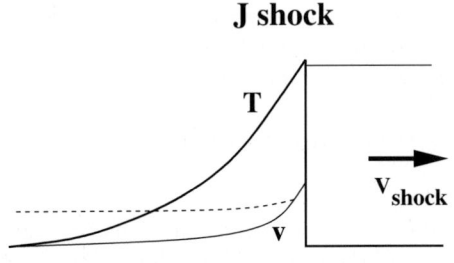
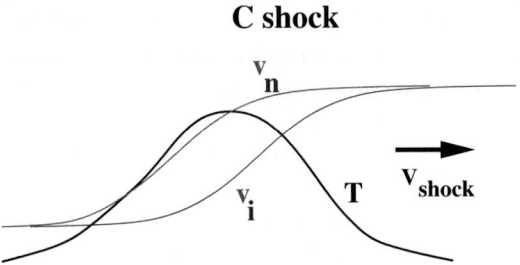

Fig. 3. Sketch, in the shock frame, of a J shock showing T and v, the dashed line corresponding to the effect of the presence of a magnetic field, and a C shock, showing the slab of gas at a relatively uniform temperature.

region where this transition occurs is the C shock. In this region the neutral gas is heated by the viscous friction, and this region radiates.

The structure of this type of shock is most easily considered using the mass and momentum conservation equations introduced in the discussion of J shocks, applied separately to the neutrals and to the magnetized ion fluid.

This time the neutral fluid has an extra force on it caused by friction with the ions, and vice versa. But in the summed momentum equation these terms cancel, and we can use the equation neglecting these forces. Generalizing the result for a magnetized one-fluid shock above by noting that flux freezing applies to the ions only, and denoting ion properties by subscript i and neutrals by n,

$$(p_n/\rho_{n,0}V_{\text{shock}}^2) + (2M_A^2\bar{v}_i^2)^{-1} + \bar{v}_n = (2M_A^2)^{-1} + 1$$

whence we find a relationship between the ion and neutral velocities.

If the cooling is such that thermal pressure is unimportant, then approximately

$$\bar{v}_n = 1 - (\bar{v}_i^{-2} - 1)/2M_A^2.$$

Far downstream where $v_i = v_n$ we have the same conditions as the cold magnetized shock discussed above.

The slip speed, that is $\bar{v}_i - \bar{v}_n$, determines the amount of frictional force and frictional heating in the flow. In the simple case here we find the maximum value (Smith, Brand & Moorhouse 1991b)

$$(\bar{v}_i - \bar{v}_n)_{\max} = 1 + 1/2 M_A^2 - 3/2 M_A^{2/3}.$$

Consider the length scales in this shock structure. The frictional body force (per unit volume) imposed on the neutrals by the ions, and *vice versa*, has magnitude $F = \rho_n \rho_i \langle \sigma v \rangle (v_n - v_i)/(m_i + m_n)$ (Draine & McKee 1993), where $\langle \sigma v \rangle$ is the ion-neutral elastic collision rate, and subscripts n, i have the obvious meaning. The equations of motion for ion fluid (neglecting ion thermal and ram pressure) and neutrals (neglecting thermal pressure) respectively are

$$\frac{d}{dz}\left(\frac{B^2}{2\mu_0}\right) = F$$

$$\frac{d}{dz}(\rho v_n^2)) = -F.$$

If we define $L_0 = V_{\text{shock}}/n_{i,0}\langle \sigma v \rangle$, roughly the initial mean free path of neutrals through the ions times Mach number, and define also the ion length $L_i = v_i/(dv_i/dz)$ and neutral length $L_n = v_n/(dv_n/dz)$, the first equation above gives $L_i/L_0 \simeq M_A^{-2}$ and we also find from the $(\bar{v}_i, \bar{v}_n)$ relation that $L_n/L_i = M_A$.

These relations for slip speed and length scales show the following results. If M_A is small the slip speed is a small fraction of the shock velocity. Such soft shocks can accelerate the neutrals without getting them very hot. The change in the ion velocity closely parallels that in the neutrals. The length of the slip zone is proportional to V_{shock}, and so can accelerate the radiating molecules to quite large velocities.

On the other hand a high value of M_A will create a flow in which the ions rapidly approach their final speed before the neutrals have been significantly decelerated, exposing the neutrals to a high rate of high velocity collisions. There are two major consequences. The heating rate is increased, and the gas may become sufficiently hot that it becomes subsonic - which engenders a hydrodynamic jump (the neglected neutral thermal pressure term in the equation above becomes important). And in these conditions at a shock velocity of 40-50 km s^{-1} the neutrals can be ionized by single collisions, leading to an ionization run-away as the increasing ion fraction reduces L_0.

In any case if the length of the drag zone is too short compared with the cooling length, the sound speed will increase and the gas may become subsonic through a hydrodynamic jump. Thus, for a limited range of conditions a C shock structure will contain a J shock which may provide much of the cooling via high excitation lines but still produce C shock-like emission of H_2 for example.

The heated region behind each kind of shock is shown schematically in Fig. 4.

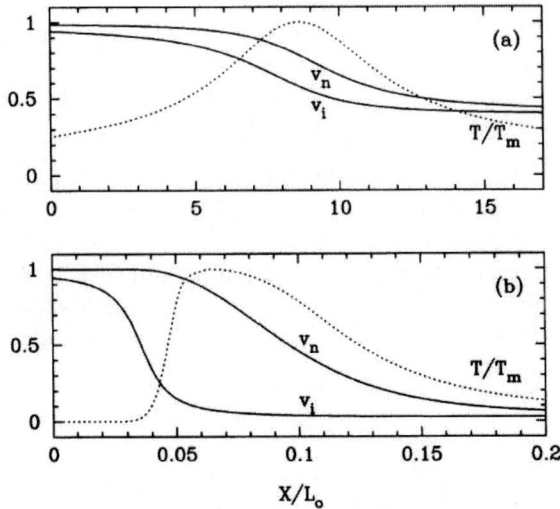

Fig. 4. (a) A low M_A C shock showing v_i and v_n changing in parallel. (b) A high M_A C shock showing the rapid change in v_i and the consequent large velocity difference (from Smith, Brand & Moorhouse 1991a).

It is evident that, in contrast to J shocks, C shocks create a zone that very roughly corresponds to a slab of gas at a given temperature, and that therefore the emission will reflect that status. That this is not shown in many of the observations is the reason for proposing bow shocks as emitters (Schwartz 1975; Hartmann & Raymond 1984; Mundt 1987).

2.3 Bow Shock Structure

There are few instances of observations of planar shocks. Usually the next best guess is that one is observing a shock driven into a bowed shape by a supersonic gas jet or supersonically moving dense clump. The first models of Herbig Haro objects were of this nature, and expose the essential complications.

The range of excitations and breadth of velocity profiles seen in quite small projected areas on the sky led to the idea that one is often observing some or indeed all of a bow shock structure. In some special cases the bow is quite discernible.

A simple but perhaps suspect way in which to create models of such regions is to use oblique planar models, and patchwork these onto the surface of a pre-decided bow geometry. This technique was pioneered by Hartigan, Raymond & Hartmann (1987) for optically emitting bows, and applied to H_2 (Smith, Brand & Moorhouse 1991a) to explain the extreme conditions at the Peak 1 in the Orion outflow.

The strongest section of the shock is clearly at the head of the bow. Here, the shock may be strong enough to ionize and dissociate the molecules, and is certainly a J-shock. Further round the side of the bow the shock may still be a J-shock but not completely dissociating the molecules, while further round, the shock becomes a C-shock (Smith & Brand 1990c).

Clearly, even if the initial magnetic field had a simple geometry, the varying angle of incidence at each part of the shock makes such a structure hard to model. The extended length scales and subtleties of structure of oblique C shocks (Wardle & Draine 1987) make this region the least straightforward part of the modelling process.

Notwithstanding, Smith, Brand & Moorhouse (1991a) were able to reproduce both the excitation (up to very high energies) and the extreme breadth of the lines with a very low Alfvén Mach number and very fast shock. The lingering problem with this explanation has always been seen to be the extremely high pre-shock magnetic field required.

The technique is to specify an axially symmetric bow shape $R = f(z)$ (z is distance along the axis; R is perpendicular distance from axis of the bow at z) supposedly created by a uniform supersonic flow, velocity V_{shock}, hitting a small obstacle, or equivalently by such an obstacle – or indeed the end of a jet – ramming supersonically into a uniform medium. If $dz/dR = \tan\psi$ the effective shock speed at a point in the bow is $V_{shock}\cos\psi$ and the shocked material has, in addition to its post-shock behaviour perpendicular to the shock, a component of velocity $V_{shock}\sin\psi$ tangential to the bow. A line of sight and a projected area around it are assumed to model an observation. At each point of the bow within that area a line intensity (per unit area of shock) through the shock flow is calculated, as are the projection factor (secant of angle between line of sight and shock normal), and the velocity component along line of sight. From this the line profile is found, the shape and intensity of the line being diagnostic of the shock properties.

3 Models and Observations

3.1 Issues

Much of the theoretical work until recently has been based on the assumption that the flow is stationary, and that timescales for chemical evolution were long, and timescales for population of internal states were short, and that upstream conditions are static. Even now the modelling of shocks, including chemistry, grain physics and electrodynamics and the effects of the dust on the chemistry and dynamics, and obliquity progresses inevitably slowly. The fact that post-shock turbulence will occur in many of the flow structures that have been measured has led to some pioneering studies of the process, but much remains to be done.

The question of how to extend a two-fluid approximation to account for the behaviour and effects of the dust grain population has been addressed several times, and because of the still uncertain properties of grains is still an open question.

Even in plane steady normal shocks several instabilities have been discovered. Are there more? Is it possible to make useful predictions from unstable shocks?

On the observational side, the ISO measurements, opening wider the near to far IR spectral region, have produced a plethora of new results. But there is still the usual urgent need of better spatial resolution (and of course signal-to-noise ratio) to compensate for our inability to see round the side of these complex regions.

Grain Charge and Drag

The picture so far of the C-shock is of a two-fluid process, but the presence of charged dust grains can significantly alter the picture, particularly in cold high density regions (Draine 1980). Several papers investigate the tricky issue of grains as a subspecies in the shock flow (Pilipp, Hartquist & Havnes 1990; Caselli, Hartquist & Havnes 1997; Wardle 1998; Flower & Pineau des Forêts 2003). Also see the chapter by Tom Hartquist and Ove Havnes.

The first issue is that the grains may be charged, and can be the dominant ionized species providing the drag on neutrals. This will occur preferentially at high densities. Pilipp, Hartquist & Havnes (1990) found that the dust in high density shocks was constrained by charge separation to move at a speed intermediate between that of the ions and of the neutrals, with the net effect of steepening the shock structure. Wardle (1998) examined the possible flows in great detail, and pointed out that the effect is accentuated by taking the out-of-plane deviations of the magnetic field into account, thus finding even steeper (and by implication hotter) shock structures. These effects are greatest at high density and in the presence of small grains (which for a given grain mass will give greater drag). Wardle (1998) noted that, since most C shocks will give temperatures between 1000K and 2000K, and molecules at that temperature will radiate most of the shock energy, there may be no dominant observed effect in changing the detailed structure of the C shock. Conversely, he pointed out that reliable diagnostics will rely on getting the physical processes correct in detail.

Flower & Pineau des Forêts (2003) addressed some of the physical and chemical detail. They show that large grains rapidly charge by electron attachment in the shock flow and become attached to the magnetic field, and that the critical velocity (at which H dissociation leads to breakdown to a J shock) is sufficiently high (Le Bourlot et al. 2002) that refractory grains can shatter. Chapman & Wardle (2006) examined the effects of obliquity of the magnetic field, and emphasized the significance of the grain physics in determining the flow structure. Ciolek, Roberge, & Mouschovias (2004) pointed out that the degree to which the grains attach to the magnetic field determines the ionic magnetoionic sound speed, and if a significant fraction of the total grain mass is thus attached, the ionic material will shock, destroying the C shock.

Grain Destruction

In all of the calculations referred to above there is the issue of grain destruction (or more positively element release) either by sputtering or shattering of dust grains.

Calculations by Draine (1995) and Flower & Pineau des Forêts (1995) have shown that C shocks can destroy dust grain mantles and cores, and Caselli, Hartquist & Havnes (1997) have shown that such shocks may release an abundance of Si consistent with observations of shocked regions, and many orders of magnitude above the average abundance. Caselli, Hartquist & Havnes (1997) emphasized that destruction is more efficient in oblique shocks than in perpendicular shocks. This process may be the dominant producer of SiO seen in high density star forming regions and indicated in the observations of Jiménez-Serra et al. (2005), for example.

Flower & Pineau des Forêts (2003) demonstrated that a specific C shock travelling at 50 km s^{-1} would shatter over half of the amorphous carbon grains present in the model.

All of these findings together imply that the behaviour of C shocks is dependent on dealing with chemistry, dust grain physics and MHD phenomena in great detail (since some of the necessary grain physics remains to be determined, this is a worry).

Chemistry and Time Dependence

Another issue which has been clarified by the refinement of shock models is that the processes determining the shocks have timescales which are commensurate with shock passage times, but also may be comparable with the time since the shock was created. This raises the possibility for the pessimist that every shock observed is unique (or rather that the number of determining parameters, now including initial conditions and time since creation, is far greater than observations can encompass). At any rate it emphasizes how important the detail can be in determining the overall effect.

Further, if the pressure driving the shock changes, or the medium into which it propagates has large gradients, or is lumpy, externally imposed time dependent effects come into play.

An example of what time-dependent driving can do is shown by Lim et al. (2002). They demonstrated that a shocked layer driven by a slowly accelerating piston can accumulate molecular gas without destroying the H$_2$, producing column densities and velocities of gas comparable with those observed in the Orion outflow in an acceleration period of tens of years, and protecting the molecular layer from the hot gas at the shock by an intermediate layer of H atoms.

Girart et al. (2005) have demonstrated that near HH2, and caused by it, a wide variety of different excitation phenomena can be created by a shock in a non-uniform medium.

Chiéze, Pineau des Forêts & Flower (1998) demonstrated that the timescale for an initial disturbance in a dark cloud to reach steady state is so long that it may exceed the lifetime of the outflow causing the shock. They showed further that during the evolution to a C shock there will be a J-shock embedded, and that the combined radiation from this region and from the rest of the (C type) flow may explain the H_2 observations in some sources where several different excitation temperatures are measured. They emphasized again the necessity of calculating in detail the degree of ionization in particular.

These more detailed studies also reveal that even if a J shock occurs all H_2 emission may not be lost! Flower et al. (2003) and Cabrit et al. (2004)) demonstrated that for fully dissociating shocks of moderate speed an H_2 emission spectrum can be produced which can account for some of the observations.

The issue of time dependence is intimately tied up with instabilities. The most marked(!) instability is the Wardle (1990) instability of plane C shocks, in which, rather like the Parker instability, ripples on the lateral magnetic field lines accumulate a higher density of ions and are dragged further downstream, trapping higher density gas which cools rapidly. The astonishing thing about this instability, which also occurs in oblique C shocks (Wardle 1991), is that it has little effect on the overall output from the shock, for reasons touched on before: the hot gas does the radiating (Mac Low & Smith 1997; Neufeld & Stone 1997).

Many of the developing models reveal one dimensional chaotic or bouncing instabilities, due for example to shortened cooling times in dissociation or ionization zones behind shocks (Lim et al. 2002; Smith & Rosen 2003; Lesaffre et al. 2004). "What will happen in three dimensional models?" is an open question.

Turbulence and Turbulent Mixing and Other Effects

Canto & Raga (1991), in a ground-breaking paper, investigated heating effects of a turbulent mixing layer. Taylor & Raga (1995) showed that such zones could mimic post-shock flows in some respects. Amongst others, Pavlovski et al. (2002), Elmegreen & Scalo (2004), and Heitsch et al. (2005) have begun to model more general cloud turbulence as a source of structure and radiation. This work is still at an exploratory stage, and in future may converge with that on shock codes.

3.2 Current Models and Observations

In the interstellar medium in our Galaxy we expect that the cooling zone in most observed shocks will be $\sim 10^{-5}$ pc in the case of J-shocks and

$\sim 10^{-2}$ pc for C-shocks in very round numbers. This means that it is unlikely that J-shocks will be resolved, while C-shocks can be resolved with current equipment.

The arrival of large masses of ISO satellite data has enlivened this area (Dishoeck 2004).

Modelling is rapidly becoming more sophisticated, with attention being paid to optimum methods (Falle 2003; Lesaffre et al. 2004), and more of the required physics (grains, chemistry, time dependence) is being included. What has become clear (Chiéze, Pineau des Forêts & Flower 1998; Flower et al. 2003; Lesaffre et al. 2004; Cabrit et al. 2004)) is that rugged predictions require very complete modelling.

Now a few examples are considered.

The extensive line list of H_2 from HH43, observed by Giannini et al. (2002), has been fit by Flower et al. (2003) by either a 25 km s^{-1} J shock *or* an 80 km s^{-1} C shock! The same modellers fit data by Wright et al. (1996) and Froebrich, Smith & Eislöffel (2002) of Ceph A West with a model of a J shock with a C precursor typical of early evolution.

O'Connell et al. (2005) modelled the H_2 emission structures seen in the HH211 outflow using C bow shocks.

Snell et al. (2005) examined the supernova remnant IC443 using the Submillimeter Wave Astronomy Satellite (SWAS) to observe H_2O lines. They concluded that the data are consistent with a combination of fast J shocks and slow (J or C) shocks, perhaps likely in a clumpy medium.

The Orion outflow, pictured in part in Fig. 1, has been a test-bed and conundrum throughout the entire development of this area.

Le Bourlot et al. (2002) produced the H_2 excitation diagram which matches the observed data of Rosenthal, Bertoldi & Drapatz (2000) by superposing two C shocks with velocities as high as 40 km s^{-1} and 60 km s^{-1}.

The discovery of the 'bullets' or fingers (Allen & Burton 1993, see also McCaughrean & Mac Low 1996) suggested an explosive event, and Stone, Xu & Mundy (1995) provided an explanation for the major structural features (in particular the fingers) by having a time variable outflow, from IRc2 or the BN object, which sweeps up a shell and then accelerates it whereupon it fragments via Rayleigh-Taylor instability.

There has been a great deal of progress in establishing the values for important rates in the microphysics and chemistry, in developing models that can cope with the subtleties of grain dynamics and time-dependent chemistry, and in the observations (particularly thanks to ISO in the mid IR); but an enormous amount remains to be done.

References

Allen, D.A., Burton, M.G. 1993 Nature **363**, 54
Beckwith, S., Persson, S.E., Neugebauer, G. 1979 Astrophys. J. **227**, 436

Beckwith, S., Persson, S.E., Neugebauer, G., Becklin, E.E. 1978 Astrophys. J. **223**, 464

Brand, P.W.J.L., Moorhouse, A., Burton, M.G., Geballe, T.R., Bird, M., Wade, R. 1988 Astrophys. J. Lett. **334**, L106

Brand, P.W.J.L., Toner, M.P., Geballe, T.R., Webster, A.S., Williams, P.M., Burton, M.G. 1989 Mon. Not. R. Astr. Soc. **236**, 929

Cabrit, S., Flower, D.R., Pineau, des Forêts, G., Bourlot, J. Le, Ceccarelli, C. 2004 Astrophys. & Sp. Sci. **292**, 501

Canto, J., Raga, A. 1991 Astrophys. J. **372**, 646

Caselli, P., Hartquist, T.W., Havnes, O. 1997 Astron. & Astrophys. **322**, 296

Chapman, J.F., Wardle, M. 2006 Mon. Not. R. Astr. Soc. **000**, 000, astro-ph/0509008

Cherneff, D., Hollenbach, D.J., McKee, C.F. 1982 Astrophys. J. Lett. **259**, L97

Chièze, J.-P., Pineau des Forêts, G., Flower, D.R. 1998 Mon. Not. R. Astr. Soc. **295**, 672

Choe, S.-U., Böhm, K.-H., Solf, J. 1985 Astrophys. J. **288**, 338

Ciolek, G.E., Roberge, W.G., Mouschovias, T.Ch. 2004 Astrophys. J. **610**, 781

Draine, B.T. 1980 Astrophys. J. **241**, 1081

Draine, B.T. 1995 Astrophys. & Sp. Sci. **233**, 111

Draine, B.T., McKee, C.F. 1993 Ann. Rev. Astr. Astrophys. **31**, 373

Draine, B.T., Roberge, W.G. 1982 Astrophys. J. Lett. **259**, L91

Draine, B.T., Roberge, W.G., Dalgarno, A. 1983 Astrophys. J. **264**, 485

Elmegreen, B.G., Scalo, J. 2004 Ann. Rev. Astr & Astrophys. **42**, 211

Falle, S.A.E.G. 2003 Mon. Not. R. Astr. Soc. **344**, 1210

Flower, D.R., Bourlot, J. Le, Pineau des Forêts, G., Cabrit, S. 2003 Mon. Not. R. Astr. Soc. **341**, 70

Flower, D.R., Pineau des Forêts, G. 1995 Mon. Not. R. Astr. Soc. **275**, 1049

Flower, D.R., Pineau des Forêts, G. 2003 Mon. Not. R. Astr. Soc. **343**, 390

Froebrich, D., Smith, M.D., Eislöffel, J. 2002 Astron. & Astrophys. **385**, 239

Gautier, T.N., Fink, U., Treffers, R.R., Larson, H.P. 1976 Astrophys. J. Lett. **207**, L129

Giannini, T., Nisini, B., Caratti, A., Lorenzetti, D. 2002 Astrophys. J. Lett. **570**, L33

Girart, J.M., Viti, S., Estalella, R., Williams, D.A. 2005 Astron. & Astrophys. **439**, 601

Graham, J.R., Wright, G.S., Hester, J.J., Longmore, A.J. 1991 Astron. J. **101**, 175

Graham, J.R., Wright, G.S., Longmore, A.J. 1987 Astrophys. J. **313**, 847

Hartigan, P., Raymond, J., Hartmann, L. 1987 Astrophys. J. **316**, 323

Hartmann, L., Raymond, J. 1984 Astrophys. J. **276**, 560

Heitsch, F., Burkhart, A., Hartmann, L., Slyz, A.D., Devriendt, J.E.G. 2005 Astrophys. J. Lett. **633**, L113

Hollenbach, D.J., Shull, M. 1977 Astrophys. J. **216**, 419

Hollenbach, D.J., McKee, C.F. 1979 Astrophys. J. Suppl. **41**, 555

Hollenbach, D.J., McKee, C.F. 1989 Astrophys. J. **342**, 306

Jiménez-Serra, I., Martíin-Pintado, J., Rodríguez-Franco, A., Martín, S. 2005 Astrophys. J. Lett. **627**, L121

Kwan, J., Scoville, N. 1976 Astrophys. J. Lett. **210**, L39

Kwan, J. 1977 Astrophys. J. **216**, 713

Le Bourlot, J., Pineau des Forêts, G., Flower, D.R., Cabrit, S. 2002 Mon. Not. R. Astr. Soc. **332**, 985
Lesaffre, P., Chièze, J.-P., Cabrit, S., Pineau des Forêts, G. 2004 Astron. & Astrophys. **427**, 157
Lim, A.J., Raga, A.C., Rawlings, J.M., Williams, D.A. 2002 Mon. Not. R. Astr. Soc. **335**, 817
London, R., McCray, R., Chu, S.-I. 1977 Astrophys. J. **217**, 442
McCaughrean, M., Mac Low, M.-M. 1996 Astron. J. **113**, 391
McKee, C.F., Hollenbach, D.J. 1980 Ann. Rev. Astr. Astrophys. **18**, 219
Mac Low, M.-M., Smith, M.D. 1997 Astrophys. J. **491**, 596
Moorhouse, A., Brand, P.W.J.L., Geballe, T.R., Burton, M.G. 1991 Mon. Not. R. Astr. Soc. **253**, 662
Mullan, D.J. 1971 Mon. Not. R. Astr. Soc. **153**, 145
Mundt, R. in *IAU Symposium 122: Circumstellar Matter* ed. by I. Appenzeller, C. Jordan, (Kluwer 1987) pp147-158
Nadeau, D., Geballe, T.R. 1979 Astrophys. J. Lett. **230**, L169
Neufeld, D.A., Stone, J.M. 1997 Astrophys. J. **487**, 283
O'Connell, B., Smith, M.D., Froebrich, D., Davis, C.J., Eislöffel, J. 2005 Astron. & Astrophys. **431** 223
Pavlovski, G., Smith, M.D., Mac Low, M.-M., Rosen, A. 2002 Mon. Not. R. Astr. Soc. **337**, 477
Pilipp, W., Hartquist, T.W., Havnes, O. 1990 Mon. Not. R. Astr. Soc. **243**, 685
Raga, A. 1986 Astrophys. J. **308**, 829
Raga, A., Böhm, K.-H. 1985 Astrophys. J. suppl. **58**, 201
Roberge, W.G., Draine, B.T. 1990 Astrophys. J. **350**, 700
Rosenthal, D., Bertoldi, F., Drapatz, S. 2000 Astron. & Astrophys. **356**, 705
Schwartz, R. 1978 Astrophys. J. **223**, 884
Schwartz, R.D. 1975 Astrophys. J. **195**, 631
Smith, M.D., Brand, P.W.J.L. 1990a Mon. Not. R. Astr. Soc. **242**, 495
Smith, M.D., Brand, P.W.J.L. 1990b Mon. Not. R. Astr. Soc. **243**, 498
Smith, M.D., Brand, P.W.J.L. 1990c Mon. Not. R. Astr. Soc. **245**, 108
Smith, M.D., Brand, P.W.J.L., Moorhouse, A. 1991a Mon. Not. R. Astr. Soc. **248**, 451
Smith, M.D., Brand, P.W.J.L., Moorhouse, A. 1991b Mon. Not. R. Astr. Soc. **248**, 730
Smith, M.D., Rosen, A. 2003 Mon. Not. R. Astr. Soc. **339**, 133
Snell, R.L., Hollenbach, D., Howe, J.E., Neufeld, D.A., Kaufman, M.F., Melnick, G.J., Bergin, E.A., Wang, Z. 2005 Astrophys. J. **620**, 758
Stone, J.M., Xu, J., Mundy, L.G. 1995 Nature **377**, 315
Taylor, S.D., Raga, A. 1995 Mon. Not. R. Astr. Soc. **296**, 823
van Dishoeck, E. 2004 Ann. Revs. Astron. & Astrophys. **43**, 119
Wardle, M. 1990 Mon. Not. R. Astr. Soc. **246**, 98
Wardle, M. 1991 Mon. Not. R. Astr. Soc. **251**, 507
Wardle, M. 1998 Mon. Not. R. Astr. Soc. **298**, 507
Wardle, M., Draine, B.T. 1987 Astrophys. J. **321**, 321
Wright, C.M., Drapatz, S., Timmermann, R., van der Werf, P.P., Katterloher, R., Graauw, Th. de 1996 Astron. & Astrophys. **315** L301
Zuckerman, B., Kuiper, T.B.H., Kuiper, E.N.R. 1976 Astrophys. J. Lett. **209**, L137

Part II

The Effects of Evolved Stars on Their Environments

Wind-Blown Bubbles around Evolved Stars

S.J. Arthur

Centro de Radioastronomía y Astrofísica, UNAM, Campus Morelia, Apartado Postal 3-72, 58090 Morelia, Michoacán, México
j.arthur@astrosmo.unam.mx

1 Introduction

Most stars will experience episodes of substantial mass loss at some point in their lives. For very massive stars, mass loss dominates their evolution, although the mass loss rates are not known exactly, particularly once the star has left the main sequence. Direct observations of the stellar winds of massive stars can give information on the current mass-loss rates, while studies of the ring nebulae and HI shells that surround many Wolf-Rayet (WR) and luminous blue variable (LBV) stars provide information on the previous mass-loss history. The evolution of the most massive stars, $M > 25 M_\odot$, essentially follows the sequence O star $\rightarrow$ LBV or red supergiant (RSG) $\rightarrow$ WR star $\rightarrow$ supernova. For a star of mass less than $\sim 25 M_\odot$ there is no final WR stage. During the main sequence and WR stages, the mass loss takes the form of highly supersonic stellar winds, which blow bubbles in the interstellar and circumstellar media. In this way, the mechanical luminosity of the stellar wind is converted into kinetic energy of the swept-up ambient material, which is important for the dynamics of the interstellar medium. In this chapter, analytic and numerical models are used to describe the hydrodynamics and energetics of wind-blown bubbles. A brief review of observations of bubbles is given, and the degree to which theory is supported by observations is discussed.

2 Classical Wind-Blown Bubbles

The effect that a stellar wind has on its surroundings can be dramatic. A massive star will first have formed an H II region around itself, and the stellar wind interacts with this ionized gas. A fast (typically 2000 km s^{-1}) stellar wind is hypersonic with respect to the ambient medium (sound speed $\sim$10 km s^{-1} in photoionized gas) and so a two-shock flow pattern forms: one shock sweeps up the ambient medium, accelerating, compressing and heating it, while the other shock decelerates the stellar wind itself, heating and compressing it

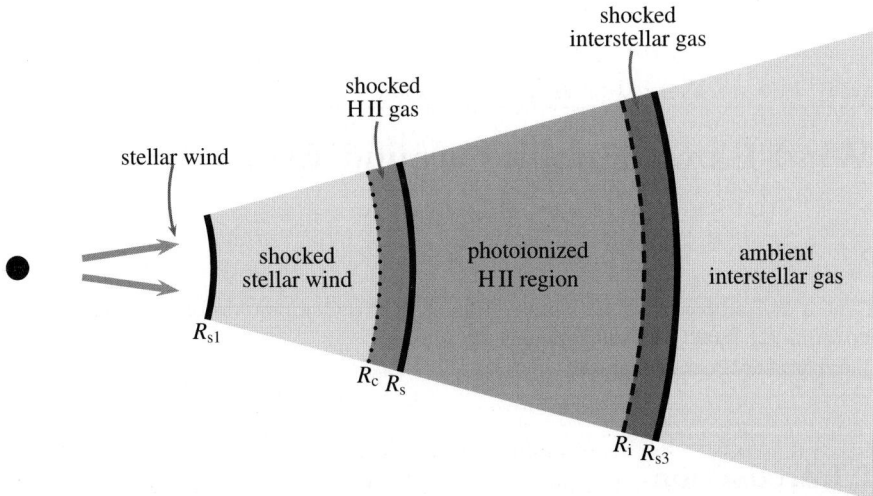

Fig. 1. Schematic of the different regions of a wind-blown bubble around a massive star. R_{s1} is the stellar wind shock, R_s is the shock set up in the ambient medium, and R_{s3} is the isothermal shock sent out ahead of the ionization front that borders the H II region, while R_c is the contact discontinuity separating shocked stellar wind material from swept-up ambient medium and R_i marks the ionization front, separating neutral and ionized gas.

(Pikel'ner 1968; Dyson & de Vries 1972; Avedisova 1972). The basic scenario is shown in Figure 1. In most treatments, the photoionized region and the neutral shell beyond the stellar wind bubble are not discussed separately, and it is assumed that the shock R_s expands into a fully ionized ambient medium.

Conditions behind the ambient medium shock, R_s, favour rapid cooling of the swept-up shell of material, which occurs once the shock velocity falls below 200 km s^{-1}, causing this region to become thin and dense (Falle 1975). The shocked stellar wind, however, can have temperature $>10^7$ K and, thus, does not cool efficiently. A hot, very low density bubble of shocked stellar wind forms, which pushes the cold, swept-up shell like a piston. Since the hot shocked wind remains adiabatic, this is known as an 'energy-driven' flow.

If the ambient medium has a radial power-law density distribution $\rho(r) = \rho_0 r^\beta$, and R_s and $V_s \equiv \dot{R}_s$, are the radius and velocity of the bubble-driven shock, respectively, then the expansion law for the hot bubble can be found by considering momentum and energy conservation, with the assumption that $R_s = R_c$, i.e., that the region of swept-up ambient medium is thin (Dyson 1984, 1989). The momentum and energy equations are

$$\frac{d}{dt}\left(\frac{4\pi}{3+\beta}\rho_0 R_s^{3+\beta} V_s\right) = 4\pi R^2 P \qquad (1)$$

$$\frac{d}{dt}\left(\frac{4\pi}{3}R_s^3\frac{P}{\gamma-1}\right) = \dot{E}_w - 4\pi R_s^2 P V_s \tag{2}$$

where P is the thermal pressure in the hot bubble, giving

$$R_s = \phi^{1/(5+\beta)}\left(\frac{2\dot{E}_w}{\rho_0}\right)^{1/(5+\beta)} t^{3/(5+\beta)}, \tag{3}$$

$$V_s = \left(\frac{3}{5+\beta}\right)\frac{R_s}{t}, \tag{4}$$

with

$$\phi = \frac{(3+\beta)(5+\beta)^3}{12\pi(11+\beta)(7+2\beta)}. \tag{5}$$

Here, the stellar wind mechanical luminosity, $\dot{E}_w \equiv \frac{1}{2}\dot{M}_w V_w$, is taken to be constant, and $\dot{M}_w$, V_w are the stellar wind mass-loss rate and velocity, respectively. A time-varying stellar wind mechanical luminosity was considered by García-Segura & Mac Low (1995a). The case $\beta = 0$ (i.e., constant density ambient medium) returns the well-known formulae of the thin shell approximation for stellar wind bubble evolution (Weaver et al. 1977). The case $\beta = -2$ would represent a bubble expanding into the density distribution left by the wind of a previous evolutionary stage of the star.

'Momentum-driven' flows occur when the shocked wind cools in a time much less than the dynamical timescale. The swept-up shell of ambient material is driven by the momentum of the wind rather than by the pressure of the shocked wind. The expansion law can be found from consideration of momentum conservation.

$$\frac{d}{dt}\left(\frac{4\pi}{3}\rho_0 R_s^{3+\beta} V_s\right) = \dot{M}_w V_w. \tag{6}$$

For constant wind momentum, $\dot{M}_w V_w$, the radius and velocity of the outer shock are given by (Dyson 1984, 1989)

$$R_s = \chi^{1/(4+\beta)}\left(\frac{\dot{M}_w V_w}{\rho_0}\right)^{1/(4+\beta)} t^{2/(4+\beta)}, \tag{7}$$

$$V_s = \left(\frac{2}{4+\beta}\right)\frac{R_s}{t}, \tag{8}$$

with

$$\chi = \frac{(3+\beta)(4+\beta)}{8\pi}. \tag{9}$$

A flow can be defined as energy or momentum driven by the thermal behaviour of the shocked stellar wind at a time, t_0, when the wind has swept up its own mass of ambient material. If the shocked wind can cool by this time then the flow will be momentum driven; otherwise it is energy driven. The cooling time in the shocked stellar wind can be estimated using Kahn's approximation to the cooling rate, $L = AT^{-1/2}n^2$ erg cm^{-3} s^{-1}, in the temperature range $10^5 < T < 2 \times 10^7$ K (Kahn 1976), where $A = 1.3 \times 10^{-19}$ is a constant and n and T are the number density and temperature, respectively, in the shocked gas. The cooling time is thus

$$t_{\rm cool} = \frac{p}{(\gamma - 1)L} \approx 4 \times 10^{-35} \frac{V_w^3}{\rho_R} , \qquad (10)$$

where a mean nucleon mass of $m = 2 \times 10^{-24}$ g has been used and it is assumed that the gas is fully ionized.[1] For a momentum-driven flow, $\rho_R = \dot{M}_w V_w / 4\pi R_{s1}^2$, where R_{s1} is assumed to be the same as R_s (see Eq. 7), since both the shocked wind and shocked swept-up ambient medium shells are thin in this approximation. Hence

$$t_{\rm cool} \simeq 1 \times 10^{-33} \chi^{2/(4+\beta)} \dot{M}_w^{-(2+\beta)/(4+\beta)} V_w^{(18+4\beta)/(4+\beta)} \rho_0^{-2/(4+\beta)} t^{2/(4+\beta)} \quad (11)$$

in this case.

When the flow is energy driven, the interior pressure of the hot bubble, P_b, is uniform and can be found through the substitution of Eqs. 3 and 4 into Eq. 1. The number density in the hot, shocked stellar wind is $n = P_b/2kT$, where T is the post-shock temperature, which depends only on the stellar wind velocity. In this case, therefore, the cooling time is given by

$$t_{\rm cool} \simeq 2 \times 10^{-35} \xi \dot{M}_w^{-(2+\beta)/(5+\beta)} V_w^{3(7+\beta)/(5+\beta)} \rho_0^{-3/(5+\beta)} t^{(4-\beta)/(5+\beta)} , \quad (12)$$

where

$$\xi = \frac{(5+\beta)^2(3+\beta)}{(7+2\beta)\phi^{2+\beta}} . \qquad (13)$$

Thus, the ratio of the cooling time to the dynamical time varies as $t^{-\beta/(4+\beta)}$ for a momentum-driven flow and $t^{-(1+2\beta)/(5+\beta)}$ for an energy-driven flow. While the cooling time is greater than the dynamical time, an initially energy-driven flow can remain energy-driven. Thus, flows can remain energy driven if $\beta \leq -\frac{1}{2}$. If $0 > \beta > -\frac{1}{2}$, energy-driven flows can become momentum driven. An initially momentum-driven flow can only be maintained if $\beta > 0$, otherwise it becomes energy driven.

Main sequence stellar winds interacting with a uniform density ambient medium ($\beta = 0$) will normally be initially energy driven, unless the wind itself is very dense and can cool immediately once it shocks. The flow will remain energy driven because of the high postshock temperatures until, eventually, this gas will begin to cool. If thermal conduction is important between the hot, shocked wind region and the cold swept-up shell, this can lower the

---

[1] The strong shock conditions $p = \frac{3}{4}\rho_R V_w^2$, $n = 4\rho_R/m$ and $T = \frac{3}{32} V_w^2 m/k$ have been used in the derivation.

temperature and raise the density in the hot bubble. This would enhance the cooling rate and so the changeover to a momentum-driven flow would occur earlier.

A Wolf-Rayet star's wind has a high mass-loss rate and could initially produce a momentum-driven flow. However, the density gradient of the red supergiant wind ($\beta = -2$) would quickly lead to a changeover to an energy-driven flow.

Observationally, it should be possible to discriminate between energy-driven and momentum-driven flows by determining the efficiency factors for the conversion of stellar wind mechanical energy and momentum into swept-up gas kinetic energy and momentum (Treffers & Chu 1982). From the analytical model, an energy-driven flow has a kinetic energy efficiency factor $\epsilon = 3(5+\beta)/(11+\beta)(7+2\beta)$ and a momentum efficiency factor $\mu = \epsilon(V_w/V_s)$, while for a momentum-driven flow the corresponding values are $\epsilon = V_s/V_w$ and $\mu = 1$. Values of $\epsilon \ll 0.1$ are taken to indicate a momentum-driven flow. In practice, however, it is difficult to estimate ϵ since the masses of both neutral and ionized components must be taken into account (Dyson & Smith 1985). Furthermore, outer shells may no longer be driven directly by the current stellar wind if the bubble has become depressurized.

2.1 Thermal Conduction

Weaver et al. (1977) obtained a self-similar solution for the temperature structure across the hot bubble under the assumptions that the pressure is constant with radius and thermal conduction from the hot shocked wind to the cold swept-up shell is important. They also assumed that radiative losses are negligible compared to the conductive and mechanical energy fluxes. In this model, the mass within the hot bubble is dominated by the evaporated mass from the shell, while this mass is negligible compared with that remaining in the shell. The self-similar solution for the temperature distribution across the hot bubble is

$$T(r) = T_b(1 - \frac{r}{R_s})^{2/5} , \qquad (14)$$

where $T_b = a\dot{E}_w^{8/35}n_0^{2/35}t^{-6/35}$ is the temperature in the inner part of the bubble and a is a constant which depends on the value of the conduction coefficient. This model has been extended to include the effect of expansion in a power-law density distribution and non-spherical (but uniform) expansion by García-Segura & Mac Low (1995a), but the same dependence on the similarity variable r/R_s is retained. It has been argued (e.g., Dyson 1981) that thermal conduction will not, in fact, be important in stellar wind bubbles since the magnetic fields in the swept-up material will suppress it. It turns out that the temperature and density profiles predicted by the self-similar model of Weaver et al. (1977) significantly overestimate the X-ray emission, as compared to what has been observed, for a given set of stellar wind bubble parameters (Wrigge et al. 2005).

3 Numerical Simulations of Wind-Blown Bubbles

Numerical simulations are a powerful tool for studying the hydrodynamics of wind-blown bubbles, complementing analytical studies since many more physical processes can be included.[2] The first, spherically symmetric, numerical studies included radiative cooling and were able to follow the transition between the initial fully adiabatic stage to the stage where the swept-up interstellar medium cools radiatively (Falle 1975). Subsequent numerical studies in two dimensions show that instabilities develop during the evolution of wind-blown bubbles, where the nature of the instability present depends on the evolutionary stage and physics being considered. Kelvin-Helmholtz, Rayleigh-Taylor, Richtmeyer-Meshkov, Vishniac, and thermal instabilities have all been reported (Rozyczka 1985; Rozyczka & Tenorio-Tagle 1985a,b; García-Segura & Mac Low 1995b; García-Segura et al. 1996a,b; Brighenti & D'Ercole 1995a,b, 1997; Strickland & Stevens 1998; Freyer et al. 2003). However, care must be taken in the identification of these instabilities, since the resolution of the numerical grid can play an important role, as can parameters such as the cutoff temperature below which radiative cooling is assumed equal to zero (see Zhekov & Myasnikov 2000b, for a more detailed discussion). Although the most widely used analytical models take into account thermal conduction (Weaver et al. 1977; García-Segura & Mac Low 1995a), very few numerical studies include this process (notable exceptions are Zhekov & Myasnikov 1998, 2000a,b). The presence of even a weak magnetic field will be enough to cause asymmetric thermal conduction and will thus lead to an asymmetric wind-blown bubble (Zhekov & Myasnikov 2000b).

In this section, the evolution of a wind-blown bubble around a $40 M_\odot$ star from the main-sequence phase, through the red supergiant phase to the final Wolf-Rayet phase is examined by means of numerical simulations. The first two stages are treated as one-dimensional (spherically symmetric) simulations, while the start of the Wolf-Rayet phase is modelled in two dimensions (cylindrical symmetry). Photoionization is included using the method described in Henney et al. (2005). Radiative cooling and photoionization heating are also taken into account. Thermal conduction and magnetic fields are not included in these models. A simple three-wind model is considered, where the stellar wind mass-loss rate and velocity is taken to be constant during each phase and the ambient medium has uniform density and temperature (cf. García-Segura & Mac Low 1995b; van Marle et al. 2005). Although a particular simulation is described here, the general features are common to all simulations of the evolution of wind-blown bubbles.

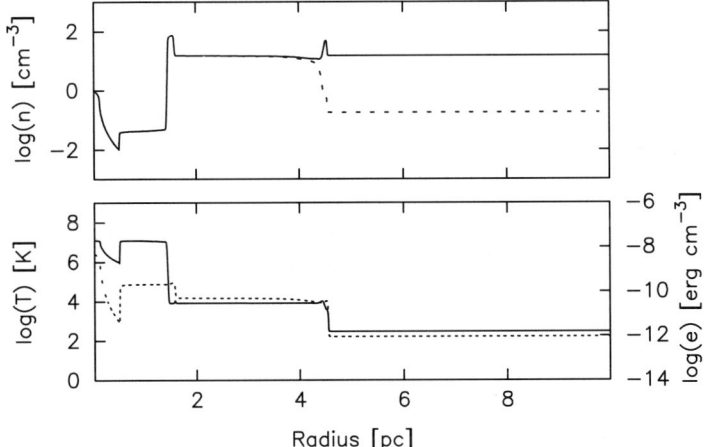

Fig. 2. Radial structure of a stellar wind bubble and H II region around a $40 M_\odot$ star at an early stage of evolution (after 40,000 years). Top panel: total number density (solid line) and ionized number density (dotted line). Bottom panel: temperature (solid line) and internal energy density ($e = p/[\gamma - 1]$ — dotted line). Stellar parameters $\dot{M}_w = 9.1 \times 10^{-7} M_\odot$ yr^{-1}, $V_w = 890$ km s^{-1}, and $S_* = 6.34 \times 10^{47}$ s^{-1} (adapted from van Marle et al. 2005).

3.1 Main Sequence Evolution

The main-sequence evolution of the ionized region (H II region) and stellar wind bubble around a $40 M_\odot$ star in a uniform, neutral ambient medium with constant stellar wind and ionizing photon rate is presented here. The initial neutral ambient medium is taken to have a hydrogen nucleus number density $n_H = 15$ cm^{-3} and temperature 300 K, while the stellar wind has a mass-loss rate of $\dot{M}_w = 9.1 \times 10^{-7} M_\odot$ yr^{-1} and wind terminal velocity $V_w = 890$ km s^{-1}, and the rate at which ionizing photons are emitted is $S_* = 6.34 \times 10^{47}$ s^{-1} (cf. Schaller et al 1992; van Marle et al. 2005).[3]

The H II region forms immediately, reaching an initial Strömgren radius of ~ 4.5 pc. A low-velocity, isothermal shock forms ahead of the ionization front in the neutral gas and the hot ($T = 10^4$ K) ionized region begins to expand. The stellar wind bubble forms inside this ionized region. Cooling in the swept up ionized gas is very efficient (see, e.g., Falle 1975) and this region quickly forms a thin, dense shell. The increase in opacity traps the ionizing photons and the main ionization front becomes confined to this swept-up shell. The

---

[2] Numerical simulations solve the full gas-dynamic equations and physical processes are generally included via source terms.

[3] In these simulations, the stellar wind is injected in the form of a thermal energy and mass source in a small volume around the star, and for this reason temperatures in the *unshocked* stellar wind region are high but do not affect the dynamics since the flow is highly supersonic in this region.

region between the shell and the outer isothermal shock is occupied by gas that belonged to the initial Strömgren sphere and is now slowly recombining (recombination time is approximately $10^5/n_{\rm H}$ years). This can be seen in Figure 2, which shows the density, ionized density, temperature and thermal energy (i.e., pressure) structure after 40,000 years of evolution. The outer ionization front is no longer sharp since gas here has started to recombine. However, the shock that was launched ahead of the ionization front continues to expand at <10 km s^{-1} into the neutral ambient medium.

In two-dimensional simulations, strong radiative cooling in the swept-up ambient medium leads to the growth of instabilities, which cause corrugation of the shell (Strickland & Stevens 1998; Freyer et al. 2003). Shadowed regions then form in the H II region beyond the stellar wind shell and recombination occurs faster here, leading to a pattern of neutral spokes and non-radial velocity fields (Freyer et al. 2003; Arthur & Hoare 2005).

Eventually, the wind bubble overtakes the outer neutral shock. By this time, the swept-up shell of material has spread out and its density has dropped. This is because the pressure in the shell is the same as that in the hot wind bubble, which falls slowly as the inner wind shock moves outwards. Since the temperature in the swept-up shell is determined by photoionization, and is, thus, a constant $\sim 10^4$ K, the density has to fall (and hence the swept-up shell must broaden) to account for the fall in pressure. The inner wind shock will continue to move outwards until the ram pressure of the stellar wind here ($\rho_w V_w^2 \equiv \dot{M}_w V_w/4\pi R_s^2$) balances the pressure of the interstellar medium. The H II region establishes itself in this, now, low-density shell, and sends a shock ahead into the neutral gas.

Figure 3 shows the radial structure of the stellar wind bubble and H II region at the end of the main sequence stage, after some 4.31×10^6 years of evolution. The inner stellar wind shock is located about 3 pc from the star. The hot ($T \sim 10^7$ K), very low density ($n_i \sim 10^{-3}$ cm^{-3}) shocked stellar wind bubble occupies the region $3 < R < 24.5$ pc. Outside of this is the shell containing the H II region. The ionized gas has density $n_i \sim 1$ cm^{-3} and temperature $T = 10^4$ K, and extends between $24.5 < R < 27.5$ pc. The broad neutral shell, $27.5 < R < 33.5$ pc, has density $n_n \sim 30$ cm^{-3} and temperature $T \sim 10^3$ K. The entire region between $3 < R < 33.5$ pc has uniform pressure and is still expanding slowly into the ambient medium.

3.2 Post-Main-Sequence Evolution

In Figure 4, the possible mass loss history of a $60 M_\odot$ star is shown (Langer et al. 1994; García-Segura et al. 1996a). A detailed study of the evolution of the wind-blown bubble around stars of initial masses 60 $M_\odot$ and 35 $M_\odot$ has been carried out by García-Segura et al. (1996a,b), for the purely hydrodynamic case, and by Freyer et al. (2003, 2005), including the radiative transfer of ionizing photons. The numerical simulations predict short-lived, observable

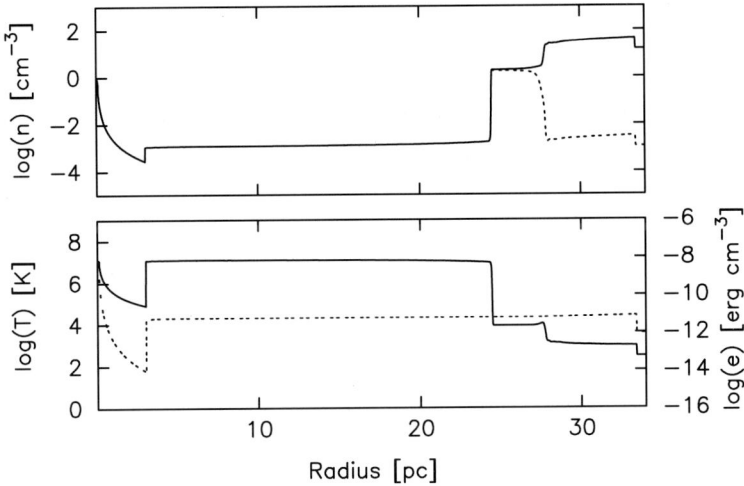

Fig. 3. Radial density and temperature structure of the stellar wind bubble around a $40 M_\odot$ star at the end of the main sequence stage, i.e. after $\sim 4.3 \times 10^6$ years. Line types are the same as for Fig. 2.

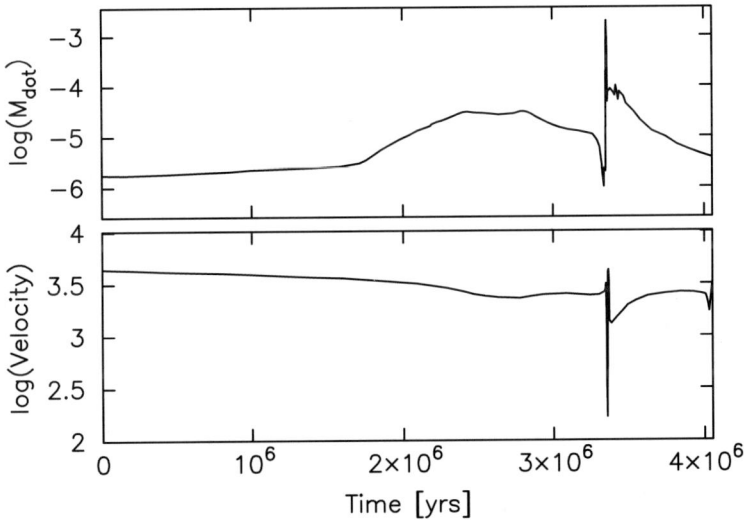

Fig. 4. Variation of mass-loss rate ($M_\odot$ yr^{-1}) and stellar wind velocity (km s^{-1}) as a function of time for a $60 M_\odot$ star (data from García-Segura et al. 1996a).

nebulae during the LBV or RSG stage and the onset of the WR stage. Instabilities are formed in the dense shells swept-up by the different stellar wind stages, and these are consistent with clumps observed in ring nebulae around some Wolf-Rayet stars. The velocity of the LBV or RSG wind is found to play a key role in the detailed structure of the ring nebulae formed during this process.

Red Supergiant Phase

During the red supergiant (RSG) phase, the ionizing photon emission rate of the 40 $M_\odot$ star drops to 3.1×10^{41} photons s^{-1}, and the stellar wind velocity drops to 15 km s^{-1} while the mass-loss rate increases to 8.3×10^{-5} $M_\odot$ yr^{-1} (van Marle et al. 2005). This phase lasts only 2×10^5 years (an LBV phase would have an even higher mass-loss rate but a much shorter timescale; see Fig. 4). The slow, dense RSG wind expands into the structure formed by the main sequence wind and a thin, dense shell of shocked RSG material forms ahead of the freely expanding wind (see Figure 5). The high thermal pressure in the hot main sequence bubble causes backflow of low-density gas towards the star, which shocks against the RSG wind. The reduction in the ionizing photon rate leads to the disappearance of the H II region as the shell of ionized gas recombines. The hot bubble remains ionized since the recombination times are long in this hot, very low-density gas. The RSG wind itself is neutral.

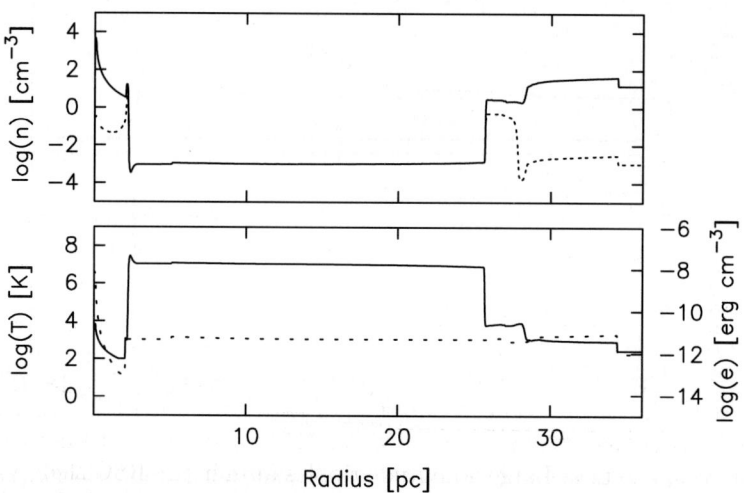

Fig. 5. Radial structure of a stellar wind bubble around a 40 $M_\odot$ star at the end of the red supergiant stage, i.e. after 4.5×10^6 years. Top panel: total number density (solid line) and ionized number density (dotted line). Bottom panel: temperature (solid line) and internal energy density ($e = p/[\gamma - 1]$ — dotted line).

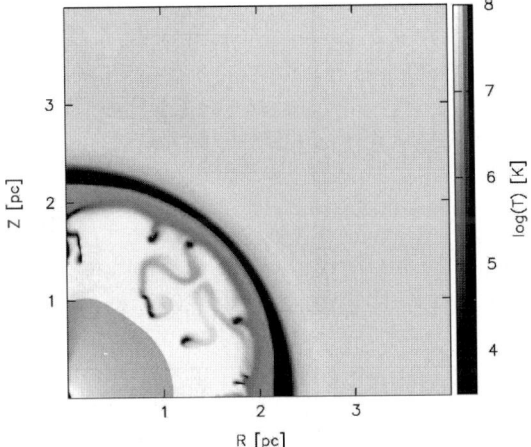

Fig. 6. Formation of Rayleigh-Taylor instabilities in the Wolf-Rayet wind shell just before the interaction with the red supergiant shell. The instabilities form as the WR wind accelerates into the r^{-2} density distribution left by the RSG wind. The temperature of the gas is shown 8000 years after the start of the Wolf-Rayet wind. Only one quadrant is shown, and the star is located at $(0,0)$.

Wolf-Rayet Phase

The Wolf-Rayet wind expands into the structure formed during the previous two phases. During the Wolf-Rayet phase, the stellar wind becomes faster ($V_w = 2160$ km s^{-1}) and remains strong ($\dot{M}_w = 4.1 \times 10^{-5}$ $M_\odot$ yrs^{-1}) and the stellar ionizing photon rate also increases ($S_* = 3.86 \times 10^{47}$ s^{-1}). The transition between the RSG and WR winds will involve a period of wind acceleration into the r^{-2} density distribution left by the RSG wind, and this situation is Rayleigh-Taylor unstable.[4] The WR shell consists of swept-up RSG wind material and shocked WR wind. It becomes Rayleigh-Taylor unstable at an early time in the simulation, well before the interaction with the RSG shell. Dense, cold clumps of material formed by the instability lag behind the main WR shell but the shell itself is not disrupted, since the instabilities cease to grow once the acceleration stage is over (see Fig. 6).

Figure 7 shows the evolutionary sequence of the interaction of the WR shell with the RSG shell. The dense, cold clumps formed due to Rayleigh-Taylor instabilities in the WR shell act as obstacles around which the hot, shocked WR wind has to flow. Large amplitude ripples form in the RSG shell, through which the WR wind eventually breaks out into the low-density bubble formed by the main-sequence wind. The dense, cold clumps formed by the Rayleigh-Taylor instabilities move outwards more slowly. These clumps will be subject

---

[4] In the numerical simulations presented here, acceleration of the wind occurs initially as a consequence of the way in which the stellar wind is included in the calculation.

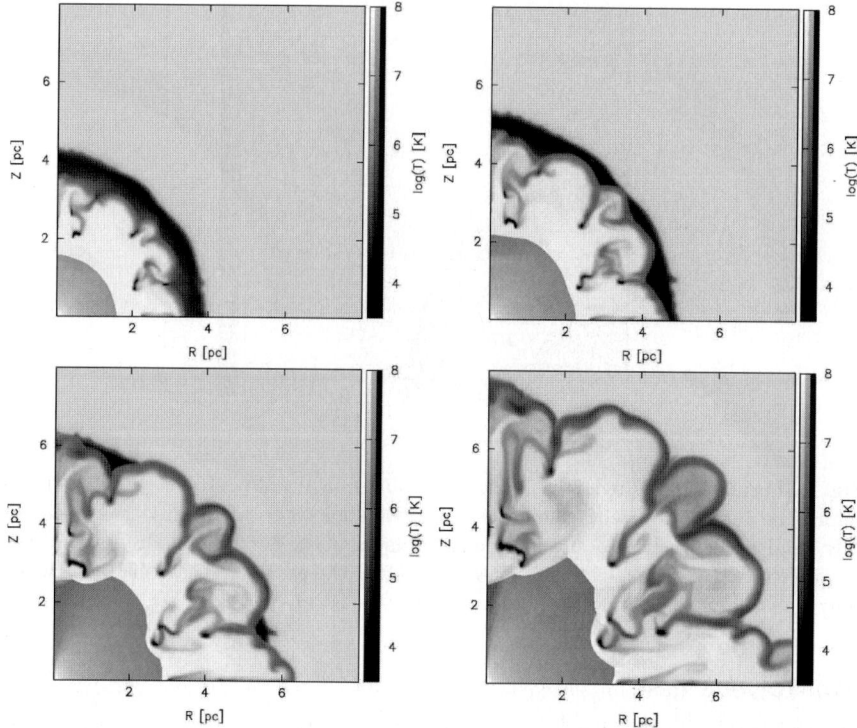

Fig. 7. Interaction of the Wolf-Rayet wind with the dense shell from the red supergiant stage. Rayleigh-Taylor instabilities form as the WR wind accelerates into the r^{-2} density distribution left by the RSG wind. The instabilities become enhanced when the WR wind collides with the dense RSG shell and ruptures it. The panels show the temperature of the gas after times (a) 15,000 years, (b) 18,000 years (c) 21,000 years and (d) 24,000 years after the start of the WR wind.

to hydrodynamic ablation and photoevaporation, but the numerical resolution of the two-dimensional simulation is not able to follow these processes. The mixing process produces gas with temperatures $\sim 10^6$ K, which should emit X-rays.

During the red supergiant phase, 16.5 $M_\odot$ of material are lost from the star. All of this material is swept into clumps and filaments by the Wolf-Rayet wind, which are potential *mass-loading* sources of the WR wind, to be discussed later.

Figure 8 shows the final radial structure of the wind bubble around the $40 M_\odot$ star just before it explodes as a supernova, from a one-dimensional (spherical symmetry) numerical simulation. The low-density bubble has been repressurized by the hot, shocked WR wind material, a new shock is being driven into the neutral shell, and the H II region is starting to reform. The

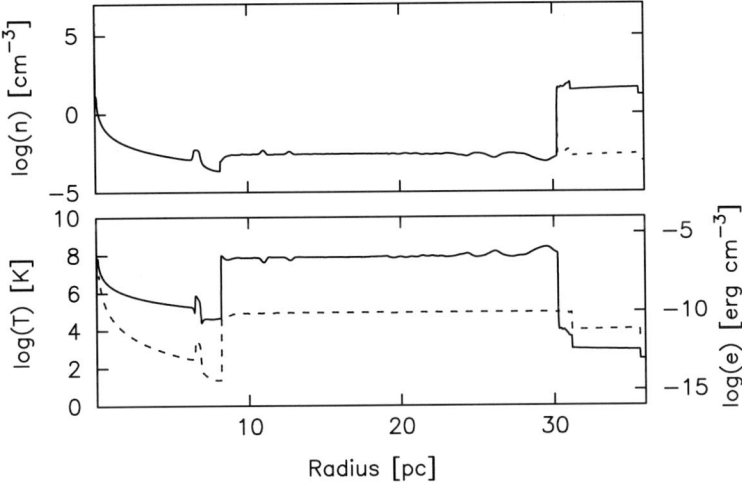

Fig. 8. Radial structure of a stellar wind bubble around a 40 $M_\odot$ star at the end of the Wolf-Rayet phase, i.e. after a total time of 4.8×10^6 years. Top panel: total number density (solid line) and ionized number density (dotted line). Bottom panel: temperature (solid line) and internal energy density ($e = p/[\gamma - 1]$ — dotted line).

one-dimensional simulation cannot show the fate of the clumps and filaments formed during the initial interaction between the WR wind and the RSG shell.

3.3 Energy Evolution in a Wind-Blown Bubble

Figure 9 shows the evolution of the neutral and ionized gas kinetic and thermal energy fractions with respect to the total stellar wind mechanical energy, $\dot{E}_w t$, resulting from the numerical simulations described above. During the main sequence stage, the initial 400,000 yrs show large variations in the relative importance of the kinetic energies in the neutral and ionized gas. This is because this period of time sees the formation and expansion of the initial H II region, which then retreats back into the shell of material swept up by the stellar wind before expanding outwards again driven by the pressure in the hot bubble. The majority of the main-sequence stage sees an energy-driven bubble in which the thermal energy of the hot bubble pushes the expansion of a neutral shell of material. The kinetic energy of the swept-up neutral shell is roughly 0.25 of the stellar wind mechanical energy for most of this stage. This is larger than the fraction $\frac{15}{77}$ predicted by the simple thin-shell model (see e.g., Dyson & Williams 1997) and can be attributed to the H II region. It is the H II region that results in the different fractions summing to more than unity, since the contribution to the thermal energy by the heating of the H II region in the shell surrounding the low-density bubble is substantial (see also Freyer et al. 2003). Observations of only the ionized gas during this phase

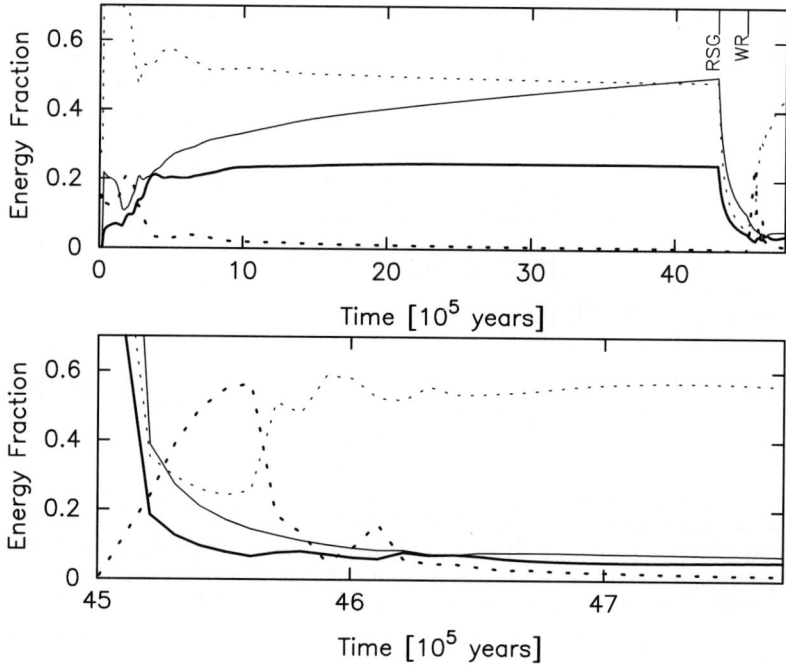

Fig. 9. Top panel: Kinetic and thermal energies of the ionized and neutral gas as fractions of the total stellar wind mechanical energy as a function of time for the numerical simulation described in this paper. Thick lines represent kinetic energy and thin lines represent thermal energy, while solid lines represent neutral gas and dotted lines represent ionized gas energy fractions. The onset of the RSG and WR are indicated. Bottom panel: Kinetic and thermal energies as fractions of the WR stellar wind mechanical energy, $\dot{E}_w(t - t_{\rm WR})$ during the WR phase.

would show $\epsilon \ll 0.1$, suggesting a momentum-driven flow, while observations of the neutral shell would give $\epsilon > 0.1$, hence an energy-driven flow.

When the red supergiant phase begins, the ionizing photons are trapped close to the star and the outer H II region recombines. Also, the driving pressure in the hot bubble drops because the RSG stellar wind is such low velocity. As a result, both the thermal and kinetic energy fractions drop. Observations of both neutral and ionized gas during this phase would give $\epsilon < 0.1$, suggesting a momentum-driven flow.

During the Wolf-Rayet phase, the RSG wind material is ionized by the high flux of ionizing photons and the high velocity WR wind forms a hot, high-pressure bubble. This is reflected in the energy fractions of the ionized gas, which rise steeply during this phase. It is a long time before the wind-blown bubble becomes pressurized out to the neutral shell, and so the neutral gas thermal and kinetic energy fractions remain low. Observations of the ionized gas during this stage would show $\epsilon > 0.1$ initially, but later in the evolution

$\epsilon \ll 0.1$ even though the flow is now energy driven. The neutral gas kinetic energy fraction corresponds to the outer swept-up shell, which is not repressurized or accelerated during the timescale we are considering. Consequently, observations of the neutral shell during this stage of evolution would indicate that $\epsilon < 0.1$.

Figure 9 shows that neither observations of only the ionized gas or only the neutral gas can reveal the full picture of the energetics of a wind-blown bubble. Observations of the outer neutral shell during this stage do not provide a true picture of what is happening with the interaction between the current stellar wind and its surroundings.

3.4 Mass Loading

The presence of dense clumps embedded in a fast flowing stellar wind and subject to a large flux of ionizing photons from the central star suggests the possibility that material from the clumps will be incorporated into the flow by one or other mass-loading processes, namely photoevaporation, conductive evaporation or hydrodynamic ablation. If the amount of mass incorporated into the flow becomes important then the dynamics of the flow can be radically altered (see the chapter by Julian Pittard). Both supersonic and subsonic flows tend to an average transonic flow, and the positions of global shock waves change (Smith 1996). Furthermore, if the mass loading is strong enough, the density in the hot shocked stellar wind can increase sufficiently that cooling becomes important in this region.

Mass loading by embedded clumps in a stellar wind was first proposed to explain the ionization potential-velocity correlation observed in the WR ring nebula RCW 58 (Smith et al. 1984; Hartquist et al. 1986). In this scenario, the mass-loading is due to hydrodynamic ablation and has two regimes: one in which the flow around the clump is subsonic and the other when it is supersonic. In the subsonic case, the mass-loading rate is proportional to $M^{4/3}$, where M is the flow Mach number. In the supersonic case, the mass-loading rate saturates and is taken to be constant. Numerical modelling of the mass-loaded flow in RCW 58 shows that the position of the stellar wind shock is fixed by the mass loading and that cooling occurs in the outer parts of the hot shocked bubble (Arthur et al. 1993, 1996).

4 Observations of Wind-Blown Bubbles

Bubbles around massive stars can be detected in a variety of ways: as optical ring-shaped nebulae (e.g., Chu 1981), as shell-shaped thermal radio continuum sources (e.g., Goss & Lozinskaya 1995), as neutral gas voids and expanding shells in the hydrogen 21 cm line-emission distribution (Cappa et al. 2005), as infrared shells (e.g., Marston 1996), and in a limited number of cases as diffuse X-ray sources (e.g., Wrigge 1999). Gamma-ray burst afterglows probe the innermost regions of the progenitor wind bubble (Chevalier et al. 2004), and Type II and Type Ib/c supernova narrow spectral absorption features can be attributed to moving shells around the massive star progenitor (Dopita et al. 1984).

Multiple concentric shells of material have been observed around many galactic WR stars (Marston 1996).[5] The outermost shells are thought to correspond to the O star phase, and are most readily observable at far infrared or 21 cm radio wavelengths, although around 8% are seen optically. These outer shells have can very large diameters (> 100 pc), depending on the ambient ISM density, and are expanding slowly (generally, $v_{\mathrm{exp}} < 10$ km s^{-1} Marston 1996). The innermost shells are easily seen in optical narrowband images, and represent dense material recently ejected from the star in a LBV or RSG phase or swept up by the stellar wind in the WR phase.

Of the ~ 150 observed Galactic WR stars about 25% are associated with optical ring-like nebulae (Heckathorn et al. 1982; Chu et al. 1983; Miller & Chu 1993; Marston et al. 1994a,b). However, only ~ 10 of these WR ring nebulae have the sharp rims and short dynamical ages (ring radius divided by expansion velocity) that suggest that they are bubbles formed by the winds of the central stars during the WR phase.[6] The remaining ring nebulae are less well defined and have dynamical ages much larger than the lifetime of a WR phase, and hence are thought to be simply stellar ejecta from the RSG or LBV stage photoionized by the WR star (like a planetary nebula), rather than dynamically shaped by its wind. Wind-blown bubbles were established as a class of object by Chu (1981) and Lozinskaya (1982).

A striking example of an optical WR ring nebula is RCW 58 (see Figure 10). This nebula consists of clumps and filaments extending radially outward from the central WN8 star, seen in Hα, with more diffuse [O III] emission extending beyond (Chu 1982; Gruendl et al. 2000). The nebula is enriched in He and N, which indicates that the material has been ejected from the star, probably during a red supergiant stage. The expansion velocity of the shell is 87 km s^{-1}, and broad linewidths indicate supersonic motions (Smith et al. 1988). The measured electron densities of the clumps are ~ 500 cm^{-3}.

Most WR wind-blown bubbles are aspherical to a lesser (e.g., RCW 58, S 308, NGC 6888) or greater (e.g., NGC 2359, NGC 3199, G2.4+1.4) degree. The asphericity is often more pronounced in Hα images than in [O III], suggesting that it arises in the RSG or LBV ejecta phase. Such morphologies could

---

[5] Ring nebulae have also been observed around WR stars in the Large Magellanic Cloud (Dopita et al. 1994) but their distance makes them less easy to study.

[6] These objects are: S 308, RCW 58, RCW 104, NGC 2359, NGC 3199, NGC 6888, Anon (MR26), G2.4+1.4, and there is kinematical evidence that the nebulae around WR 116 and WR 133 have expansion velocities consistent with wind-blown bubbles (Esteban & Rosado 1995). Anon(WR 128) and Anon(WR 134) are possibly wind-blown bubbles, too (Gruendl et al. 2000). There are also wind-blown bubbles around two other evolved stars: NGC 6164-5 (around an O6.5fp star), NGC 7635 (around an O6.5III star).

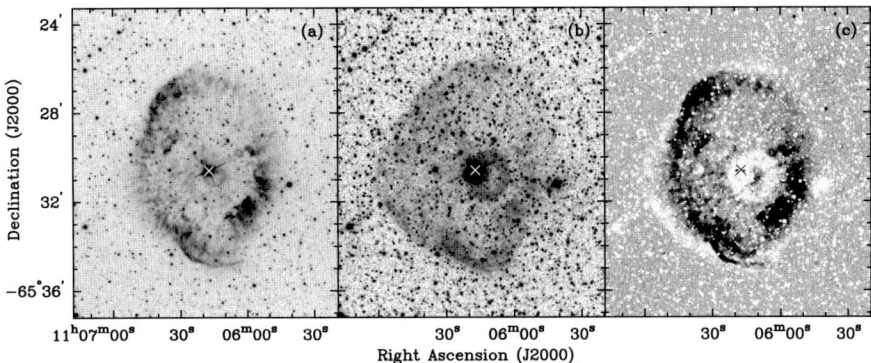

Fig. 10. (a) Optical Hα and (b) [O III] λ5007 images of the ring nebula RCW 58, together with (c) an image of their difference (Gruendl et al. 2000). (Reproduced with permission of the AAS).

also be due to the environment: for example, NGC 2359 and NGC 3199 are each bounded on one side by molecular clouds (Schneps et al. 1981; Marston 2001) and their roughly parabolic shapes in Hα images could be the result of the wind-bubble blowing out from near the surface of the dense cloud and expanding down a strong density gradient, as has been suggested in the case of G2.4+1.4 (Dopita & Lozinskaya 1990). An alternative explanation for this sort of morphology could be that the star is moving supersonically through the ambient medium and forms a bowshock ahead of it (this was suggested as an explanation for G2.4+1.4 by Brighenti & D'Ercole 1995b).

Four of the 8 WR wind-blown bubbles have been observed with X-ray satellites but only two of these, NGC 6888 and S 308, have been detected in diffuse X-rays. Both were detected with the *ROSAT* PSPC in the energy range 0.1–2.4 keV and spectral analysis shows that the hot gas in these bubbles is dominated by the component at $\sim 1.5 \times 10^6$ K (Bochkarev 1988; Wrigge et al. 1994; Wrigge 1999), although this could be partially contaminated by higher energy point source emission. *ASCA* SIS observations of NGC 6888 suggests that there is an additional component of hot gas at a temperature of 8×10^6 K (Wrigge et al. 1998, 2005), while *XMM-Newton* EPIC observations of S 308 indicate that the spectrum is very soft, suggesting plasma temperatures of only 1×10^6 K (Chu et al. 2003). Figures 11 and 12 show the optical [O III] λ5007 and X-ray images of S 308 and NGC 6888, taken from Chu et al. (2003). The S 308 images clearly show that the X-ray emission is limb-brightened, and NGC 6888 appears to be limb-brightened also.[7] In both cases the X-ray emission is interior to the [O III] emission, which marks the position of the main shock wave.

---

[7] Although not clear from the *ROSAT* image, this is evident in recent *Chandra* images: http://www.chandra.harvard.edu/photo/2003/ngc6888.

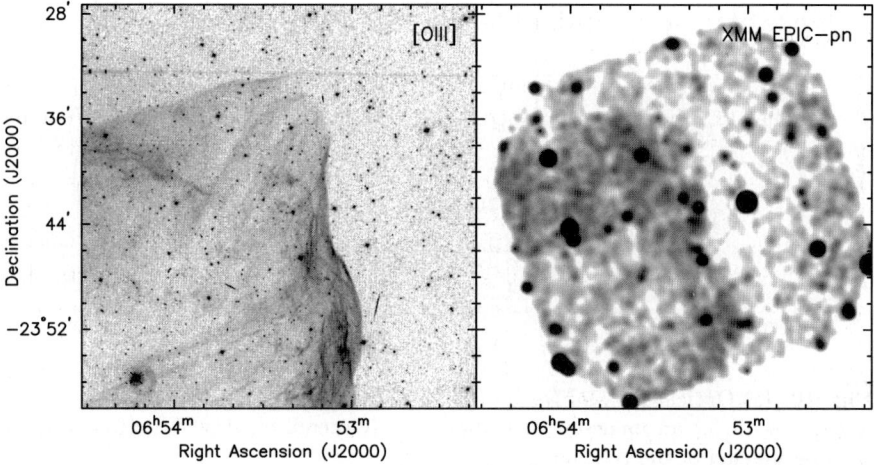

Fig. 11. Optical [O III] λ5007 image (left) and *XMM-Newton* EPIC image (right) of the northwest quadrant of S 308 (Chu et al. 2003, figure reproduced with permission from *Revista Mexicana de Astrononomía y Astrofísica*).

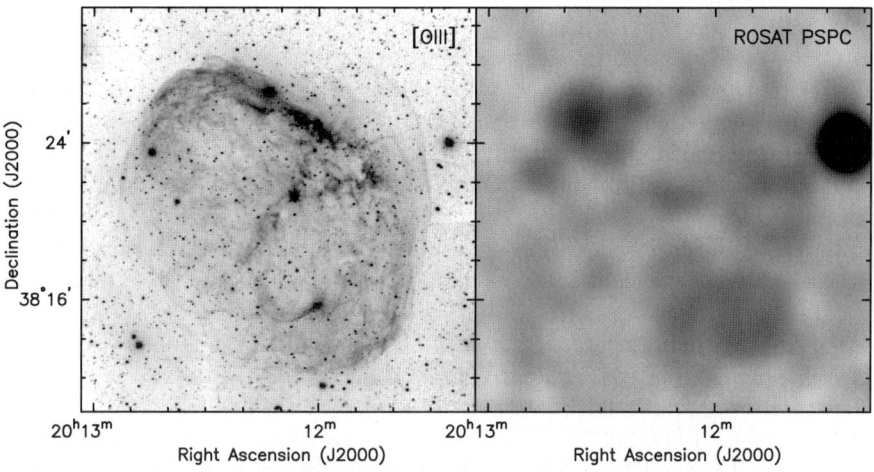

Fig. 12. Optical [O III] λ5007 image (left) and *ROSAT* PSPC image (right) of NGC 6888 (Chu et al. 2003, figure reproduced with permission from *Revista Mexicana de Astronomía y Astrofísica*).

5 Observations Confront Theory

It has been proposed that nebulae around massive stars begin with wind-blown shells in the main-sequence stage, evolve into amorphous H II regions, then ring-like H II regions, then nebulae with stellar ejecta in the RSG or LBV phase before the star explodes as a supernova (McKee et al. 1984). The bubbles blown by WR winds occur within the circumstellar material of the RSG/LBV phase. In practise, wind-blown bubbles have rarely been observed around main-sequence O stars (although kinematic evidence suggests that wind-blown bubbles do exist inside H II regions, Nazé et al. 2001). Ring-like H II regions may be too faint to observe but expanding HI shells with masses up to $10^4 M_\odot$ should occur around massive stars, as illustrated in Figure 3. In the later stages of massive star evolution and, indeed, around some supernova remnants, expanding HI shells are detected (Cappa et al. 2005) that must have originated in the main-sequence stage.

The fact that so few of the known WR stars are surrounded by wind-blown bubbles suggests that this is a short-lived phase of evolution (Chu 1981). This is borne out by numerical simulations, which predict observable bubbles only during the early interaction of the WR wind with the LBV or RSG nebula (Freyer et al. 2003, 2005). More common are amorphous nebulae, which are observed around many WR stars, indicating a photoionization rather than wind-blown origin, even when there are clear indications that the central star possesses a strong stellar wind. These cases could be the result of the environment in which the nebula is found: if a strong density gradient is present the stellar wind will blow out in the direction of decreasing density rather than form a pressure-driven expanding bubble (Arthur & Hoare 2005, discuss similar models in the context of cometary H II regions). Clumps and filaments observed in nebulae such as RCW 58 are formed naturally in numerical simulations by Rayleigh-Taylor and other instabilities (García-Segura et al. 1996a,b).

Theoretical estimates of X-ray emission as predicted by analytic theory fail to match the reality. Models that do not include thermal conduction have very high temperatures in the shocked stellar wind and do not produce the soft X-ray spectrum that has been detected in NGC 6888 and S 308. On the other hand, the self-similar models of Weaver et al. (1977) and García-Segura & Mac Low (1995a), which include thermal conduction, predict much higher values of the soft X-ray luminosities than are observed. It appears that some sort of dissipative process, though not the classical thermal conduction proposed by Weaver et al. (1977), is necessary to produce the soft X-rays. This argument presupposes that the X-ray emission is produced in the shocked Wolf-Rayet wind material. Two-dimensional numerical simulations of the wind-blown bubble produced by a $35 M_\odot$ star Freyer et al. (2005), on the other hand, show that the X-ray emission actually arises in the shocked, dense RSG wind material. This can be appreciated in Figure 7, where gas temperatures of $\sim 10^6$ K, characteristic of soft X-rays are present in the shocked

RSG shell material. Also, it could be expected that material photoevaporated or ablated from the dense clumps formed by the instabilities could augment the soft X-ray luminosity in the interaction region.

6 Final Remarks

Although the general characteristics of the formation and evolution of wind-blown bubbles around evolved stars are well understood, there remain a number of important features to explain. Most obviously, it is interesting to know whether the asphericity of nebulae such as NGC 6888 and RCW 58 is due to the non-spherical ejection of material in the RSG or LBV stage, or whether it is due to some direction-dependent physical process, such as thermal conduction, in the WR bubble phase. Photoionization is clearly important for the thermal balance and energetics of wind-blown bubbles and should not be neglected in numerical calculations. The role of instabilities in producing the clumps and filaments observed in some (but not all) bubbles should be examined in more detail. Cooling within the clumps can lead to the formation of very dense, neutral structures within the hot, ionized bubble. The ablation, either by photoevaporation, thermal conduction or hydrodynamic ablation, of this neutral material will modify the physical properties and chemical abundances of the hot, shocked wind and such processes should, therefore, be studied further. Theories should be tested with real data and new observations will always be needed, at all wavelengths, to provide more detailed kinematical information, abundance data, and improved stellar wind parameters.

References

Arthur, S.J., Dyson, J.E., Hartquist, T.W. 1993 MNRAS **261**, 425
Arthur, S.J., Henney, W.J., Dyson, J.E. 1996 A&A **313**, 897
Arthur, S.J., Hoare, M.G. 2005 ArXiv Astrophysics e-prints, arXiv:astro-ph/0511035
Avedisova, V.S. 1972 Soviet Astronomy **15**, 708
Bochkarev, N.G. 1988 Nature **332**, 518
Brighenti, F., D'Ercole, A. 1995a, MNRAS **277**, 53
Brighenti, F., D'Ercole, A. 1995b, MNRAS **273**, 443
Brighenti, F., D'Ercole, A. 1997 MNRAS **285**, 387
Cappa, C., Niemela, V.S., Martín, M.C., McClure-Griffiths, N.M. 2005 A&A **436**, 155
Chevalier, R.A., Li, Z.-Y., Fransson, C. 2004 ApJ **606**, 369
Chu, Y.-H. 1981 ApJ **249**, 195
Chu, Y.-H. 1982 ApJ **254**, 578
Chu, Y.-H., Gruendl, R.A., Guerrero, M.A. 2003 Revista Mexicana de Astronomia y Astrofisica Conference Series **15**, 62
Chu, Y.-H., Guerrero, M.A., Gruendl, R.A., García-Segura, G., Wendker, H.J. 2003 ApJ **599**, 1189

Chu, Y.-H., Treffers, R.R., Kwitter, K.B. 1983 ApJS **53**, 937
Dopita, M.A., Cohen, M., Schwartz, R.D., Evans, R. 1984 ApJ **287**, L69
Dopita, M.A., Lozinskaya, T.A. 1990 ApJ **359**, 419
Dopita, M.A., Bell, J.F., Chu, Y.-H., Lozinskaya, T.A. 1994 ApJS **93**, 455
Dyson, J.E. 1981 Investigating the Universe, ASSL **91**, 125
Dyson, J.E. 1984 Ap&SS **106**, 181
Dyson, J.E. 1989 Structure and Dynamics of the Interstellar Medium LNP **350**, 137
Dyson, J.E., de Vries, J. 1972 A&A **20**, 223
Dyson, J.E., Smith, L.J. 1985 Cosmical Gas Dynamics, 173
Dyson, J.E., Williams, D.A. 1997 The physics of the interstellar medium. Edition: 2nd ed. Publisher: Bristol: Institute of Physics Publishing, 1997. Edited by J.E. Dyson and D.A. Williams. Series: The graduate series in astronomy. ISBN: 0750303069
Esteban, C., Rosado, M. 1995 A&A **304**, 491
Falle, S.A.E.G. 1975 A&A **43**, 323
Freyer, T., Hensler, G., Yorke, H.W. 2003 ApJ **594**, 888
Freyer, T., Hensler, G., Yorke, H.W. 2005 ArXiv Astrophysics e-prints, arXiv:astro-ph/0512110
García-Segura, G., Mac Low, M.-M. 1995a ApJ **455**, 145
García-Segura, G., Mac Low, M.-M. 1995b ApJ **455**, 160
García-Segura, G., Mac Low, M.-M., Langer, N. 1996a A&A **305**, 229
García-Segura, G., Langer, N., Mac Low, M.-M. 1996b A&A **316**, 133
Goss, W.M., Lozinskaya, T.A. 1995 ApJ **439**, 637
Gruendl, R.A., Chu, Y.-H., Dunne, B.C., Points, S.D. 2000 AJ **120**, 2670
Hartquist, T.W., Dyson, J.E., Pettini, M., Smith, L.J. 1986 MNRAS **221**, 715
Heckathorn, J.N., Bruhweiler, F.C., Gull, T.R. 1982 ApJ **252**, 230
Henney, W.J., Arthur, S.J., García-Díaz, M.T. 2005 ApJ **627**, 813
Kahn, F.D. 1976 A&A **50**, 145
Langer, N., Hamann, W.-R., Lennon, M., Najarro, F., Pauldrach, A.W.A., Puls, J. 1994 A&A **290**, 819
Lozinskaya, T.A. 1982 Ap&SS **87**, 313
Marston, A.P., Chu, Y.-H., García-Segura, G. 1994a ApJS **93**, 229
Marston, A.P., Yocum, D.R., García-Segura, G., Chu, Y.-H. 1994 ApJS **95**, 151
Marston, A.P. 1996 AJ **112**, 2828
Marston, A.P. 2001 ApJ **563**, 875
McKee, C.F., van Buren, D., Lazareff, B. 1984 ApJ **278**, L115
Miller, G.J., Chu, Y.-H. 1993 ApJS **85**, 137
Nazé, Y., Chu, Y.-H., Points, S.D., Danforth, C.W., Rosado, M., Chen, C.-H.R. 2001 AJ **122**, 921
Pikel'ner, S.B. 1968 Astrophysical Letters **2**, 97
Rozyczka, M. 1985 A&A **143**, 59
Rozyczka, M., Tenorio-Tagle, G. 1985a A&A **147**, 202
Rozyczka, M., Tenorio-Tagle, G. 1985b A&A **147**, 209
Schaller, G., Schaerer, D., Meynet, G., Maeder, A. 1992 A&AS **96**, 269
Schneps, M.H., Haschick, A.D., Wright, E.L., Barrett, A.H. 1981 ApJ **243**, 184
Smith, S.J. 1996 ApJ **473**, 773
Smith, L.J., Pettini, M., Dyson, J.E., Hartquist, T.W. 1984 MNRAS, 211, 679
Smith, L.J., Pettini, M., Dyson, J.E., Hartquist, T.W. 1988 MNRAS, 234, 625
Strickland, D.K., Stevens, I.R. 1998 MNRAS **297**, 747

Treffers, R.R., Chu, Y.-H. 1982 ApJ **254**, 569
van Marle, A.J., Langer, N., García-Segura, G. 2005 A&A **444**, 837
Weaver, R., McCray, R., Castor, J., Shapiro, P., Moore, R. 1977 ApJ **218**, 377
Wrigge, M., Wendker, H.J., Wisotzki, L. 1994 A&A **286**, 219
Wrigge, M. 1999 A&A **343**, 599
Wrigge, M., Chu, Y.-H., Magnier, E.A., Kamata, Y. 1998 The Local Bubble and Beyond LNP **506**, 425
Wrigge, M., Chu, Y.-H., Magnier, E.A., Wendker, H.J. 2005 ApJ **633**, 248
Wrigge, M., Wendker, H.J. 2002 A&A **391**, 287
Zhekov, S.A., Myasnikov, A.V. 1998 New Astronomy **3**, 57
Zhekov, S.A., Myasnikov, A.V. 2000a ApJ **543**, L53
Zhekov, S.A., Myasnikov, A.V. 2000b Ap&SS **274**, 243

Do Fast Winds Dominate the Dynamics of Planetary Nebulae?

J. Meaburn[1]

Instituto de Astronomia, UNAM, Apdo. Postal 877, Ensenada, BC 22800, México.
jm@ast.man.ac.uk

1 Introduction

A review of recent observations of the kinematics of six objects that represent the broad range of phenomena called planetary nebulae (PNe) is presented. It is demonstrated that Hubble–type ouflows are predominant, consequently it is argued that ballistic ejections from the central stars could have dominated the dynamical effects of the fast winds in several, and perhaps all, of these objects. An alternative possibility, which involves an extension to the Interacting Winds model, is considered to explain the dynamics of evolved planetary nebulae.

A consensus has been established (e.g. Kastner et al. 2003) about the basic processes for the creation of a planetary nebula (PN): a star with an initial mass of roughly 1–8 $M_\odot$ loses mass in its Asymptotic Giant Branch (AGB) phase at $\leq 10^{-4}$ $M_\odot$ yr^{-1} by emitting a 'superwind' flowing at 10 - 20 km s^{-1} over $\leq 10^5$ yr (depending on the initial stellar mass). The star eventually becomes an 0.5 - 1.0 $M_\odot$ White Dwarf (WD) which produces enough Lyman photons to ionise a substantial fraction of the circumstellar envelope recognisable as the expanding PN. The whole structure can be enveloped as well in a prior, similarly slow, low density Red Giant (RG) wind. In the transition from the AGB to WD phase the outflow mass loss rate declines to 10^{-8} $M_\odot$ yr^{-1} but the wind speed increases dramatically to several 1000 km s^{-1} to blow as a fast wind for an as yet unknown period. As the WD star evolves further, the fast wind declines.

Obviously, the real story is hugely more complicated in detail and variable between objects (Balick & Frank, 2002). Morphologies range from simple, spherical shells to complex poly-polar structures (e.g. NGC 2440, López et al. 1998) probably around close binary systems. The ejected, dusty, post–AGB, molecular material is often very clumpy (e.g. the cometary globules of NGC 7293, Huggins et al. 1992; Meaburn et al. 1992; O'Dell & Handron 1996; Meaburn et al. 1998). High–speed jets (e.g. IRAS17423-1755, Riera et al. 1995) and 'bullets' (e.g. MyCn 18, Bryce et al. 1997; O'Connor et al. 2000) are found. Shell or lobe expansion velocities can range from 20 km s^{-1} to $\geq$ 500 km s^{-1} (e.g. He2–111, Meaburn & Walsh 1989 and NGC 6302, Meaburn et al. 2005c). Sometimes many of these distinctly separate phenomena occur in one object reflecting the separate stages of its complex evolution. The ages of observed PNe range from the initial proto-PN stage to that of a well–evolved PN $\geq 10^4$ yr later, around an ageing WD star.

In current dynamical theories of PNe much emphasis is placed on the importance, even dominance, of the fast wind. In the elegant interacting winds (IWs) model and its variants (Kwock, Purton & Fitzgerald 1978; Kahn & West 1985; Chu et al. 1993; Mellema 1995, 1997; Balick & Frank 2002) the shocked ($10^6 - 10^8$ K) fast wind can form an energy–conserving, pressure–driven 'bubble' in the preceding smooth AGB wind whose density declines as the inverse square of the distance from the star (see the chapter by Jane Arthur for an overview of wind-blown bubbles). This bubble is similar in principle to that considered in pioneering work by by Dyson & de Vries (1972). The characteristic shell of a simple PN is consequently formed between the shocks in the fast wind and AGB wind and, being pressure–driven by the superheated gas, is expanding faster than this ambient AGB outflow. A variant would have the momentum of the isotropic fast wind simply sweeping up the AGB outflow and accelerating an expanding shell. For the creation of a bi-polar PN, Cantó (1978) and Barral & Cantó (1981) considered something similar. Here a fast wind from a star embedded in a dense circumstellar disk forms cavities on either side of it which are delineated by stationary shocks across which the fast wind refracts to form bi–polar, momentum–conserving, outflows parallel to the cavity walls. Again, energy conserving, elongated 'bubbles', pressure-driven by the shocked wind on either side of this disk, would also form expanding bi-polar lobes. Steffen & López (2004) examined the effects of the fast wind on a clumpy AGB wind which is more realistic than the smooth density distributions usually considered in the IWs models.

There is now an abundance of observational evidence that fast winds exist within PNe and some evidence that they interact significantly with the circumstellar envelopes. Patriarchi & Perinotto (1991) discovered that 60 percent of central stars of PNe emit particle winds of 600–3500 km s^{-1}. However, direct observational evidence of their interaction with the circumstellar medium is more limited. Collimated and truncated ablated flows, where the fast wind has mixed with, and is slowed by, photoionised gas evaporating from dense, stationary, globules (Hartquist et al. 1986; Dyson et al. 1989a; Dyson, Hartquist & Biro 1993) are detected in the hydrogen–deficient PNe A30 (Borkowski et al. 1995; Meaburn & López,1996) and A78 (Meaburn et al. 1998). Also, diffuse X–ray emission is found in the cores of five PNe (NGC 7009, Hen 3–1475, BD+30deg 3639, NGC 6543 & NGC 2392 – Chu et al. 2001; Gruendl et al. 2001 Guerrero, Chu & Gruendl 2004; Chu et al. 2004; Guerrero et al. 2005). Similarly, Kastner et al. (2003) observed such diffuse X–ray emission in the core of the bi–polar PN Menzel 3 (Mz 3) and Montez et al. (2005) inside the main shell of NGC 40 which is a PN generated by a WR–type star. All of these authors interpret the X–ray emissions to be the consequences of the collisions of the fast winds with the slower moving surrounding AGB winds. They suggest that conductive cooling is occurring for the temperatures of the hot gases emitting the X–rays are far lower than if simply generated by shocks in the fast winds at their measured speeds. Nonetheless, they imply that over–pressured 'bubbles' of super-heated gases are forming and driving the expansions of the ionised PNe shells as predicted by the IWs model.

The principal purpose of the present chapter is to examine, on the basis of observations made recently with the two Manchester echelle spectrometers (MES - Meaburn et al. 1984 and 2003), the part played by the fast winds in the creation of the well–evolved PNe NGC 6853 (Dumbbell) and 7293 (Helix), the young PN NGC 6543, and the outer lobes of the bi–polar (poly–polar) 'PNe' NGC 6302, Mz 3 and MyCn 18, for these are all recently observed examples of the range of circumstellar phenomena broadly designated as PNe.

2 The Evolved PNe NGC 6853 and 7293

Cerruti–Sola & Perinotto (1985) and Patriarchi & Perinotto (1991) failed to detect any fast winds of $\geq 10^{-10}$ $M_\odot$ yr^{-1} from either of the central stars of NGC 6853 and 7293 in their IUE observations. This is not surprising for both stars are well into their WD phases with surface temperatures $\approx 10^5$ K (Górny, Stasińska & Tylenda 1997; Napiwotzki 1999) and well past the transitions from their AGB to WD phases during which periods the fast winds are expected to blow. The question is, do the present morphologies and kinematics of these PNe depend critically on the previous emissions of fast winds if they in fact occurred? Observations of NGC 6853 were reported by Meaburn et al. (2005a) and of NGC 7293 by Meaburn et al. (2005b) in an attempt to throw light on this question. These should be combined with an appreciation of the complementary imaging for NGC 7293 by O'Dell (1998) and O'Dell, McCullough & Meixner (2004).

Some aspects of the ionisation stratification of NGC 6853 and 7293 can be appreciated in Figs. 1, 2 & 3 and of the corresponding velocity structure in Figs. 3 and 4a & b. Highly excited gas emitting He II lines is expanding slowly around both exciting stars. These central volumes are themselves both surrounded by faster shells of lower excitation emitting the [O III] lines. All are contained within outer and even faster expanding lowly ionized shells emitting the [N II] lines. For NGC 6853 the central He II volume (0.38 x 0.33 pc^2) is expanding at ≤ 7 km s^{-1}, the inner [O III] shell at 13 km s^{-1} and the outer ellipsoidal (0.50 x 0.67 pc^2) [N II] shell at 35 km s^{-1}. For NGC 7293 the central He II volume (0.21 pc diam.) is expanding at ≤ 11 km s^{-1}, the inner [O III] shell (0.25 pc. diam.) at 12 km s^{-1} and the outer [N II] structure (0.64 pc. across and shown in Meaburn et al. 2005b to be bipolar with an axis tilted at 37° to the sight line to give the characteristic helical appearance of NGC 7293) is expanding at 25 km s^{-1}. The deep images in Figs. 1 and 2 show that

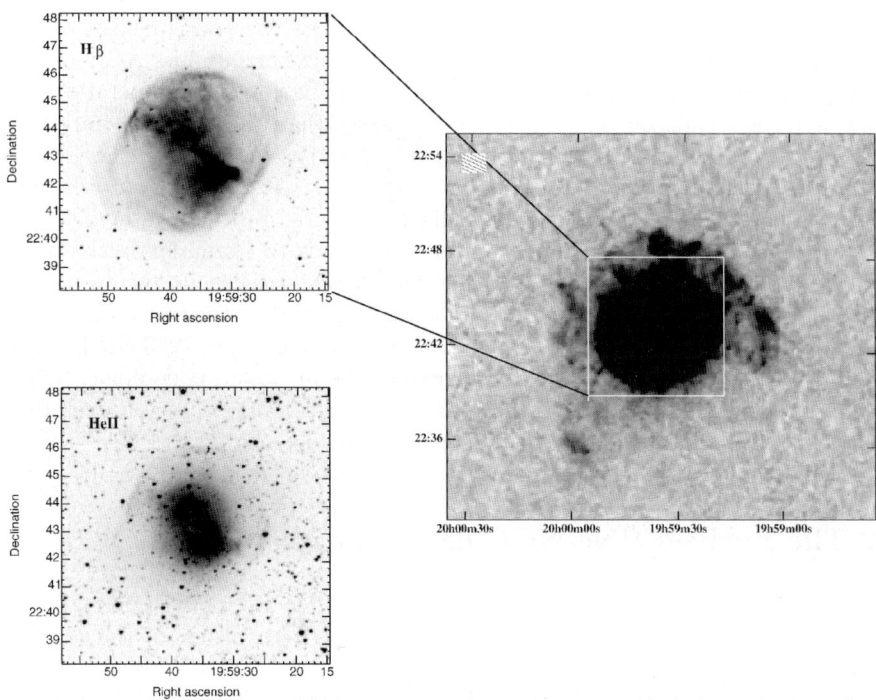

Fig. 1. A deep Hα + [N II] image of NGC 6853 taken by Panos Boumis is compared with Hβ and He II 4686 Å images (compliments of Bob O'Dell) of the familiar 'Dumbbell' nebula (coords throughout are for J2000).

the bright regions of both nebulae are surrounded by clumpy haloes which could be, within the IWs model, the as yet unaccelerated AGB wind but could alternatively be the prior RG wind.

Even if fast winds were present in NGC 6853 and 7293, they could not reach the outer shells unless they percolated as mass–loaded flows through clumpy inner He II and [O III] emitting regions (Meaburn & White, 1982). However, with no current fast wind observed, the IWs model could then only apply in the earliest post–AGB stages of the formation of these PNe; i.e. the fast winds initially formed expanding bubbles then switched off. A somewhat elaborate consequence within the IWs model for an evolved PN could be the acceleration inwards of the inside surface of the outer shell, for it would no longer be supported by the over-pressure of the super–heated shocked gas, to cause the slower moving inner regions reported in Meaburn et al. (2005 a & b). The gap between the outer edge of the [O III] inner shell and the inside edge of the [O III] and [N II] outer shell (Fig. 2) is not simply explained by this inward acceleration model.

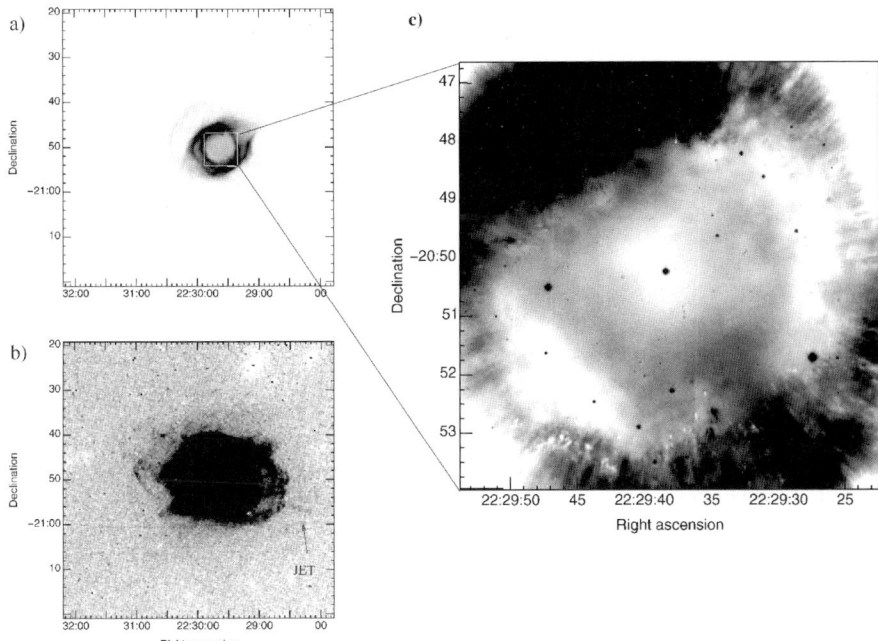

Fig. 2. The Hα + [N II] image in a) shows the familar, bright helical structure which gives the nebula its name 'Helix'. It is shown in b) that this is enveloped in a complex outer halo that may even include a jet. The NTT (La Silla) [O III] 5007 Å image in c) reveals particularly well the inner [O III] 5007 Å shell.

Alternatively, the fast wind never existed or was ineffective over the long term. The expanding shells could then simply be ejections of AGB wind that are pulsed or of higher speed for unknown reasons. The dynamical ages of the central He II volume, the inner [O III] and the outer [O III] plus [N II] NGC 7293 shells are all $\approx 10^4$ yr and those of the central He II volume and outer [N II] shell of NGC 6853 $\approx$ 8000 yr. Within this picture, in each PN, all of these emitting regions would have been ejected at about the same time but with decreasing velocities i.e. Hubble–type outflows. The haloes of NGC 6853 and 7293 could still be the prior, but lower speed, AGB or even RG winds emitted over $\geq 10^5$ yr.

A comment must be made about the tails of the cometary knots in NGC 7293. As the fast wind is currently not observed in this PN it cannot be the cause of these radial tails. An alternative remains that dense knots in the AGB (or even RG wind) seen initially as SiO maser spots (Dyson et al. 1989b) are overrun by a pulse of AGB wind to draw these tails out.

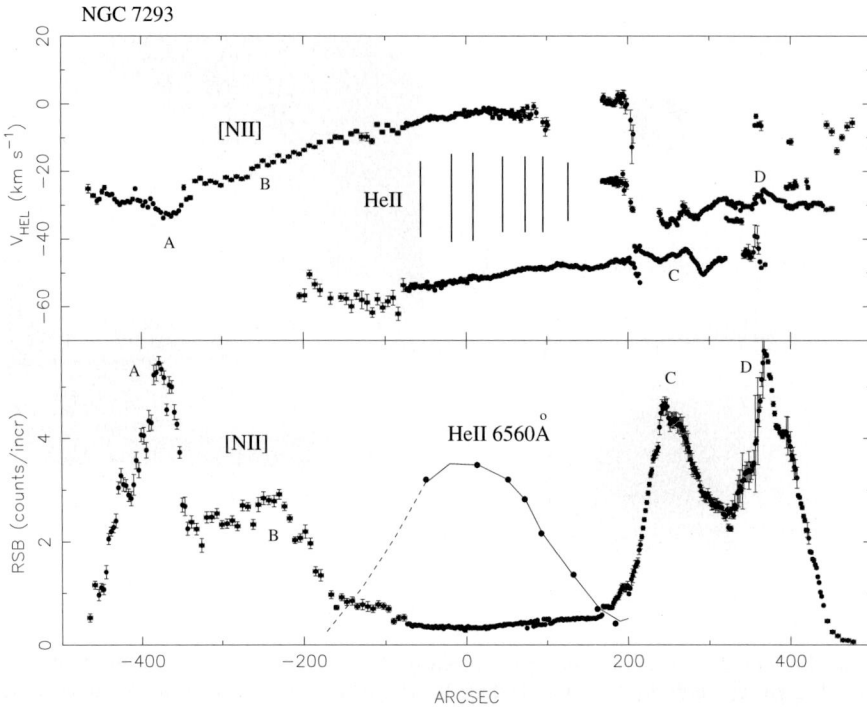

Fig. 3. In the top panel the radial velocities of separate velocity components in the [N II] 6584 Å profiles along an EW cut through the central star and over the helical structure in Fig. 2a are compared with the widths of single Gaussians that simulate the He II 6560 Å profiles around the nebular core. The latter have been corrected for instrumental broadenings. The relative surface brightness (RSB) variations of the [N II] 6584 Å and He II 6560 Å profiles are shown in the bottom panel along the same cut. The brightness peaks A-D over the helical structure are marked.

3 The Young PN NGC 6543

The 'Cat's Eye' nebula, NGC 6543, is a young PN photoionised by an O7+WR–type star which emits a high–speed particle wind at 1900 km s^{-1} (Patriarchi & Perinotto, 1991). Its bright filamentary core is complex but within an overall 25 arcsec × 17 arcsec ellipse ($\equiv$ 0.12 pc × 0.08 pc for a distance of 1001 ± 269 pc as given by Reed et al. 1999). This core is surrounded by a highly filamentary structure with a 330 arcsec diam. ($\equiv$ 1.6 pc and see the image by Romani Corradi in Mitchell et al. 2005). Chu et al. (2001) have shown conclusively that diffuse X–ray emission is confined within an 'inner' ellipse of optical line emission, with a minor axis of 8 arcsec across, which is itself embedded in the larger bright core. Miranda and Solf (1992) had shown that this inner elliptical feature is expanding at 16 km s^{-1} to give a dynamical age of 2400 yr. The fast wind must therefore be confined to this small inner

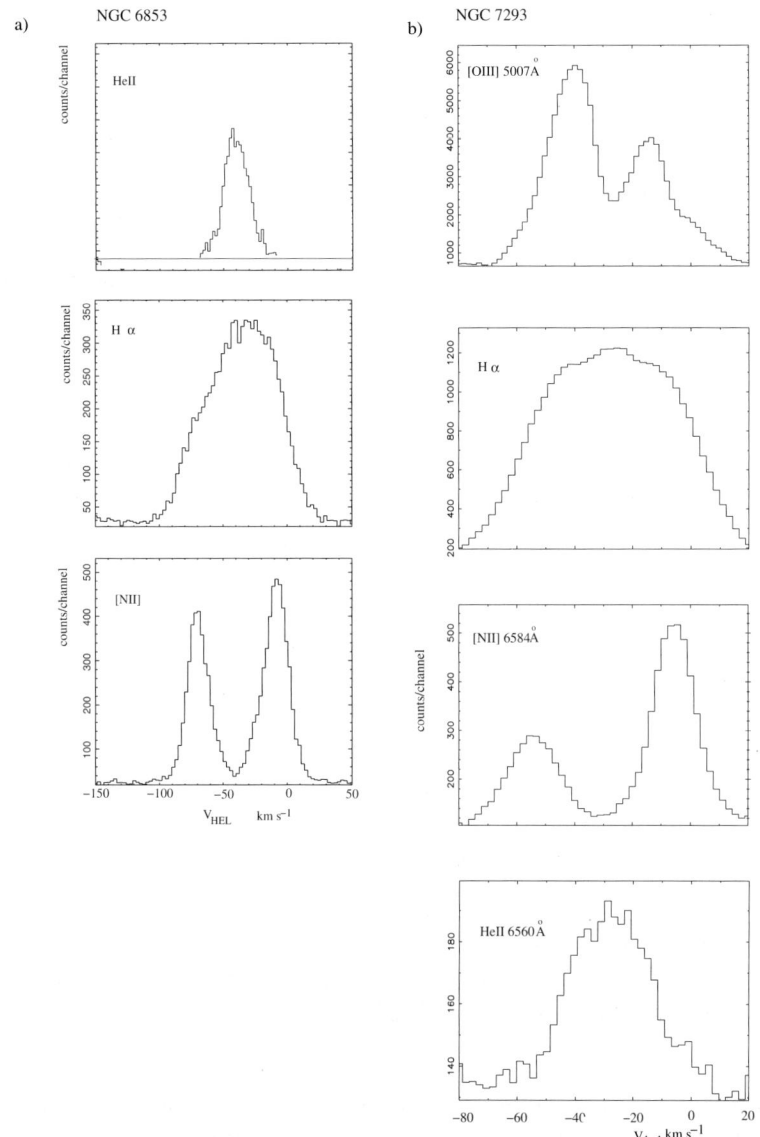

Fig. 4. a) Line profiles over the core of NGC 6853 are compared. b) As for a) but for NGC 7293.

region. The larger filamentary features in the core, which surround this inner ellipse, and the bi–polar jets, seem independent of the presence of the fast wind and are most likely ejecta.

Bryce et al. (1992) have demonstrated that the outer, high excitation halo is very inert. It is expanding globally at 4.5 km s^{-1} which gives it a dynamical age of 1.7×10^5 yr. Mitchell et al. (2005) showed that all of the flows off globules in this halo occur at around the sound speed and, therefore, argued that they are solely consequences of ionisation fronts created by photoionisation. There is no evidence of any interaction with the fast wind which must be confined to the inner ellipsoidal shell within the nebular core.

It is concluded that the post–AGB phase started only around 2400 yr ago and that a small shell, predicted by the IWs model is being driven by the fast wind into an extremely clumpy AGB wind. The outer halo may even be the slow moving relic of the most recent RG wind. It is difficult to visualise the eventual creation of a large expanding shell in such a clumpy outer halo even if the fast wind were to blow for 10 times its current age.

4 The Bi–polar PNe NGC 6302, Mz 3 and MyCn 18

These bi–polar nebulae must have more complicated stellar systems than NGC 6853, 7293 and 6543 and have circumstellar disks as well. Here, following Bains et al. (2004), Smith (2003) and Smith & Gehrz (2005), they are designated PNe for they have, arguably, post–AGB elements in their natures although the central stars are most likely close binaries.

4.1 NGC 6302 – A High Excitation Poly–polar PN

NGC 6302 (PN G349.5+01.0) is a poly-polar planetary nebula (PN), which was described and drawn as early as 1907 by Barnard. It is in the highest excitation class of PNe with a central O VI–Type White Dwarf and possible binary companion (Feibelman 2001). This stellar system is heavily obscured by a dense circumstellar disk (Matsuura et al. 2005).

The kinematics of the prominant NW lobe of NGC 6302 (Fig. 5) have been determined in detail by Meaburn & Walsh (1980) and most recently by Meaburn et al. (2005c) – see examples from the latter in Fig. 6. Meaburn & Walsh (1980) showed 'velocity ellipses' in the position–velocity (PV) arrays of line profiles across the diameter of the lobe (as sketched in Fig. 7b and see Meaburn et al. 2005c) showing that this outflow is Hubble–type reaching V = 600 km s^{-1} at the extremities of the lobe (Figs. 6b & c). A 'spot' value of V = 263 km s^{-1} at position A' in Fig. 5 (1.71 arcmin from the star) is shown in Fig. 7c. No fast wind has been directly observed (Feibelman 2001) from the O VI–Type WD star and its possible companion. For an expansion–proper motion distance of 1.04 ± 0.16 kpc (Meaburn et al. 2005c) the dynamical age of the lobe is 1900 yr.

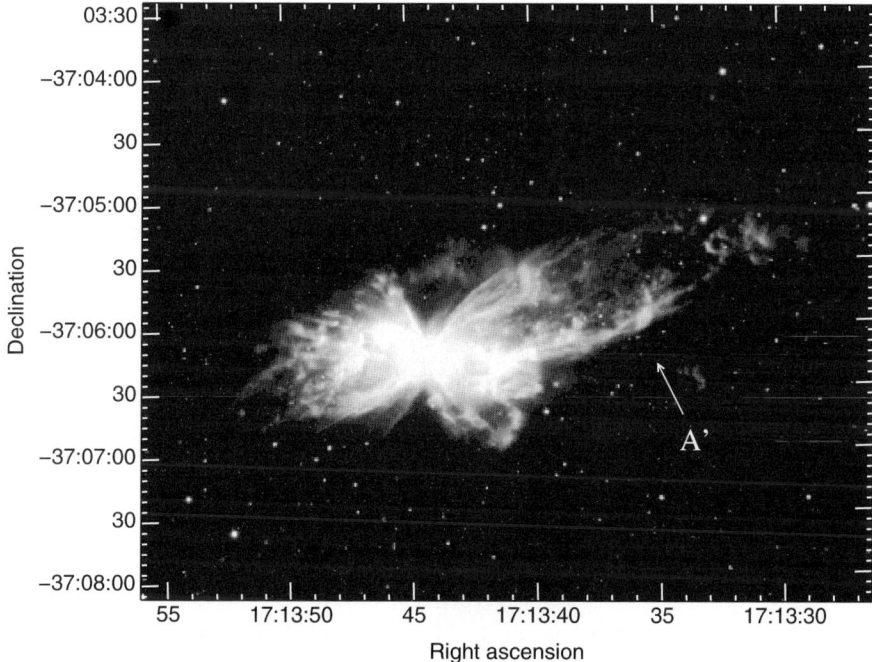

Fig. 5. An Hα + [N II] image of NGC 6302 taken by Romani Corradi with the 3.6–m La Silla telescope. The cut A' shown in Fig. 7 is marked.

Meaburn et al. (2005c) concluded that an eruptive event 1,900 yr ago created the prominent NW lobe and possibly many of the other lobes.

4.2 Mz 3 – a symbiotic PN

Bains et al. (2004) and Smith (2003) suggested that the central stellar system of the PN Mz 3 (Fig. 8) is a symbiotic binary. López & Meaburn (1983) had shown that the bright central bi–polar shells of Mz 3 (the 9 arcsec diam. N shell and 14 arcsec diam. S shell in Fig. 8) are in spherical expansion at 40 km s^{-1} and 55 km s^{-1} respectively. For a distance to Mz 3 of 1.3 kpc (see Bains et al. 2004 for a review of possible distances) a mean dynamical age of 1435 yr for these inner shells is implied. These shells are on either side of a dense disk (Meaburn & Walsh 1985) which obscures the central stellar system.

Furthermore, Meaburn & Walsh (1985) revealed that the N and S lobes in Fig. 8 had circular sections, for velocity ellipses occurred in the PV arrays of line profiles over their diameters, and that their outflows at velocity V along vectors directed away from the central star (as for NGC 6302 in Fig. 7 though with different angles) are very similar to those of NGC 6302 and similarly Hubble–type. Spot values of V at 20 arcsec N and S of the central star are given by Meaburn & Walsh (1985) as 90 and 93 km s^{-1} respectively. As for

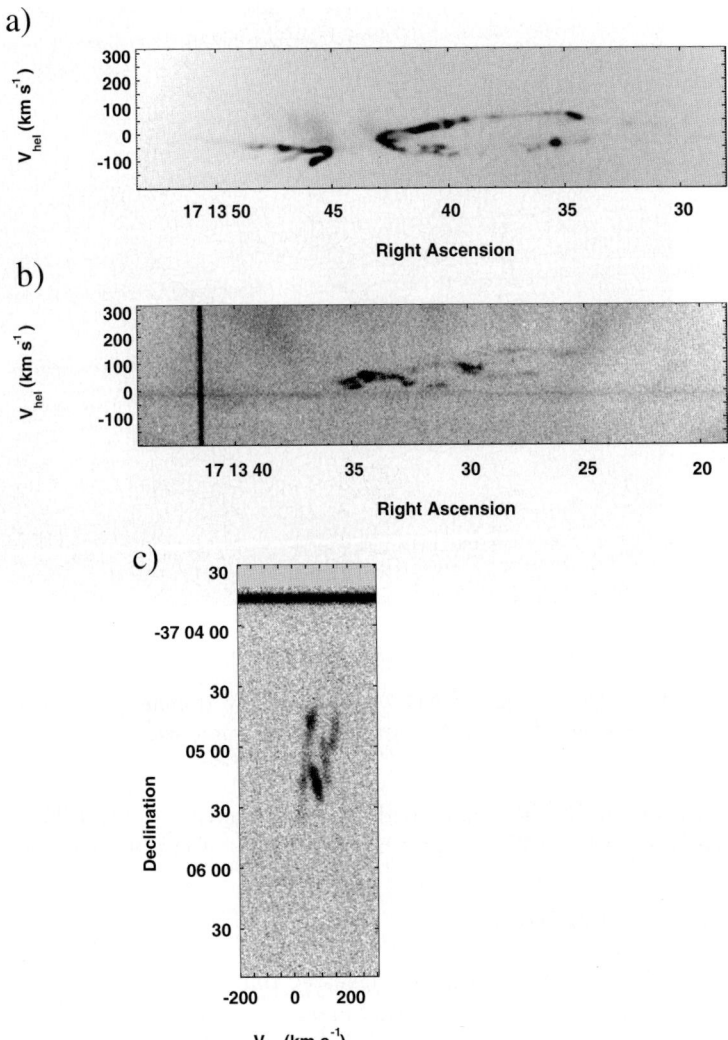

Fig. 6. Sample PV arrays of [N II] 6584 Å line profiles over NGC 6302 are shown. Those in a) and b) are for an EW slit centred on DECs -37 05 50 and -37 05 02 respectively. That in c) is a NS slit centred on RA 17 13 28. By comparison with the image in Fig. 5 it can be seen that the arrays in b) and c) cover the extremities of the prominant NW lobe of NGC 6302.

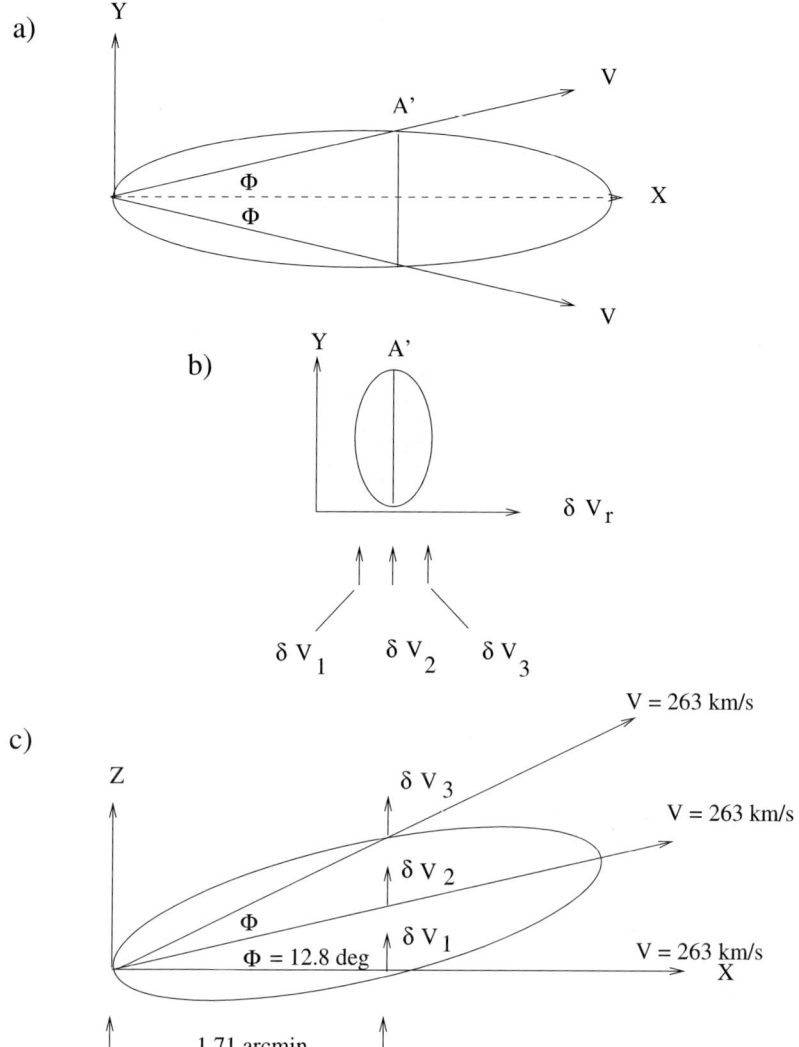

Fig. 7. The image of the NW lobe of NGC 6302 is shown schematically as an ellipsoidal structure with circular section in a) where X and Y are in the plane of the sky. The expansion velocity V is shown to be Hubble–type. The velocity ellipse found along cut A' (1.71 arcmin from the central star) in Fig. 5 is from Meaburn & Walsh (1980) and gives the parameters shown in c) where the Z dimension is perpendicular to the plane of the sky (the observer is below c). With Hubble–type expansion, $V \approx 600$ km s^{-1} at the extremity of the NW lobe (see Fig. 6b).

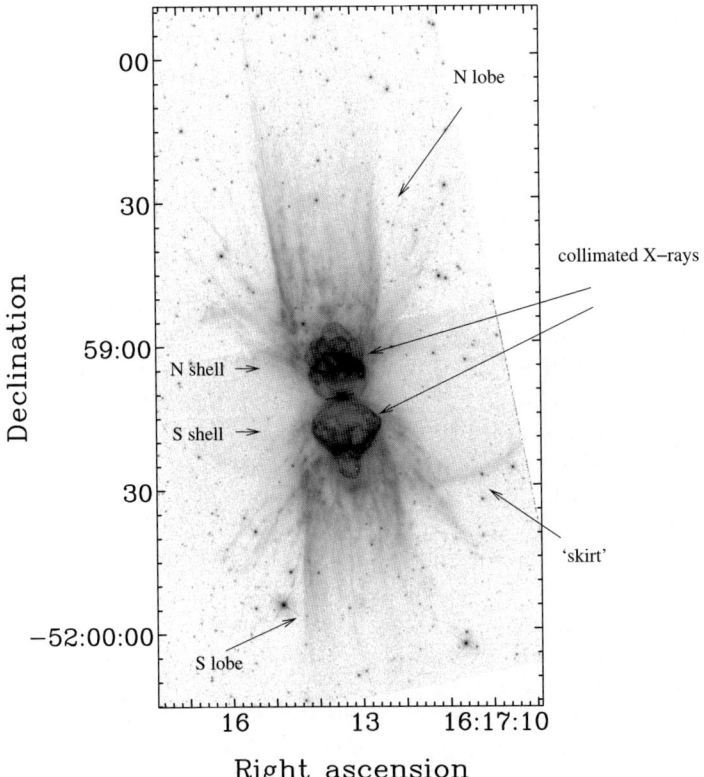

Fig. 8. An HST image of Mz–3 in the light of Hα + [N II]. The N and S lobes exhibit Hubble–type outflows and the central bright shells, N and S shell, contain diffuse X–ray emission though collimated along the lobe axis. The high speed 'skirt' identified by Santander–Garcia et al. (2004) is marked.

NGC 6302 (Sect. 3.1) the Hubble–type nature of the outflows implies that at the limits of detection of these lobes (82 arcsec N and 52 arcsec S) V reaches 369 km s^{-1} and 242 km s^{-1} respectively, as confirmed by Santander-Gracia et al. (2004). The corresponding dynamical ages for the N and S lobes of Mz 3 are consequently 1317 and 1273 yr respectively which are remarkably similar to those of the N and S shells.

The implication is that *all* of the ouflows of the N and S shells and the N and S lobes are within a general Hubble–type velocity system which reinforces the possibility that they are all the consequence of ejections at closely similar times but with different ejection speeds. In any case, the fast wind must be contained within the N and S shells to shield the N and S lobes from any direct interaction. Furthermore, the X–ray emission (Kastner et al. 2003) suggests that this fast wind is collimated along the bi–polar axis of Mz 3 and is causing the secondary protusions at the apices of the N and S shells evident in Fig. 8.

This would preclude the N and S shells themselves from being 'bubbles' driven only by this fast wind.

Incidentally, the high–speed skirt (Fig. 8) has been properly identified by Santander-Garcia et al. (2004) as the origin of the high–speed 'velocity ellipse' in the PV arrays of Meaburn and Walsh (1985) and the high–speed feature described by Redman et al. (2000).

4.3 MyCn 18 – A Nova–like PN

MyCn 18 is also aptly known as the 'Engraved Hourglass' nebula due to the visually dramatic bipolar appearance of its bright core (Sahai et al. 1999 and references therein). However, interest in MyCn 18 has recently been further heightened by the discovery of the knots of ionized gas flowing in both directions along its bipolar axis at speeds of up to 660 km s^{-1}(Bryce et al. 1997; O'Connor et al. 2000). These can be seen in the continuum subtracted image in Fig. 9a. Corradi & Schwarz (1993) had previously investigated the bright core of MyCn18 and concluded that it is a young PN. The presence of a dusty, molecular, equatorial waist region, suspected by Sahai et al. (1999) on the basis of an excess in the stellar K-band photospheric flux, substantiated this young age. The radio thermal emission map of Bains & Bryce (1997) reveals the ionised inside surface of the dense waist region to be very bright in comparison to emission from polar directions.

The Hubble-type nature of the knotty outflow is clear in Fig. 9b (from O'Connor et al. 2000). It is notable that the best fit straight lines are significantly displaced from the systemic radial velocity near the nebular core. O'Connor et al. (2000) showed that these knots were ejected over a 300 yr period with a dynamical age of 1250 yr (Bryce et al. 1997). They also concluded that dynamically, the most plausible explanation seems to be that the high speed knotty outflow from MyCn18 is the result of a (possibly recurrent) nova–like ejection from a central binary system. This is in harmony with the considerations of Sahai et al. (1999) who favoured a close binary system to generate the morphology of the very innermost regions of MyCn 18. In this picture the knots would be the manifestation of ballistic, dense bullets ejected with a range of speeds and not the consequence of acceleration by a fast wind.

5 Conclusions

The IWs model certainly seems applicable to the very innermost shell of the young PN NGC 6543 (and very clearly to the main shell of NGC 40 – see Sect. 1). However, considerable modification of this theory is needed to explain the current state of the evolved PNe NGC 6853 and 7293. The fast wind could have switched off well before the 10^4 yr age of their expanding shells generating an inward acceleration of their inside surfaces. Alternatively, it remains possible that eruptive events over 10^4 yr have dominated any effects

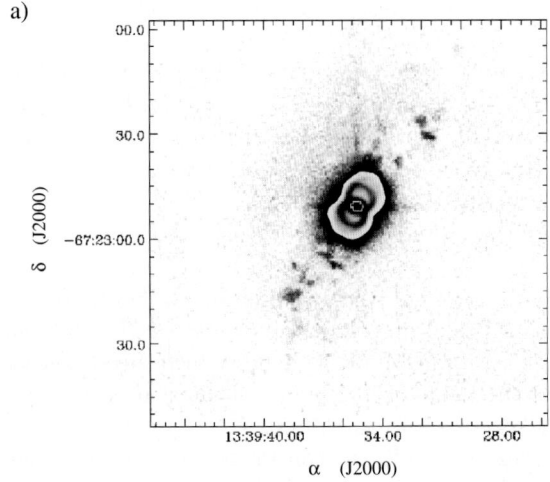

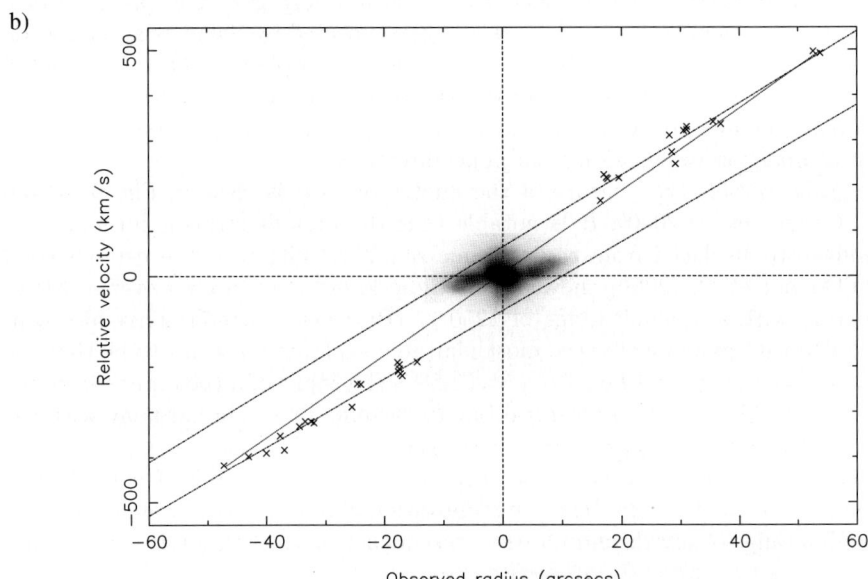

Fig. 9. In a) the image of the bright core of MyCn 18 taken with the AAT (Bryce et al. 1997 and O'Connor et al. 2000) is shown compared with the faint knots along the lobe axis. In the PV diagram in b) the relative radial velocities of these knots are shown as crosses. Very high–speed, Hubble–type, bi–polar motions are indicated for the radial velocities follow straight lines.

of the fast winds, active only for a few thousand years in these PNe, to create the Hubble–type expansions throughout their volumes that are currently observed.

The preponderance of Hubble–type outflows of the lobes of the bi–polar PNe NGC 6302, Mz 3 and MyCn 18 (and see Corradi 2004 for other examples) invites the simplest interpretation: that they are consequences of ejections over short periods of time but of material with different speeds. Even when a fast wind is currently present (e.g. Mz 3) the outer high–speed lobes are shielded from it by inner shells.

The present chapter has been deliberately limited to the consideration of the dynamics of a small number of objects whose motions have been well observed. It seems clear in this small sample that simple ballistic ejections, maybe involving close binary systems in some cases, could dominate the dynamical effects of the fast winds.

Acknowledgements

The author is grateful to Bob O'Dell for providing the $H\beta$ and He II 4686 Å images in Fig. 1 and to Romani Corradi for the image in Fig. 5.

References

Balick, B., Frank, A. 2002 Annu. Review Astron. & Astrophys. **40**, 439
Bains, I., Redman, M.P., Bryce, M., Meaburn, J. 2004 MNRAS **354**, 549
Bains, I., Bryce, M. 1997 Acta Cosmologica **23**, 107
Barral, J.F., Cantó, J.F. 1981 Rev. Mex. A & A **5**, 101
Borkowski, K.J., Harrington, J.P., Tsvetanov, Z.I. 1995 ApJ **449**, L143
Bryce, M., Meaburn, J., Walsh, J.R., Clegg, R.E.S. 1992 MNRAS **254**, 477
Bryce, M., López, J.A., Holloway, A.J., Meaburn, J. 1997 ApJ **487**, L161
Cantó, J. 1978 A&A **70**, 111
Cerruti–Sola, M., Perinotto, M. 1985 ApJ **291**, 237
Corradi, R.L.M., Schwarz, H.E. 1993 A&A, 268, 714
Corradi, R.L.M. 2004 Asymmetric PNe III, ASP Conference Series, eds. M. Meixner, J. Kastner, B. Balick & N. Soker **313**, 148
Chu, Y.-H., Kwitter, K.B., Kaler, J.B. 1993 AJ **106**, 650
Chu, Y.H., Guerrero, M.A., Gruendl, R.A., Williams, R.M., Kaler, J.B. 2001 ApJ **553**, L69
Chu, Y.H., Guerrero, M.A., Gruendl, R.A., Webbink, R.F. 2004 AJ **127**, 477
Dyson, J.E., de Vries, J. 1972 A&A **20**, 223
Dyson, J.E., Hartquist, T.W., Pettini, M., Smith, L.J. 1989a MNRAS **241**, 430
Dyson, J.E., Hartquist, T.W., Pettini, M., Smith, L.J. 1989b MNRAS **241**, 625
Dyson, J.E., Hartquist, T.W., Biro S. 1993 MNRAS **261**, 430
Feibelman, W.A. 2001 ApJ **550**, 785
Górny, S.K., Stasińska, G., Tylenda, R. 1997 A&A **318**, 256
Gruendl, R.A., Chu, Y.H., O'Dwyer, Guerrrero, M.A. 2001 AJ **122**, 308

Guerreo, M.A., Chu, Y.H., Gruendl, R.A. 2004 Asymmetric PNe III, ASP Conference Series, eds. M. Meixner, J. Kastner, B. Balick & N. Soker **313**, 259
Guerreo, M.A., Chu, Y.H., Gruendl, R.A., Meixner, J. 2005 A&A **430**, L69
Hartquist, T., Dyson, J.E., Pettini, M., Smith, L. 1986 MNRAS **221**, 715
Huggins, P.J., Bachiller, R., Cox, P., Forveille, T. 1992 ApJ **401**, L43
Kahn, F., West, K. 1985 MNRAS **212**, 837
Kastner, J.H., Balick, B., Blackman, E.G., Frank, A., Soker, N., Vrtilek, D.S., Li, J. 2003 ApJ **591**, L37
Kwok, S., Purton, C.R., Fitzgerald, P.M. 1978 ApJ **219**, L125
López, J.A., Meaburn, J. 1983 MNRAS **204**, 203
López, J.A., Meaburn, J., Bryce, M., Holloway, A.J. 1998 ApJ **493**, 803
López, J.A., Meaburn, J., Palmer, J.W. 1993 ApJ **415**, L135
López, J.A., Meaburn, J., Bryce, M., Holloway, A.J. 1998 ApJ **493**, 80
Matsuura, M, Zijlstra, A.A., Molster, F.G., Waters, L.B.F.M., Nomura, H., Sahai, R., Hoare, M.G. 2005 MNRAS **359**, 383
Meaburn, J., Walsh, J.A. 1980 MNRAS **193**, 631
Meaburn, J., Walsh, J.R. 1985 MNRAS **215**, 761
Meaburn, J., White, N.J.,1982, Astrophys. Space Sci. **82**, 423
Meaburn, J., Walsh, J.R. 1989 A&A **223**, 277
Meaburn, J., Blundell, B., Carling, R., Gregory, D.F., Keir, D., Wynne, C. 1984 MNRAS **210**, 463
Meaburn, J., Walsh, J.R., Clegg, R.E.S., Taylor, D & Berry D.S. 1992, MNRAS **255**, 177
Meaburn, J., Clayton, C.A., Bryce, M., Walsh, J.R., Holloway, A.J., Steffen, W. 1998 MNRAS **294**, 201
Meaburn, J., López, J.A. 1996 ApJ **472**, L45
Meaburn, J., López, J.A., Bryce, M., Redman, M.P. 1998 A&A **334**, 670
Meaburn, J., López, J.A., Gutiérrez, L., Quiróz, F., Murillo, J.M., Valdéz, J., Pedrayez, M. 2003 Rev Mex AA **39**, 185
Meaburn, J. Boumis, P., Christopoulou, P.E., Goudis, C.D., Bryce, M., López, J.A. 2005a Rev. Mex. A., A. **41**, 109
Meaburn, J., Boumis, P., López, J.A., Harman, D.J., Bryce, M., Redman, M.P., Mavromatakis, F. 2005b MNRAS **360**, 963
Meaburn, J., López, J.A., Steffen, W., Graham, M.F., Holloway, A.J. 2005c AJ in press
Mellema, G. 1995 MNRAS **277**, 173
Mellema, G. 1997 A&A **321**, L29
Miranda, L.F., Solf, J. 1992 A&A **260**, 397
Mitchell, D.L., Bryce, M., Meaburn, J., Lopez, J.A., Redman, M.P., Harman, D., Richer, M.G., Riesgo, H. 2005 MNRAS **326**, 1286
Montez, R., Kastner, J.H., de Marco, O., Soker, N. 2005 ApJ, in press
Napiwotzki, R. 1999 A&A **350**, 101
O'Connor, J.A., Redman, M.P., Holloway, A.J., Bryce, M., López, J.A., Meaburn, J. 2000 ApJ **531**, 336
O'Dell, C.R., Handron, K.D. 1996 AJ **111**, 1630
O'Dell, C.R. 1998 AJ **116**, 1346
O'Dell, C.R., McCullough P., Meixner M. 2004 AJ **128**, 2339
Patriarchi, P., Perinotto, M. 1991 ApJS **91**, 325

Reed, D.S., Balick, B., Hajian, A.R., Klayton, T.L., Giovanardi, S., Casertano, S., Panagia, N., Terzian, Y. 1999 AJ **118**, 2430
Riera, A., Garcia–Lario, P., Manchado, A., Pottasch, S.R., Raga, A.C. 1995 A&A **302**, 137
Redman, M.P., O'Connor, J.A., Holloway, A.J., Bryce, M., Meaburn, J. 2000 MNRAS **312**, L23
Sahai R., Dayal A., Watson A. M., et al. 1999 ApJ **118**, 468
Santander–Garcia, M., Corradi, R.L.M., Balick, B. and Mampaso, A. 2004 A&A **426**, 185
Smith, N. 2003 MNRAS **342**, 383
Smith, N., Gehrz, R.D. 2005 AJ **129**, 969
Steffen, W., López, J.A. 2004 ApJ **612**, 319
Wilson, O.C. 1950 ApJ **111**, 279

Spectral Studies of Supernova Remnants

J.C. Raymond[1]

Center for Astrophysics, 60 Garden St., Cambridge, MA 02176
jraymond@cfa.harvard.edu

1 Introduction

Supernova Remnants (SNRs) are the main sources of kinetic energy in the interstellar medium, the primary injection sites of metals and probably the main sources of cosmic rays. Spectral studies of supernova remnants are the basic tools for understanding the dynamics and evolution of SNR, the composition of both SN ejecta and ambient ISM material, and the physics of collisionless astrophysical shocks. The general properties of shocks are reviewed by McKee & Hollenbach (1980) and Draine & McKee (1993). The present chapter is meant more as a summary of useful spectral techniques than a complete list of important results. Some are applicable to shocks occurring in other sorts of objects as well, including Herbig-Haro objects (see the chapter by Tom Ray) and wind-blown bubbles (see the chapter by Jane Arthur).

Doppler shift measurements are perhaps the most direct application of spectroscopy, providing line-of-sight velocities for reconstruction of the 3-dimensional shapes of SNRs (e.g. Clark et al. 1983; Shull & Hippelein 1991; Reed et al. 1995; Flanagan et al. 2004; Lozinskaya et al. 2005). This method requires only the line centroids at many positions in the SNR. Line centroids as a function of position across a single filament can also be used to derive the local shock structure (Raymond et al. 1988).

Line profiles provide the next level of detail. These can reveal the thermal velocities of shocked particles, which in turn can be used to derive shock speeds (Chevalier & Raymond 1978; Ghavamian et al. 2001) and important aspects of the energy partition among different particle species in the shocked plasma. In other cases they provide information about the level of turbulence or the curvature of the shock front (Shull et al. 1982; Greidanus & Strom 1992; Raymond 2003). Absorption line profiles can also be used to infer temperatures and ion column densities (e.g. Jenkins, Silk & Wallerstein 1976, 1984).

Relative line intensities can be obtained from lower resolution data. The relative intensities of lines of different ions can be used to infer the electron temperature, which is often used to estimate shock speeds, and the relative intensities of lines of different elements indicate the abundance pattern (e.g. Cox 1972a; Dopita, Dodorico & Benvenuti 1980).

At lower resolution, X-ray observations provide the electron temperature behind the blast wave, and IR observations provide the temperature of dust grains. Synchrotron emission has been observed at radio, IR, optical, and X-ray wavelengths, providing information about energetic electrons, magnetic fields and the history of the shocked plasma. Gamma ray emission detections are rare, but they constrain the high energy electrons and cosmic rays (Aharonian et al. 2005).

Because the spectral signatures are quite different, we will divide the shocks into radiative and non-radiative classes (McKee & Hollenbach 1980). The former effectively convert the thermal energy of the shocked plasma into radiation, and they tend to be bright at optical and UV wavelengths. The latter have radiative cooling time scales much longer than the relevant dynamical times. They produce the X-ray emission seen in non-plerionic SNRs, and if they are detected at optical or UV wavelengths it is in emission lines produced in a narrow ionization zone close to the shock front. In the Crab nebula and similar remnants, filaments of plasma photoionized by the synchrotron emission are detected, and models of photoionized shocks are required.

This chapter concerns mainly moderate to high resolution spectroscopy, capable of measuring individual atomic or molecular lines. Therefore, its focus is on spectral lines rather than continuum radiation, and it covers thermal rather than non-thermal emission from SNRs. Fesen & Hurford (1996) provide a catalog of over 250 UV, optical and near-IR emission lines observed in supernova remnants, and dozens more are seen at X-ray and IR wavelengths.

2 Non-Radiative Shocks

Non-radiative shocks are those whose dynamical time scale is much shorter than their radiative cooling times (McKee & Hollenbach 1980). Therefore, only a small fraction of the energy dissipated in the shock is converted to radiation. Nearly all X-ray observations of SNRs pertain to non-radiative shocks, and understanding these shocks is a requirement for interpretation of the X-ray data. Non-radiative shocks have been observed at optical wavelengths in about a dozen SNRs and several pulsar-driven bow shocks (Raymond 1991). Optical and UV emission arises from a narrow ionization zone just behind the shock, so that Coulomb collisions do not have time to bring the different particle species into equilibrium. Therefore, it is possible to infer some of the properties of collisionless shock waves from the observations.

Collisionless shocks are those having a thickness much smaller than the collisional mean free path of the particles involved. They have long been studied in the solar wind (see Li, Zank & Russell 2005 for recent results). They can accelerate particles to relativistic speeds and convert some of the kinetic energy of the shocked plasma into turbulent magnetic energy. Because of the lack of collisions, there is no reason to expect that electrons, protons and ions will reach the same temperature, or in fact any Maxwellian distribution at all. The Coulomb collision time increases with $T_e^{3/2}$ (Spitzer 1978), and it would take 10^6 years for electrons to reach the same temperature as protons in the 1000 year old remnant SN1006 through Coulomb collisions alone. There are strong fields and many modes of plasma waves in collisionless shocks, however. Consequently, some transfer of energy among particle species is expected. For shocks slower than a few hundred km/s, Coulomb collisions may be able to bring the plasma into thermal equilibrium within a dynamical time scale.

2.1 X-ray Spectra

Electron Temperature

The electron temperature, T_e, is the parameter most directly extracted from X-ray spectra, because the dominant continuum emission is usually thermal bremsstrahlung with a Boltzmann factor energy dependence $e^{-h\nu/kT_e}$. Here, h is Planck's constant, ν is the photon frequency and k is the Boltzmann constant. Even if the spectrum is dominated by emission lines, the Boltzmann factor is in all the excitation rates and its very strong temperature dependence means that reliable temperatures can be derived in spite of any uncertainties in composition or atomic rates. Only under unusual circumstances can recombination to excited levels or charge transfer could alter some of the line ratios enough to make them ambiguous. In general, the derived temperatures range from about 10^6 K to a few 10^7 K. Lower temperatures would be hard to detect, because they produce photons too soft to penetrate the ISM. Especially at the low end of the temperature range there will be some ambiguity between temperature and absorbing column density.

There is evidence for differences between proton and electron temperatures in a few cases. In 1E 0102.2-7219, Tycho's SNR and SN1006, the shock speed can be inferred from the proper motion of the expanding X-ray emission, and the electron temperature is far below the value that would be found if electron and ion temperatures were equal (Hughes et al. 2000; Hwang et al. 2002; Long et al. 2003). The kinetic temperatures of the ions could in principle be measured from the line widths, but so far this has only proven possible for the O VII emission from a knot in SN1006, where the oxygen temperature was found to be 528 keV (Vink et al. 2003), far higher than T_e or T_p.

Ionization State

Just as Coulomb collisions cannot bring the plasma into thermal equilibrium behind a fast shock, collisional ionization requires a long time to bring the plasma into ionization equilibrium. An element such as oxygen may enter the shock in its neutral or singly ionized state and be quickly ionized through the

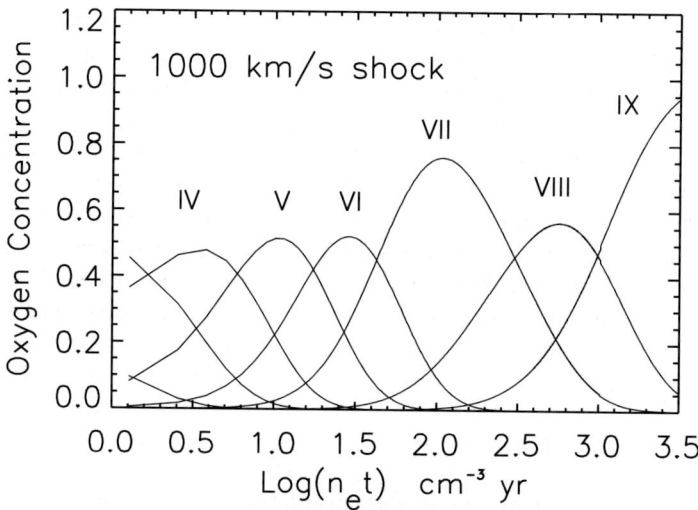

Fig. 1. Variation of oxygen ion concentrations behind a 1000 km/s shock, for a pre-shock density of 1 cm^{-3} and equal electron and ion temperatures.

lower ionization states. However, the ionization cross sections for He-like and H-like ions are relatively small, so the oxygen may be stuck in those ionization states for a long time. Figure 1 shows the concentrations of different ions of oxygen behind a 1000 km/s shock. For simplicity, for this model it was assumed that $T_e = T_i$. In ionization equilibrium, 99% of the oxygen would be totally stripped. As can be seen from the X-axis of Figure 1, the ionization state depends on the product $n_e t$ of density and time. Thus, if T_e can be derived from the spectral shape, the value of $n_e t$ can be derived from line ratios that correspond to the ionization state. Vedder et al. (1986) applied this idea to the Cygnus Loop, using the forbidden to resonance line ratio of O VII as a temperature indicator and the ratio of O VIII to O VII to infer $n_e t$. Flanagan et al. (2004) have more recently applied it to the shocked SN ejecta in 1E 0102.2-7219, and they were able to discern the ionization length scale behind the reverse shock by comparing images in hydrogenic and helium-like lines. Depending on which quantities are the most uncertain, one can then estimate the density or the time since the shock encountered the gas that is now emitting (e.g. Long et al. 2003; Vink et al. 2003).

Abundances

Given a measurement of the temperature and some idea of the ionization state or $n_e t$, it is possible to extract elemental abundances. In many cases, the H-like and He-like ions are observed, and they dominate the ionization balance, so that the correction for unobserved ionization states is small. The excitation rates of the strong lines contain the Boltzmann factor $e^{-h\nu/kT_e}$, but so does the bremsstrahlung emissivity that provides the nearby continuum. Therefore, the equivalent width given by the line to continuum ratio is relatively insensitive to the temperature, and derived abundances should be reliable.

There are a number of complications, however. Under some circumstances inner shell excitation or dielectronic recombination satellite lines can be important, and a lower energy can replace $h\nu$ in the Boltzmann factor (e.g. Gabriel 1972; Vink et al. 2003). It is also possible to get a significant contribution to both lines and continuum from radiative recombination. This is most important if the plasma has cooled rapidly due to adiabatic expansion or to radiative cooling in gas enriched in metals. Because the recombination rate tends to be smaller than the collisional excitation rates of strong lines, such supercooled regions tend not to contribute strongly to the total emission. The biggest caveat is probably the presence of plasma spanning a range of electron temperatures (e.g. Fabbiano et al. 2004). A single temperature fit will often underestimate the abundance of an element whose observable lines peak near the fitted temperature and overestimate the abundances of lines whose observed lines peak away from that temperature. Of course, if a non-thermal continuum is present, that must also be properly taken into account.

While these are serious caveats, difficulties that they potentially present can be overcome with an adequate model if the data quality is good. Warren & Hughes (2004) presented a detailed analysis of the Chandra spectrum of the LMC remnant 0509-67.5. They found enhancements of the elements Si through Ca that not only imply a Type Ia supernova, but also favor a Delayed Detonation model for the explosion. The long Chandra observation of Cas A shows distinct knots having very different silicon and iron abundances (Hwang et al. 2004), and Flanagan et al. (2004) derived abundances in 1E 0102.2-7219.

2.2 Optical/UV Spectra

Non-radiative shocks are sometimes seen at visible wavelengths as faint filaments of essentially pure Balmer line emission. Figure 2 shows an Hα image of the NW section of SN1006, along with the Balmer line profiles of the Hα filament.

These filaments are the result of the excitation of neutral H atoms swept up by the shock (Chevalier & Raymond 1978; see review in Raymond 1991). Though the H atoms are quickly ionized, 20-25% of them will be excited to produce an Hα photon before that happens. The number of photons per hydrogen atom passing through the shock is given by the ratio of the excitation rate to the ionization rate, so that the surface brightness in any line is just

$$I_\lambda = n(Z) V \frac{q_{ex}}{q_i} / 4\pi \text{ photons}/(\text{cm}^2 \text{ s sr}) \qquad (1)$$

where n(Z) is the density of element Z in ionization states below or equal to that producing the line in question, V is the shock speed, q_{ex} is the excitation

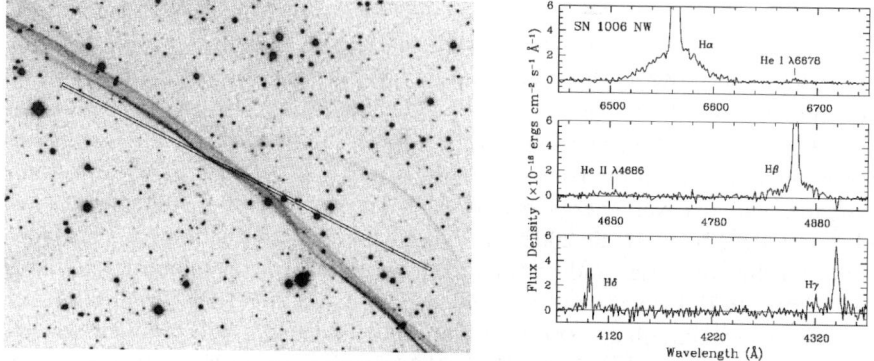

Fig. 2. Left: Hα filaments in the northwest region of SN1006. Right: Balmer line profiles from the spectrum at the slit position shown on the left (Ghavamian et al. 2002).

rate for line λ and q_i is the ionization rate of the ion that produces the line. The forbidden lines that dominate the optical spectra of radiative shocks and other nebulae have small values of q_{ex} as well as elemental abundances far below that of neutral H in partially ionized phases of the ISM, so those lines are much fainter than the Balmer lines.

The most remarkable property of these shocks is associated with the Balmer line profiles. Some of the atoms passing through the shock are unaffected until they are excited. They do not feel the electric or magnetic fields of the collisionless shock, and they retain their pre-shock velocity distribution. Other atoms undergo charge transfer with protons in the post-shock plasma, acquiring the velocity distribution of the shocked protons. In the case of the SN1006 spectrum shown in Figure 2, the Hα FWHM is 2300 km/s (Ghavamian et al. 2002). Thus a measurement of the Hα line profile gives the pre-shock and post-shock ion temperatures quite directly from the widths of the narrow and broad components, respectively.

The atomic rates that go into models of non-radiative shocks are very basic, but there are several subtleties. First, the charge transfer cross section drops rapidly with energy above about 2 keV because the interaction ceases to be resonant, but at about the same energy charge transfer into excited levels can produce a significant number of photons (Chevalier, Kirshner & Raymond 1980). Second, radiative transfer in the Lyman lines must be taken into account for the narrow component. Each time a Lyβ photon is absorbed, it stands a 12% chance of being converted to Hα and a two-photon pair (Chevalier et al. 1980; Laming 1996). Third, excitation and ionization by protons becomes important in the faster shocks, because protons at 10 keV energies mimic electrons at energies of 10s of eV (Raymond et al. 1995; Laming et al. 1996).

An important result from the line profile is obtained from the intensity ratio of the broad and narrow components, since the ratio depends upon the ratio of charge transfer and ionization rates (Ghavamian et al. 2001). Because the post-shock ion temperature can be derived from the broad component width, the ratio T_e/T_p can be obtained. Ghavamian et al. found that T_e/T_p decreases from >0.7 for the relatively slow 350 km/s shock in the Cygnus Loop to ≃0.5 for 600 km/s shocks in RCW86 to < 0.2 in a 2100 km/s shock of Tycho's SNR. Ghavamian et al. (2002) extended this to $T_e/T_p < 0.07$ in the 3000 km/s shock of SN1006, and analysis of DEM L71 confirms the trend (Rakowski et al. 2003).

The profiles of ultraviolet emission lines can be used to measure the kinetic temperatures of other elements in non-radiative shocks. He II, C IV, N V and O VI lines were detected in SN1006 (Raymond et al. 1995; Laming et al. 1996), and their differing sensitivities to electron and proton excitation can be used to derive limits on T_e/T_p comparable to that obtained from the broad-to-narrow component intensity ratio. Korreck et al. (2004) measured the O VI profile in SN1006 with FUSE. They found temperatures well above the proton temperature, with $T_O \simeq 8T_p$. In the 350 km/s shock in the Cygnus Loop, FUSE observations of the O VI profile show $T_O < 2.7T_p$ (Raymond et al. 2003). Thus, both electron and ion temperatures indicate that the different particle species are not far from thermal equilibrium with each other in the slow shock, but they are far from equilibrium in the faster shocks (see Rakowski 2005 for a review).

Another important aspect of the non-radiative shock profiles is the width of the narrow component. This is generally larger than expected for 10^4 K pre-shock gas, and higher temperatures would imply that H would be fully ionized in equilibrium. Thus, the hydrogen seems to be heated to $\simeq 4 \times 10^4$ K in a narrow precursor region ahead of the shock (Smith et al. 1994; Hester et al. 1994). This could be a result of the precursor required by models of particle acceleration in shocks, or it could be attributed to fast neutrals leaking out through the shock into the pre-shock plasma, though the latter idea is not supported by the model calculations of Lim & Raga (1996). The fastest shock observed, that in SN1006, is the only one that does not show evidence for heating in a precursor (Sollerman et al. 2003). At first glance, this would seem to favor the cosmic ray precursor interpretation, because so much energy would be available for each neutral that overtakes the shock. However, the charge transfer cross section falls rapidly at speeds above 2000 km/s, so that head-on collisions are much less likely to lead to charge transfer than are overtaking collisions. This may drastically reduce the number of fast neutrals that overtake the shock at the fastest shock speeds.

In addition to the physics of collisionless shocks, the Balmer filaments are useful because they provide shock speeds that can be combined with proper motions to obtain distances to Galactic SNRs (e.g. Chevalier & Raymond 1978; Blair et al. 1999; Winkler, Gupta & Long 2003 for Tycho, the Cygnus Loop and SN1006, respectively).

2.3 Other Optical Lines

He I and He II lines are barely detected in some non-radiative shocks, and they can be used to determine the pre-shock ionization state (Ghavamian et al. 2002). [Fe X] and [Fe XIV] emission is detected farther behind the shock in a few SNR, including the Cygnus Loop (Ballet et al. 1989; Teske 1990) and Puppis A (Teske & Petre 1987). These lines help to pin down the temperature of the X-ray emitting gas.

3 Radiative Shocks

3.1 Nature of the Flow

In a radiative shock, radiative cooling is able to convert the thermal energy of the shocked gas into photons. As the gas cools, it is compressed, and because the cooling rate increases with density, the temperature drops suddenly. In a steady flow situation, the gas is compressed by up to a factor of 4 at the shock, and it moves away from the shock front at $V_s/4$. It cools almost in pressure equilibrium, and as the density increases, the speed with respect to the shock approaches zero. For typical ISM parameters, the compression is likely to be halted by magnetic pressure by the time the gas cools to 10,000 K.

When the gas is suddenly heated by the shock, it is far from ionization equilibrium. As shown in Figure 1, it approaches equilibrium after some time, but at the same time the gas begins to cool. Radiative shocks in SNRs are typically slower than about 300 km/s, because the cooling time decreases rapidly with shock speed. These shocks produce temperatures of about 10^6 K or less. At those temperatures the gas cools rapidly enough that recombination cannot keep up, and the gas becomes overionized (Cox 1972a). Therefore, the interpretation of shock spectra requires models of the flow that include time-dependent ionization.

Radiative shocks faster than about 100 km/s produce enough ionizing radiation to form a photoionized zone of compressed gas, where most of the optical emission is produced (Cox 1972a). The ionizing radiation also ionizes the pre-shock gas, and this can drastically affect the emission spectrum in two ways (Shull & McKee 1979; Raymond 1979; Hartigan et al. 1987; Dopita & Sutherland 1996). First, a shock having $V_s \sim 100$ km/s in partially neutral gas will rapidly use a substantial fraction of its thermal energy to ionize and excite H I. The rapid cooling means that the gas will not attain as high an ionization state as it would if the pre-shock hydrogen were ionized (Figure 3). Second, a significant amount of optical radiation can be produced in the pre-shock medium. For galactic SNR filaments this is generally spatially resolved from the post-shock emission, but for AGNs, for instance, the pre-shock and post-shock emission cannot be distinguished and the pre-shock contribution must be taken into account. The importance of photoionization requires that

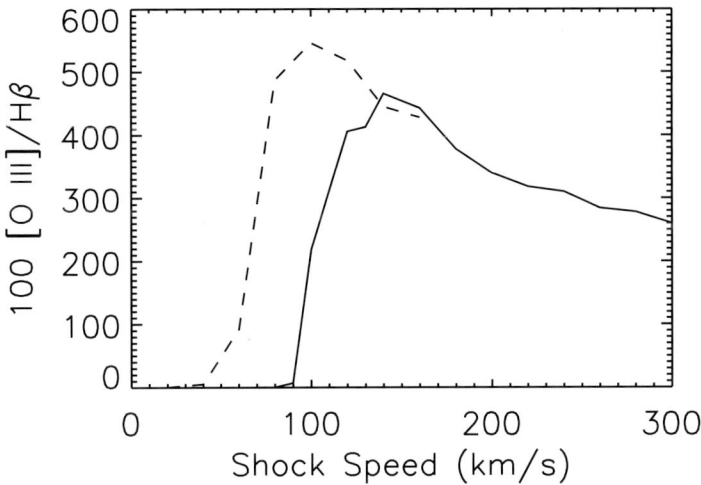

Fig. 3. Ratio of [O III] λ5007 to Hβ normalized to Hβ=100 (Hartigan et al. 1987). Solid line shows models for which pre-shock ionization state is in equilibrium with photons produced by the shock. Dashed line shows models with singly ionized pre-shock gas.

shock models include a radiative transfer calculation and computation of the heating by photoionization.

Tabulated sets of shock models generally are for steady flows, but radiative shocks faster than about 150 km/s are thermally unstable (Binette, Dopita & Tuohy 1985; Gaetz, Edgar & Chevalier 1988). The radiative cooling rate is

$$P = n_e\, n_H \Lambda(T,X) \quad \mathrm{erg/(cm^3\, s)} \qquad (2)$$

where X is the ionization state. Thus a small region of higher density experiences rapid cooling because the cooling rate scales as $n_e n_H$, while the thermal energy content scales as $n_e + n_H$. At temperatures above about 2×10^5 K and below about 2×10^7 K, the cooling rate coefficient $\Lambda(T,X)$ scales roughly as $T^{-1/2}$. These factors combine so that the higher density region experiences runaway cooling. Under some circumstances, a region large enough that the sound crossing time exceeds the radiative cooling time can undergo catastrophic cooling and drop out of pressure equilibrium. This leads to secondary shocks and a very complicated flow and thermal pattern (e.g. Innes, Giddings & Falle 1987). Innes (1992) showed that this affects emission line ratios at the factor of 2 level and generates complex line profiles. Sutherland, Bicknell & Dopita (2003) computed similar models in 2 dimensions, in which they find possibly fractal structure in the post-shock gas and a significant conversion of shock energy into turbulence. In section 5 of their chapter Tom Hartquist, Alex Wagner and Sam Falle have addressed the possible modification of the instability due to the presence of cosmic rays.

Another major problem with the real world is that density inhomogeneities play a major role in shock wave evolution. Both models and observations show the importance of the interaction of blast waves with dense clouds (e.g. Klein, McKee & Collela 1994; Levenson et al. 1996; Patnaude et al. 2002), but the roles of thermal conduction and turbulent stripping are not well enough understood for quantitative spectral predictions for specific shock-cloud interactions. Julian Pittard has considered the rates of thermal conduction and hydrodynamic ablation on clump evaporation in his chapter. The role of magnetic fields in limiting the turbulent cascade and thermal conduction is especially important (Mac Low et al. 1994).

3.2 SNR Ejecta Morphology

SNR ejecta seen at optical wavelengths are generally moving more or less in free expansion. Therefore, their position along the line of sight is simply given by their velocity times the time since the explosion. In SNRs such as Cas A, N132D and 1E 0102.2-7219, the shock that lights up an optical knot causes some deceleration, but it is small (e.g. Fesen et al. 2005). This is because the internal shock speed is of order 100 km/s, while the free expansion speed is of order 5000 km/s. In the case of the Crab, the optical filaments arise when shocks driven by the pulsar wind nebula compress the photoionized plasma, making it bright enough to observe (Sankrit & Hester 1997). Again, the shock speeds are small compared to the expansion speed. Therefore, the positions of the optical knots in 3 dimensions can be derived from the Doppler shifts and proper motions. Clark et al. (1983) and Lawrence et al. (1995) showed that the optical filaments of the Crab are located on an elliptical shell that is open at the ends. Reed et al. (1995) and Lawrence et al. (1995) showed that the optical knots of Cas A are located mostly on large rings on the surface of a shell. As demonstrated by these papers, either a spatial scan with a long slit spectrograph or a series of Fabry-Perot maps at different velocities can produce excellent results.

3.3 Identifying Shocks

The first question that often arises is whether a newly observed nebula is excited by shocks or photoionization. The optical spectra of both shocks and H II regions are dominated by the hydrogen Balmer lines and the forbidden lines of [O I], [O II], [O III], [N II] and [S II]. A shock generally produces [O III] by collisional ionization, so temperatures of order 10^5 K must be present. Thus the [O III] I(4363)/I(5007) ratio typically indicates temperatures of at least 30,000 K in shocks, compared with 10,000 K in H II regions and planetary nebulae (Osterbrock & Ferland 2006). Even better temperature-sensitive line ratios are obtained by comparing optical and UV lines, such as [O III] I(1664)/I(5007). A second indication, and one that is often easier to obtain, is the presence of a broad range of ionization states and the strength of the lowest ionization lines. While H II region spectra typically show I([S II])/I(Hα) < 0.2, SNR shocks are usually characterized by I([S II])/I(Hα) > 0.4 (e.g., Fesen et al. 1985). There are exceptions to this rule in cases where the shock serves mainly to compress gas within a photoionizing field and increase its emissivity, as has been found for filaments in the Crab Nebula (Sankrit & Hester 1997) and for HH Objects in the Orion Nebula (Reipurth et al. 1998). There are also incomplete shocks in which the the gas has not had time to cool to the temperatures at which the [S II] emissivity peaks (Raymond et al. 1988; see below).

3.4 Shock Parameters

Shock Speed

The most straightforward measurement is that of the Doppler shift. For an expanding spherical shell, the velocities will trace an ellipse as a function of distance from the center of the SNR. This can be used to estimate the SNR expansion speed (e.g. Lozinskaya et al. 2005). Extending that to the SNR age is dangerous, in that the radiative shocks often pertain to slower shocks in relatively dense clouds. The speed derived for the bright optical filaments is typically less than 150 km/s, while speeds 2 or 3 times higher are required to match the X-ray spectra produced by the blast wave. Faster radiative shocks are sometimes observed in the shells of young SNRs. In particular, speeds above 400 km/s were seen in IC443 (Meaburn et al. 1990) and in N 49 (Chu & Kennicut 1988; Vancura et al. 1992).

The structures of SNRs are often quite complex, so that while the velocity ellipse method may work in a statistical sense, it can lead to erroneous results for individual filaments. This is made worse by the fact that a sheet of gas seen edge-on will appear as a bright filament, and will therefore be an attractive target for observation, but its motion will be transverse to the line of sight, so its Doppler velocity will be near zero.

An alternative method for estimating the shock speed is the measurement of the intensity ratios of bright spectral lines. Figure 3 shows the ratio of [O III] $\lambda 5007$ to Hβ (normalized to Hβ=100) as a function of shock speed. The solid curve shows models from Hartigan et al. (1987) with the pre-shock ionization state in equilibrium with the ionizing photon flux produced by the shock, while the dashed curve shows predictions for fully ionized pre-shock gas. The latter could occur as a result of ambient starlight, but it is also expected for normal SNR evolution, because as the shock slows down from speeds of order 300 km/s, it produces intense ionizing radiation capable of ionizing a thick surrounding region. By the time it reaches speeds below 100 km/s, it may still be moving in this pre-ionized medium, which will not have had time to recombine at typical ISM densities (Cox 1972b).

Figure 3 demonstrates that a given ion typically appears at a characteristic shock speed when the temperature is high enough to ionize plasma to that state, and its ratio to the Balmer lines slowly declines for higher speeds. This occurs because the plasma must always cool through that ionization state, but the Balmer line emission slowly increases as more ionizing photons are produced. Figure 3 also demonstrates the importance of the pre-shock ionization state for shock speeds below about 120 km/s. Faster shocks are much less sensitive, but they still include a collisionally excited component that would be absent in the pre-ionized case (Cox & Raymond 1985). Thus, a small [O III]/Hβ ratio can prove that the shock is slow, but a moderate [O III]/Hβ ratio by itself is ambiguous.

For shock speeds above about 120 km/s, the [O III]/Hβ is relatively insensitive to V_s. The ultraviolet emission of N V has been observed with IUE and HST, and it is sensitive to shock speeds up to about 150 km/s. The [Ne V] λ3425 line and the O VI $\lambda\lambda$1032, 1037 doublet provide diagnostics for shock speeds up to about 200 km/s (Hartigan et al. 1987). The [Ne V] line has been observed by Fesen & Itoh (1985) and Szentgyorgyi et al. (2000), but both the atmospheric cutoff and the lack of blue response of most CCDs make the observations difficult. The O VI doublet has been observed with HUT and FUSE (e.g. Raymond et al. 2001; Sankrit et al. 2004). In addition to extinction by dust, the the UV lines can be affected by resonance scattering along the line of sight and within the emitting sheet of shocked gas (Long et al. 1992; Cornett et al. 1992; Raymond et al. 2003). Nevertheless, they are so sensitive to shock conditions that they are very useful as diagnostics. Rocket measurements and Voyager observations of O VI with low spatial resolution can be used to determine the UV and EUV luminosities of SNRs (e.g. Rasmussen & Martin 1992; Vancura et al. 1993). When UV and optical lines are compared, it often turns out that a broad range of shock speeds is required to match the spectrum (e.g. Vancura et al. 1992; Raymond et al. 2001).

Density

The densities of the shocks seen in the optical can be obtained from one of the classical density-sensitive line ratios of [O II] I(3729)/I(3726) or [S II] I(6717)/I(6730). These ratios are useful in the range 100 to 10,000 cm^{-3} (e.g. Osterbrock 1989). Both pertain to gas at temperatures near 10,000 K. It is often assumed that this density is related to the pre-shock density by a factor of 4 density jump at the shock and subsequent cooling from the post-shock temperature to 10,000 K at nearly constant pressure. This is generally wrong, because pressure support by the magnetic field and/or cosmic rays is almost certain to dominate by the time the shocked plasma has radiated away most of its internal energy (e.g. Raymond et al. 1988). Thus, the density derived from [O II] or [S II] yields only a lower limit to the pre-shock density even when the shock speed has been determined.

An alternate method for estimating densities is to divide a quantity that scales with density and depth along the line of sight as n^2L by one that scales as nL, where n is the post-shock density, or 4 times the pre-shock density. An example of the former is the X-ray emission measure. The latter is provided by The brightness of any emission line (except for a few forbidden lines that are quenched at high density) from a radiative shock, as well as the the brightness of an optical or UV line from a non-radiative shock, which scales as nLV_s. This technique has been used to separate pre-shock density from line-of-sight depth by comparing [Ne V] and X-ray observations (Szentgyorgyi et al. 2000) and by comparing O VI with X-rays (Raymond et al. 2003) at positions in the Cygnus Loop. This method has the advantages that it is independent of any assumptions about magnetic or cosmic ray pressure. It does require that the X-ray and optical observations pertain to the same gas. This is likely for clean, well-defined filaments, but difficult to prove.

A third method, which avoids these difficulties, can be applied to shocks seen at least partially face-on. Most studies focus on the brightest, most easily observed filaments, but these are sheets of gas viewed edge-on (Hester 1987). Shocked gas that is oblique to the line-of-sight will produce a Doppler shift equal to $V_s cos\theta$, where θ is the angle between the flow direction and the line of sight, since gas that has cooled has a speed approaching the shock speed. The surface brightness of the shock scales as $n_0 I_0(V)/cos\theta$, where $I_0(V)$ is the surface brightness produced by a face-on shock of velocity V in a medium of density 1 cm^{-3}. Thus, once the shock speed has been determined, the product of Doppler shift times surface brightness is proportional to n_0. This method was applied to shocks in the Cygnus Loop (Raymond et al., 1988) and Vela SNR (Raymond et al. 1997). In addition to the pre-shock density, it can provide a 3D picture of the filament. Even more directly, it provides the ram pressure of the shock, which can be compared with the pressure of the X-ray emitting plasma and with the thermal pressure determined from [O II] or [S II] line ratios.

Composition

SNR shocks provide a means to measure the elemental abundances in the ISM. As compared to H II regions, they have the advantage that grain destruction in the shock allows one to measure the complete chemical composition instead of the depleted composition of the gas phase alone. Jones, Tielens & Hollenbach (1996) found that even shocks as slow as 100 km/s return much of their material to the gas phase. Shock wave determinations of abundances also do not depend on the spectral shape of an illuminating stellar radiation field. As in the H II region or planetary nebula composition determinations, the relative strengths of the emission lines in an SNR shock give a first order idea of the elemental abundances, but detailed models of the shock emission are required for a reliable abundance determination. In both cases uncertainties in the atomic rates enter, and radiative transfer and thermal equilibrium must be handled with care. In one interesting example, the radial abundance gradients of oxygen and nitrogen in M33 determined from SNRs and H II regions agree quite closely, but the absolute N abundances differ by a factor of 2 (Smith et al. 1993). The Galactic abundance gradient has also been explored (Binette et al. 1982).

SNR shocks have the additional advantage that UV emission lines are relatively bright, providing a larger number of observable ions of several elements. This is especially important for C, Mg and Si, which have few lines in the optical range. A good example of the use of combined optical and UV spectra to obtain abundances of the ISM in the LMC is the analysis of N49 by Vancura et al. (1992), which agrees with H II region determinations of elements such as N, O and S, but is more reliable for the elements best observed in the UV, such as C, Mg and Si. Analysis of a shocked interstellar cloud at the periphery of the LMC remnant N132D yielded somewhat smaller abundances (Blair et al. 2000).

Abundances of SN ejecta are most often determined from X-ray spectra of non-radiative shocks (section 2.1), primarily because the optical knots represent a very small fraction of the ejected mass. However, the optical knots provide a means of measuring the small scale variations in composition. In some cases the optical knots also provide measurements of elements not observable in the X-ray due to large absorbing columns (e.g. Cas A). Shocks in hydrogen-depleted gas differ from those in the normal ISM in two major ways. First, the mean molecular weight, μ, of the gas is considerably higher, especially if the gas is only partially ionized. Therefore, a given shock speed will produce a much higher temperature since kT $\sim$ $3\mu V_s^2/16$. Second, the radiative cooling rate at temperatures above 50,000 K is higher by several orders of magnitude. It is so fast that the cooling time can be comparable with the time scale for sharing of energy between electrons and ions. Therefore, a situation can arise in which $T_e << T_i$ and the radiative losses of the electrons are balanced by the energy transferred from the ions by Coulomb collisions.

The first models of shocks of this sort were computed by Itoh (1981; 1988) for pure oxygen, and those models were extended to include other elements by Sutherland & Dopita (1995) and Blair et al. (2000). They showed the low electron temperature described above, and they showed precipitous cooling when the thermal energy of the ions is exhausted. For instance, a shock that produces $T_e \sim 10^5$ K ionizes most of the oxygen to O III. The electron temperature decreases slowly until the ions cool, and then both temperatures plunge to less than 10^3 K before the oxygen can recombine. Therefore, unlike in the ordinary ISM shocks, little [O II] or [O I] emission is produced. Observed spectra generally show emission from several ionization stages, unlike the spectra of these models. A range of shock speeds may be present in even the highest spatial resolution spectra, with the consequence that the relative line strengths reflect the fractions of shock surface moving at different speeds. It is also possible, however, that the interaction of a dense ejecta knot with the SNR reverse shock produces a turbulent flow that cannot be modeled as the sum of planar shocks. Another suggestion is that photoionization of ejecta gas that has not yet reached the shock contributes a significant fraction of the forbidden line emission (Sutherland & Dopita 1995).

Another feature of the Itoh (1988) models was the O I recombination line at 7774 Å due to photoionization of the cold, compressed gas behind the shock. The predicted strength was much larger than that observed. A possible explanation is that the ejecta knots are so small that the shock travels through a knot in a short time, such as the 10 year typical lifetime of Cas A fast moving knots. In that case, there may not be enough column density to absorb the ionizing radiation until the shock has traversed the knot and the photoionizing flux drops.

The most ambitious attempt so far to interpret the optical and UV spectra of shocked SN ejecta was the analysis of HST spectra of N132D and 1E 0102-7219 by Blair et al. (2000). They used the FOS and also the velocity structure to isolate individual features. Unlike Cas A, the ejecta of these remnants showed only C, O, Ne and Mg, with none of the oxygen-burning products expected for Type II SN. This led Blair et al. to suggest that Type Ib supernovae produced these remnants.

Optical knots in the ejecta of the Crab require a different sort of model. While a shock seems to be important for compressing the gas and producing UV emission lines (Sankrit & Hester 1997), most of the optical emission results from photoionization and heating of the compressed gas by synchrotron emission from the nebula. Pure photoionization models can match the optical spectra (e.g. MacAlpine et al. 1996), but models of shocks in a background photoionizing field are also available (Sankrit & Hester 1997). The models for the Crab filaments indicate elemental abundances expected for a relatively low mass SN (MacAlpine et al. 1996).

Ejecta abundances can also be determined from absorption lines if a suitable background star is present. So far, SN1006 is the only SNR that can be studied in detail (Wu et al. 1983; Blair et al. 1996; Hamilton et al. 1997; Winkler et al. 2005). Broad absorption lines of Si, II, Fe II and Fe III probe both cold, photoionized ejecta in free expansion and material that has recently passed through the reverse shock. The observations show less iron that expected from SN Ia models, along with a substantial asymmetry along the line of sight.

Incomplete Shocks

Filaments with [O III] emission much stronger than predicted by the tabulated shock models are found in many SNRs (e.g. Raymond et al. 1988; Boumis et al. 2005). This is because the tabulated models generally are based on the assumption of a steady flow in which the gas cools to 1000 K or below, so that there is time for hydrogen to recombine and produce strong Balmer lines. If a filament is actually a shock that has been propagating through a dense region for less than the recombination time ($\sim 10^5/n_e$ years), the cool part of the predicted flow is missing, and the lines of H, [N II], [O I], [O II] and [S II] are anomalously weak compared to [O III]. In many cases, these incomplete shocks show where the SNR blastwave has recently encountered a dense cloud, but in some cases, such as G65.3+5.7 and CTA 1 (Fesen, Gull & Ketelsen 1983), very large coherent incomplete shocks are seen.

3.5 Shocks in Dense Clouds

When a SNR blastwave encounters a dense molecular cloud, the approximate constancy of ram pressure, ρv^2, implies that the shock in the cloud gas will be very slow. This, combined with the high extinction typical of molecular clouds, means that these shocks are best observed in the IR or radio. Only a few SNRs have been observed in the IR at spectral resolution capable of showing emission lines so far, but the Spitzer Space Telescope is expected to provide many more. IR and radio observations can address questions about the nature of the slow shocks, the medium they encounter and the strength of the magnetic field. These observations have the advantage of penetrating galactic dust obscuration, and they are also important for understanding observations of star formation in other galaxies.

The IR range includes the fine structure lines of the neutral and singly ionized species of several of the most abundance elements. Among the brightest are [C II] 156μm, [O I] 63μm, [Ne II] 12.8μm, [Si II] 34.8μm and [Fe II] 1.64, 25μm. These ions can reveal the evolution of shocked gas after it cools below 1000 K, the temperature at which it becomes invisible in the optical emission lines. Burton et al. (1990) found that [O I] 63 μm emission accounted for 40-75% of the IRAS 60 micron emission in IC443. Reach & Rho (1996) found that the bright [O I] emission in W44 and 3C391 accounted for 10-20% of the IR emission, the remainder being attributed to heated dust. [O I] can arise either from a J-shock in moderate density gas or a slower C shock in much denser material. The filamentary structure seen in [Fe II] emission is reminiscent of the SNR filaments seen in the optical, suggesting that the ionic lines are produced by J-shocks when the gas cools below a few thousand K. The strength of the [Fe II] emission suggests efficient grain destruction (e.g. Hollenbach & McKee 1989; Reach et al. 2002).

Molecular hydrogen emission can be produced by molecules that reform as the gas cools behind an ordinary (J-) shock (e.g. Hollenbach & McKee 1989) or by molecules present in the gas encountered by a continuous (C-) shock (see the review by Draine & McKee 1993 and section 2.2 of the chapter by Peter Brand). Because many H_2 lines are accessible, it is possible to derive an excitation temperature and to find departures from a thermal distribution. Richter, Graham & Wright (1995) measured over 20 H_2 lines in IC443, and they find remarkably similar excitation conditions, $T_{exc} \sim 1940$ K. The lack of Brγ emission is difficult to reconcile with bow shock models, leading Richter et al. to favor a partially dissociative J-shock model (Brand et al.

1988; Moorhouse et al. 1991). Faint emission regions show evidence for formation pumping of the higher levels. The H_2 emission in SNRs is generally very clumpy (Richter et al. 1995; Reach et al. 2002) and spatially separate from ionic emission (Oliva et al. 1999), leading to the picture that the H_2 emission comes from shocks in dense cores engulfed by an ionic shock in lower density gas.

Radio observations have revealed other molecules, including CO, CS and HCO^+ (Reach, Rho & Jarrett 2005). CO line widths indicate 20-30 km/s shock speeds, and the brightness and morphologies indicate that they move through dense cores. Observations of OH masers provide another handle on the slow C-type shocks in dense clouds (e.g. Yusef-Zadeh et al. 1999). These masers also make it possible to derive magnetic field strengths through Zeeman splitting (e.g. Hoffman et al. 2005). The remnants for which magnetic fields have been estimated from masers show field strengths of order 1 mG. This presumably results from compression of the already high fields in dense clouds by fairly slow shocks.

4 Summary

Spectra of SNRs reveal the shock speeds, compositions, molecular content and in some cases the degree of ion-electron thermal equilibration. Interpretation of these spectra involves some straightforward line ratio diagnostics, but more often sophisticated models involving time-dependent ionization and radiative transfer. Spitzer observations promise to greatly increase the spectral information available, especially for remnants in the Galactic plane.

This work was supported by NASA grants NNG05GD94G and NAG5-12446 to the Smithsonian Astrophysical Observatory.

References

Aharonian, F. et al. 2005 A&A **437**, L7
Arendt, R.G., Dwek, E., Moseley, S.H. 1999 ApJ **521**, 234
Ballet, J., Rothenflug, R., Dubreuil, D., Soutoul, A., Caplan, J. 1989 A&A **211**, 217
Binette, L., Dopita, M.A., Dodorico, S., Benvenuti, P. 1982 A&A **115**, 315
Binette, L., Dopita, M.A., Tuohy, I.R. 1985 ApJ **297**, 476
Blair, W.P., Morse, J.A., Raymond, J.C., Kirshner, R.P., Hughes, J.P., Dopita, M.A., Sutherland, R.S. 2000 ApJ **538**, 61
Blair, W.P., Long, K.S., Raymond, J.C. 1996 ApJ **468**, 871
Blair, W.P., Sankrit, R., Raymond, J.C., Long, K.S. 1999 AJ **118**, 942
Boumis, P., Mavromatakis, F., Xilouris, E.M., Alikakos, J., Redman, M.P., Goudis, C.D. 2005 A&A **443**, 175
Brand, P.W.J.L., Moorhouse, A., Burton, M.G., Geballe, T.R., Bird, M. Wade, R. 1988 ApJL **334**, L103
Burton, M.G., Hollenbach, D.J., Haas, M.R., Erickson, E.F. 1990 ApJ **355**, 197

Chevalier, R.A., Raymond, J.C. 1978 ApJ **225**, L27
Chevalier, R.A., Kirshner, R.P., Raymond, J.C. 1980 ApJ **235**, 186
Chu, Y.-H., Kennicut, R.C. Jr. 1988 AJ **95**, 1111
Clark, D.H., Murdin, P., Wood, R., Gilmozzi, R., Danziger, J., Furr, A.W. 1983 MNRAS **204**, 415
Cornett, R.H. et al. 1992 ApJL **359**, L9
Cox, D.P., 1972a ApJ **178**, 143
Cox, D.P., 1972b ApJ **178**, 159
Cox, D.P., Raymond, J.C. 1985 ApJ **298**, 651
Dopita, M.A., Dodorico, S., Benvenuti, P. 1980 ApJ **236**, 628
Dopita, M.A., Sutherland, R.S. 1996 ApJS **102**, 161
Draine, B.T., McKee, C.F. 1993 ARA&A **31**, 373
Fabbiano, G., Baldi, A., King, A.R., Ponman, T.J., Raymond, J., Read, A., Rots, A., Schweizer, F., Zezas, A. 2004 ApJL **605**, 21
Fesen, R.A., Blair, W.P., Kirshner, R.P. 1985 ApJ **292**, 29
Fesen, R.A., Hurford, A.P. 1996 ApJS **106**, 563
Fesen, R.A., Itoh, H. 1985 ApJ **295**, 43
Fesen, R.A., Gull, T.R., Ketelsen, D.A. 1983 ApJS **51**, 337
Fesen, R.A., Hammell, M.C., Morse, J., Chevalier, R.A., Borkowski, K.J., Dopita, M.A., Gerardy, C.L., Lawrence, S.S., Raymond, J.C., van den Bergh, S. 2005 submitted to ApJ
Flanagan, K.A., Canizares, C.R., Dewey, D., Houck, J.C., Fredericks, A.C., Schattenburg, M.L., Markert, T.H., Davis, D.S. 2004 ApJ **605**, 230
Gabriel, A.H. 1972 MNRAS **160**, 99
Gaetz, T.J., Edgar, R.J., Chevalier, R.A. 1988 ApJ **329**, 927
Ghavamian, P., Raymond, J., Smith, R.C., Hartigan, P. 2001 ApJ **547**, 995
Ghavamian, P., Winkler, P.F., Raymond, J.C., Long, K.S. 2002 ApJ **572**, 888
Greidanus, H., Strom, R.G. 1992 A&A **257**, 265
Hamilton, A.J.S., Fesen, R.A., Wu, C.-C., Crenshaw, D.M., Sarazin, C.L. 1997 ApJ **481**, 838
Hartigan, P., Raymond, J., Hartmann, L. 1987 ApJ **316**, 323
Hester, J.J. 1987 ApJ **314**, 187
Hester, J.J., John Raymond, C., William Blair P., 1994 ApJ **420**, 721
Hoffman, I.M., Goss, W.M., Brogan, C.L., Claussen, M.J. 2005 ApJ **627**, 803
Hollenbach, D., McKee, C.F. 1989 ApJ **342**, 306
Hughes, J.P., Rakowski, C.E., Decourchelle, A. 2000 ApJL **543**, L61
Hwang, U., Decourchelle, A., Holt, S.S., Petre, R. 2002 ApJ **581**, 1101
Hwang, U. et al. 2004 ApJ **615**, L117
Innes, D.E. A&A 256, 660
Innes, D.E., Giddings, J.R., Falle, S.A.E.G. 1987 MNRAS **227**, 1021
Itoh, H. 1981 PASJ **33**, 1
Itoh, H. 1988 PASJ **38**, 717
Jenkins, E.B., Wallerstein, G., Silk, J. 1976 ApJS **32**, 681
Jenkins, E.B., Wallerstein, G., Silk, J. 1984 ApJ **278**, 649
Jones, A.P., Tielens, A.G.G.M., Hollenbach, D.J. 1996 ApJ **469**, 740
Klein, R.I., McKee, C.F., Collela, P. 1994 ApJ **420**, 213
Korreck, K.E., Raymond, J.C., Zurbuchen, T.H., Ghavamian, P. 2004 ApJ **615**, 280
Laming, J.M., Raymond, J.C., McLaughlin, B.M., Blair, W.P. 1996 ApJ **472**, 267

Lawrence, S.S., MacAlpine, G.M., Uomoto, A., Woodgate, B.E., Brown, L.W., Oliversen, R.J., Lowenthal, J.D., Liu, C. 1995 ApJ **109**, 2635
Levenson, N.R., Graham, J.R., Hester, J.J., Petre, R. 1996 ApJ **468**, 323
Li, G., Zank, G.P., Russell, C.T. 2005 *The Physics of Collisionless Shocks*, (Melville, NY: AIP)
Lim, A.J., Raga, A.C. 1996 MNRAS **280**, 103
Long, K.S., Blair, W.P., Vancura, O., Bowers, C.W., Davidsen, A.F., Raymond, J.C. 1992 ApJ **400**, 214
Long, K.S., Reynolds, S.P., Raymond, J.C., Windler, P.F., Dyer, K.K., Petre, R. 2003 ApJ **586**, 1162
Lozinskaya, T.A., Komarova, V.N., Moiseev, A.V., Blinnikov, S.I. 2005 Ast. Lett **31**, 243
MacAlpine, G.M., Lawrence, S.S., Sears, R.L., Sosin, M.S., Henry, R.B.C. 1996 ApJ **463**, 650
Mac Low, M.-M., McKee, C.F., Klein, R.I., Stone, J.M., Norman, M.L. 1994 ApJ **433**, 757
McKee, C.F., Hollenbach, D.J. 1980 Ann. Revs. A&A **18**, 219
Meaburn, J., Whitehead, M.J., Raymond, J.C., Clayton, C.A., Marston, A.P. 1990 A&A **227**, 191
Moorhouse, A., Brand, P.W.J.L., Geballe, T.R., Burton, M.G. 1991 MNRAS **253**, 662
Oliva, E., Moorwood, A.F.M., Drapatz, S., Lutz, D., Sturm, E. 1999 A&A **343**, 943
Osterbrock, D.E., Ferland, G.J. 2006 1989 *Astrophysics of Gaseous Nebulae and Active Galactic Nuclei (2nd Edition)*, (Sausalito: University Science Books)
Patnaude, D.J., Fesen, R.A., Raymond, J.C., Levenson, N.A., Graham, J.R., Wallace, D.J. 2002 AJ **124**, 2118
Rakowski, C.E. 2005 Ad. Sp. Res. **35**, 1017
Rakowski, C.E., Ghavamian, P., Hughes, J.P. 2003 ApJ **590**, 846
Rasmussen, A., Martin, C. 1992 ApJL **396**, L103
Raymond, J.C. 1979 ApJ Suppl **39**, 1
Raymond, J.C. 2003 RMAA, Serie de Conferencias **15**, 258
Raymond, J.C., Blair, W.P., Fesen, R.A., Gull, T.R. 1983 ApJ **275**, 636
Raymond, J.C., Hester, J.J., Cox, D.P., Blair, W.P., Fesen, R.A., Gull, T.R. 1988 ApJ **324**, 869
Raymond, J.C., Blair, W.P., Long, K.S. 1995 ApJL **454**, L31
Raymond, J.C., Blair, W.P., Long, K.S., Vancura, O., Edgar, R.J., Morse, J., Hartigan, P., Sanders, W.T. 1997 ApJ **482**, 881
Raymond, J.C., Li, J., Blair, W.P., Cornett, R.H. 2001 ApJ **560**, 763
Raymond, J.C., Ghavamian, P., Sankrit, R., Blair, W.P., Curiel, S. 2003 ApJ **584**, 770
Reach, W.T., Rho, J. 1996 A&A **315**, L277
Reach, W.T., Rho, J., Jarrett, T.H., Lagage, P.-O. 2002 ApJ **564**, 302
Reach, W.T., Rho, J., Jarrett, T.H. 2005 ApJ **618**, 297
Reed, J.E., Hester, J.J., Fabian, A.C., Winkler, P.F. 1995 ApJ **440**, 706
Reipurth, B., Bally, J., Fesen, R.A., Devine, D. 1998 Nature **396**, 343
Richter, M.J., Graham, J.R., Wright, G.S. 1995 ApJ **454**, 277
Sankrit, R., Hester, J.J. 1997 ApJ **491**, 796
Sankrit, R., Blair, W.P., Raymond, J.C. 2004 AJ **128**, 1615
Shull, J.M., McKee, C.F. 1979 ApJ **227**, 131

Shull, P. Jr., Hippelein, H. 1991 ApJ **383**, 714
Shull, P. Jr., Dufour, R.J., Parker, R.A.R., Gull, T.R. 1982 ApJ **253**, 682
Smith, R.C., Kirshner, R.P., Blair, W.P., Long, K.S., Winkler, P.F. 1993 ApJ **407**, 564
Smith, R.C., Raymond, J.C., Martin Laming, J. 1994 ApJ **420**, 286
Sollerman, J., Ghavamian, P., Lundqvist, P., Smith, R.C. 2003 A&A **407**, 249
Spitzer, L. Jr. 1978 *Physical Processes in the Interstellar Medium*, (New York: Wiley)
Sutherland, R.S., Dopita, M.A. 1995 ApJ **439**, 381
Sutherland, R.S., Bicknell, G.V., Dopita, M.A. 2003 ApJ **591**, 238
Szentgyorgyi, A.H., Raymond, J.C., Hester, J.J., Curiel, S. 2000 ApJ **529**, 279
Teske, R.A. 1990 ApJ **365**, 256
Teske, R.A., Petre, R. 1987 ApJ **314**, 673
Vancura, O., Blair, W.P., Long, K.S., Raymond, J.C. 1992 ApJ **394**, 158
Vancura, O., Blair, W.P., Long, K.S., Raymond, J.C., Holberg, J.B. 1993 ApJ **417**, 663
Vedder, P.W., Canizares, C.R., Markert, T.H., Pradhan, A. 1986 ApJ **307**, 269
Vink, J., Laming, J.M., Gu, M.F., Rasmussen, A., Kaastra, J.S. 2003 ApJL **578**, 31
Warren, J.S., Hughes, J. 2004 ApJ **608**, 261
Winkler, P.F., Gupta, G., Long, K.S. 2003 ApJ **585**, 324
Winkler, P.F., Long, K.S., Hamilton, A.J.S., Fesen, R.A. 2005 ApJ **624**, 189
Wu, C.-C., Leventhal, M., Sarazin, C.L., Gull, T.R. 1983 ApJL **269**, L5
Yusef-Zadeh, F., Goss, W.M., Roberts, D.A., Robertson, B., Frail, D.A. 1999 ApJ **527**, 172

Part III

Multicomponent Flows and Cosmic Rays

Mass-Loaded Flows

J.M. Pittard

School of Physics and Astronomy, The University of Leeds, Leeds, LS2 9JT, UK
jmp@ast.leeds.ac.uk.

1 Introduction

A key process within astronomy is the exchange of mass, momentum, and energy between diffuse plasmas in many types of astronomical sources (including planetary nebulae (PNe), wind-blown bubbles (WBBs), supernova remnants (SNRs), starburst superwinds, and the intracluster medium) and dense, embedded clouds or clumps. This transfer affects the large scale flows of the diffuse plasmas as well as the evolution of the clumps (e.g., Fig. 1). While in much theoretical work this interaction has been ignored, its consequences can be fundamental, as a growing body of literature now shows. Indeed, the standard model of the interstellar medium is based on such exchanges (McKee & Ostriker 1977), which occur through, for example, conduction, ablation and photoevaporation. The injection and mixing of mass from condensations into a surrounding supersonic medium induces shocks, increasing the pressure of the flowing medium (e.g., Pittard et al. 2005). This can lead to clump crushing and the reduction of the Jeans mass causing star formation, and is likely to play a role in sequential star formation (e.g., Elmegreen & Lada 1977), and may allow a starburst to develop (Hartquist, Dyson & Williams 1997). Radiative cooling is one way in which a starburst might be regulated, as it acts to reduce the pressure of the ambient medium once the mass injection rate becomes too high.

Three lengthscales characterize the entrainment of material from a clump into a surrounding flow (Hartquist & Dyson 1993). The smallest lengthscale is associated with the turbulent boundary layer around the clump. On intermediate scales the material injected into the flow forms a cometary-like tail, such as those seen around clumps in PNe. On the largest scales, the material is completely mixed into the flow and becomes indistinguishable from it. Unfortunately, and despite huge effort, the effectiveness of the physical processes in controlling the interchange of dense and diffuse material remains uncertain, in part because of the complexity of the turbulent boundary layers which exist between them. In addition, the microphysics which may drive some global

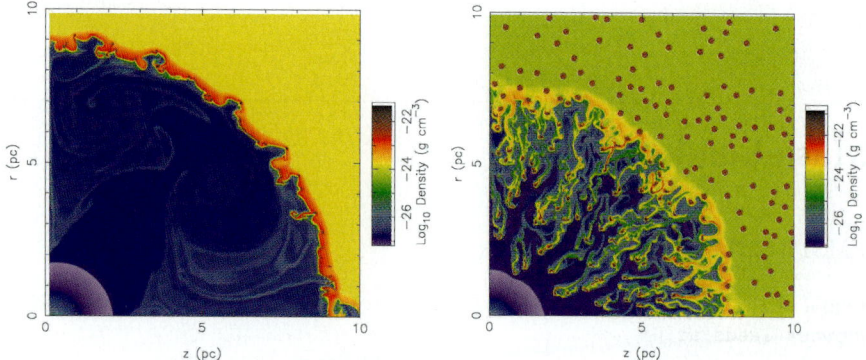

Fig. 1. The effect of mass-loading on a 'cluster-wind' from a group of early-type stars. In the left panel the wind expands into a smooth medium, while in the right panel the ambient medium is clumpy. The entrainment of mass from the clumps increases the density of the diffuse gas in the bubble interior, reduces its temperature, and slows its expansion.

processes is poorly understood. For example, magnetic reconnection, which may be necessary in order for clump and diffuse material to fully mix, is a difficult subject.

The influence of John Dyson in the field of mass-loaded flows cannot be overemphasized. He was one of the first to study the process of photoevaporation, he was involved in the development of the widely used analytical theory for ablatively-driven mass-loading, and in many subsequent works he has investigated the effect of mass-loading on a wide variety of astrophysical sources. I am grateful that I have had the opportunity to work with him in this field.

In Sect. 2 I review our current understanding of mass-injection processes. Section 3 focuses on intermediate-scale structure, while Sect. 4 examines the global effect of mass-loading on a flow. Section 5 concerns the mass-loading of a variety of diffuse sources. For an excellent summary of existing theoretical and observational studies on the interface between clouds and their surroundings see Hartquist & Dyson (1993).

2 Mass Exchange Processes

Consider a cloud of radius r_c, density ρ_c, and mass $M_c = 4/3\pi\rho_c r_c^3$, embedded in a medium of temperature T, density ρ, velocity v, and pressure $P = \rho kT/\mu m_H$. Let c_c and c be the isothermal sound speed in the cold cloud and in the hotter surroundings, respectively, and $\mathcal{M}$ be the Mach number of the flow relative to the cloud. Mass can be lost from the cloud and entrained into the surroundings through three main mechanisms, as discussed in the following subsections. I describe our current understanding of each process, detail analytical estimates of the rate of mass-loss, and highlight current uncertainties. The mass-loss rates driven by each process are then compared for clumps in a variety of different situations.

2.1 Hydrodynamic Ablation

Numerical simulations of the interaction of a supersonic wind or a strong shock with a single cloud have been presented many times. The evolution for the case of an adiabatic cloud can be broken into 4 consecutive stages: an initial transient stage when the shock first strikes the cloud, a compression stage, a re-expansion stage, and finally a destruction stage. During the initial interaction, a bow shock forms around the cloud, while a shock is driven into the cloud with velocity $v_c \approx \chi^{-1/2} v_s$, where χ is the density ratio between the cloud and its initial (e.g., pre-shock) surroundings, and v_s is the velocity of the shock through the ambient medium. The characteristic timescale for the cloud to be crushed by the transmitted shock is $\tau_{cc} = r_c/v_c \approx \chi^{1/2} r_c/v_s$. When the transmitted shock reaches the back of the cloud, a strong rarefraction is reflected back into the cloud, causing its subsequent re-expansion downstream. This is accompanied by a lateral expansion driven by the high pressure in the cloud and the lower pressure in the surrounding medium at its sides. The cloud is disrupted by the action of both Kelvin-Hemlholtz (KH) and Rayleigh-Taylor (RT) instabilities, with the Richtmyer-Meshkov instability playing a minor role unless the surface of the cloud is irregular. Destruction occurs after several crushing times, with the cloud material expanding and diffusing into the ambient flow (Klein, McKee & Colella 1994). In 3-D simulations, instabilities drive a richer structure (Stone & Norman 1992; Xu & Stone 1995). Recent laser experiments confirm that the vortex ring which forms at the back of the cloud is broken up by the action of azimuthal bending mode instabilities (Klein et al. 2003). In contrast, radiative clouds break up into numerous dense cold fragments which survive for many dynamical timescales (Mellema, Kurk & Röttgering 2002; Fragile et al. 2004). Self-gravity can become dynamically important in the dense fragments behind the compression shock. External magnetic fields generally increase the compression of the cloud and enhance radiative cooling, while magnetic fields internal to the cloud resist compression (see Fragile et al. 2005, and references therein). An example of a numerical calculation of the time evolution of a cold cloud interacting with a supersonic wind is shown in Fig. 2.

An analytical theory for the hydrodynamic ablation of material from dense clumps into the surrounding flow was presented by Hartquist et al. (1986). First, consider a clump embedded in a subsonic flow. The magnitude of the pressure variations over the surface of the clump, created by the well-known Bernoulli effect, is (Landau & Lifshitz 1959)

$$\Delta P \approx P_s \left[1 - \left\{ 1 + \frac{\gamma - 1}{2} \mathcal{M}^2 \right\}^{-\gamma/(\gamma-1)} \right], \quad (1)$$

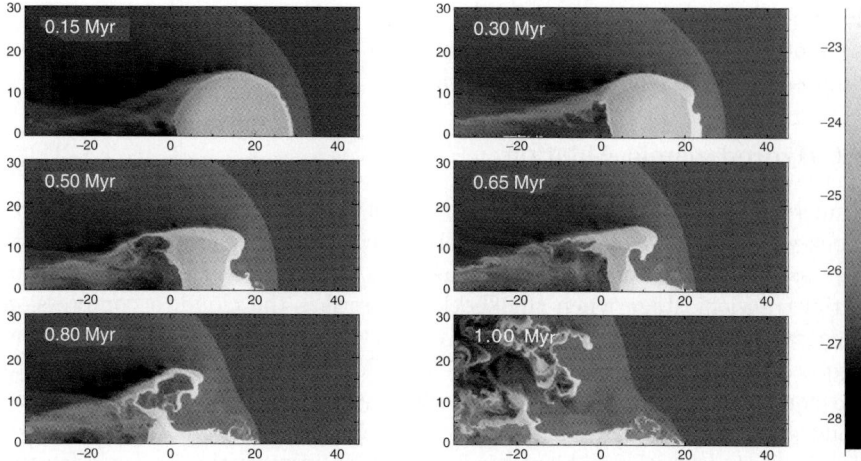

Fig. 2. The destruction of a cold cloud in a supersonic flow by hydrodynamic ablation. The time evolution of the logarithm of mass density is shown (in units of $\mathrm{g\,cm^{-3}}$), with distances given in pc. At the beginning of the simulation the cloud center is at $z = 20$ pc. $\chi = 500$ and $\mathcal{M} = 3$. There is no thermal conduction or photoevaporation, magnetic fields, or self-gravity, and the simulations are performed using 2D axisymmetry. Radiative cooling is included. (From Marcolini et al. 2005).

where the stagnation pressure, $P_s = P + \rho v^2$. In the small Mach number regime, $\Delta P \approx \mathcal{M}^2 P_s$. As the flow is fastest at the sides of the clump, the pressure is reduced, and the clump expands in directions normal to the flow at a speed

$$v_{\mathrm{exp}} \approx \frac{\gamma c_c}{\gamma - 1} \ln\left\{1 + \frac{\gamma - 1}{2}\mathcal{M}^2\right\}. \qquad (2)$$

For small Mach numbers, $v_{\mathrm{exp}} \approx (\gamma/2)c_c\mathcal{M}^2 \approx c_c\mathcal{M}^2$. Mixing between the cloud material and the flow occurs within a region of size l which is set by the requirement that the rate of mass-loss from the clump, $\dot{M}_{\mathrm{ab}}$, is comparable to the mass-flux of the ambient flow through this region, $\dot{M}_s$. Since $\dot{M}_{\mathrm{ab}} \sim M_c/t \sim M_c v_{\mathrm{exp}}/l$, and $\dot{M}_s = \rho v l^2$, $l \approx (\mathcal{M}^2 M_c C_c/\rho v)^{1/3}$. In contrast, if the flow is supersonic, mixing occurs largely as a result of a low pressure region over the reverse face of the clump, 'shadowed' from the flow. Since the mass cannot leave the clump faster than its sound speed, $v_{\mathrm{exp}} \sim c_c$, and in this case $l \approx (M_c c_c/\rho v)^{1/3}$.

The rate of mass-loss from the clump, $\dot{M}_{\mathrm{ab}} \approx l^2 \rho_l v_{\mathrm{exp}}$, where ρ_l is the characteristic density of ablated material at distance l. Momentum conservation requires that $\rho_l v_{\mathrm{exp}} = \rho v$, so in subsonic flows $\dot{M}_{\mathrm{ab}} \approx \mathcal{M}^{4/3}(M_c c_c)^{2/3}(\rho v)^{1/3}$, while in supersonic flow $\dot{M}_{\mathrm{ab}} \approx (M_c c_c)^{2/3}(\rho v)^{1/3}$ (i.e. independent of $\mathcal{M}$). These estimates have received some limited support from the numerical simulations calculated by Klein et al. (1994), although the predicted scaling with the flow parameters remains to be confirmed.

An alternative approach based on 'mixing-length' theory has been presented by Cantó & Raga (1991) (see also Arthur & Lizano 1997). While the boundary layer around the cloud is likely to be turbulent, even if the cloud and the surroundings are magnetized (Hartquist & Dyson 1988), such theories are complicated by the unknown degree to which clump gas and the tenuous plasma physically mix, and I do not discuss them further here.

Finally, it is unclear whether the ablation process by itself can merge the stripped material with the global flow in the sense that its temperature, velocity, and density approach those of the surrounding tenuous material. It may therefore be necessary to invoke another process, such as the transfer of heat by thermal conductivity, for the stripped material to acquire the physical state of the surrounding medium. Thermal conduction can accomplish this phase transition without microscopic mixing, and acceleration to the global flow speed is effected by the response of stripped material to pressure gradients and viscous coupling, which may arise from a host of mechanisms including turbulence.

2.2 Conductively-Driven Thermal Evaporation

Cold clouds embedded in a hot medium may also lose mass to their surroundings as hot electrons deposit energy in the surface regions of the clump. This process is referred to as thermal evaporation. The rate of mass loss is dependent on many factors, including the temperature of the hot phase, the clump radius, whether the conductivity is saturated, the presence of magnetic fields and plasma instabilities, and whether there is a velocity difference between the ambient medium and the cloud (e.g., Cowie & McKee 1977; McKee & Cowie 1977). Nonspherical clumps may be treated in an approximate way by adopting half the largest dimension as the radius of the clump (Cowie & Songalia 1977).

If the mean-free-path for electron-electron collisions, λ_{ee}, is approximately less than the temperature scale-height, $T/|\nabla T|$, then the heat flux into the cloud, q, is given by the classical theory of thermal conduction i.e. $q = q_{cl} = -\kappa \nabla T$. The mean-free-path $\lambda_{ee} = t_{ee}(3kT/m_e)^{1/2}$, where the electron-electron equipartition time is given by (Spitzer 1962)

$$t_{ee} = \frac{3m_e^{1/2}(kT)^{3/2}}{4\pi^{1/2}n_e e^4 \ln \Lambda}, \qquad (3)$$

where $\ln \Lambda = 29.7 + \ln (T/10^6 \sqrt{n_e})$ is the Coulomb logarithm and the other symbols have their usual meaning. I have implicitly assumed that $T_e = T$. The thermal conductivity, κ, in a fully ionized hydrogen plasma is (see, e.g., Spitzer 1962; Cowie & McKee 1977)

$$\kappa = 1.84 \times 10^{-5} \frac{T^{5/2}}{\ln \Lambda} \text{ erg s}^{-1} \text{ K}^{-1} \text{ cm}^{-1}, \qquad (4)$$

(the zero current requirement reduces the effective coefficient of conductivity by a factor ≈ 0.4 - see Spitzer & Härm 1953). The evaporative mass-loss rate from a single clump (Cowie & McKee 1977) is then

$$\dot{M}_{\rm con} = \frac{16\pi\mu\kappa\omega r_{\rm c}}{25k} = 2.75 \times 10^{19}\, \omega\, r_{\rm pc}\, T_6^{5/2} \; {\rm g\, s}^{-1} \qquad (5)$$

where $r_{\rm pc}$ is the clump radius in parsecs, and $T_6 = T/10^6$ K. For classical evaporation, $\omega = 1$. As conductively driven evaporation has a very temperature sensitive rate, ablation is likely to regulate clump dispersal in lower temperature media.

When $\lambda_{\rm ee} \gtrsim T/|\nabla T|$, the classical theory of thermal conduction, which is based on a diffusion approximation, may no longer be used. Instead the heat flux reaches a limiting value; i.e. it becomes saturated. The approach to saturated conduction is still partly empirical, and it is common practice to take a flux-limited form for the heat flux: $q_{\rm sat} = 5\phi\rho c^3$ (Cowie & McKee 1977), where ρ and c are in the hot phase, and ϕ is a parameter of order unity which describes the uncertainty in the numerical value of the saturated heat flux. Observations of the highly saturated solar wind, laboratory plasma experiments, and Fokker-Planck calculations suggest that $0.3 \lesssim \phi \lesssim 1.1$ (Giuliani 1984). Balbus & McKee (1982) conjectured that the effective heat flux, q, can be approximated by the harmonic mean of $q_{\rm sat}$ and $q_{\rm cl}$,

$$\frac{1}{q} \approx \frac{1}{q_{\rm sat}} + \frac{1}{q_{\rm cl}}. \qquad (6)$$

The resulting heat flux reduces to the smaller of the two conduction forms when there is a large disparity between them, and has the convenient property of a smooth transition from diffusive to flux-limited transport. The ratio of classical to saturated heat flux is

$$\sigma = \frac{q_{\rm cl}}{q_{\rm sat}} = \frac{\kappa}{5\phi\rho c^3}\frac{dT}{dr}, \qquad (7)$$

where the last expression explicitly assumes spherical symmetry i.e. $\nabla T = dT/dr$. The heat flux is then

$$q = \frac{\kappa}{1+\sigma}\frac{dT}{dr}, \qquad (8)$$

where σ is the *local* stauration parameter. This expression becomes the diffusive flux when $\sigma \ll 1$, and the saturated flux when $\sigma \gg 1$.

Cowie & McKee (1977) also defined a *global* saturation parameter,

$$\sigma_0 \equiv \frac{2\kappa T}{25\phi\rho c^3 r_{\rm c}} = \left(\frac{T}{1.54 \times 10^7}\right)^2 \frac{1}{nr_{\rm pc}\phi}, \qquad (9)$$

where $\ln \Lambda = 30$ has been assumed. Whereas σ measures the saturation locally, σ_0 measures global scales and allows a quick assessment of the importance of saturation effects (σ_0 is essentially the local saturation parameter, σ, evaluated at the ambient conditions with $dT/dr = T/r_c$). For $\sigma_0 \lesssim 0.03/\phi$, radiative losses quench the evaporation, and the clump grows in mass as surrounding material condenses onto it (McKee & Cowie 1977). This, of course, may make the clump gravitationally unstable, and initiate new star formation. For $0.03/\phi \lesssim \sigma_0 \lesssim 1$, the clump is evaporated at the classical rate. The onset of saturation occurs when σ_0 is of order unity, with highly saturated flows having $\sigma_0 \gg 1$. The mass-loss rate in the saturated regime is specified with $\omega \approx (1 + \sigma_0)^{-0.7}$ (Giuliani 1984). Since the onset of saturation is dependent on the radius of the clump for a specified hot phase, larger clumps will tend to evaporate in the classical limit, while the evaporation of mass from smaller clumps will tend towards saturation. If the temperature and density of the hot phase is evolving (e.g., because it is the interior of a WBB or SNR), the radius of clumps which are just at the onset of saturation will also change.

The dynamics of the evaporation process have been analyzed by McKee & Cowie (1975) and Cowie & McKee (1977), and an analogy with ionization fronts can be made. For classical evaporation (i.e. $0.03 \lesssim \sigma_0/\phi \lesssim 1$), the velocity of the conduction front, $v_{\text{cond}} \sim 2\sigma_0 \phi c_c^2/c$. If $\sigma_0 \gtrsim 0.25/\phi$, the conduction front drives a shock into the cloud. Otherwise, the cloud evaporates subsonically. Conduction fronts in the saturated regime have a velocity $v_{\text{cond}} \sim 1.12\sigma_0^{1/8} c_c^2/c_w$ (for $\phi = 1$), and always drive a shock into the cold cloud ahead of the conduction front itself.

Time-dependent hydrodynamical simulations of clouds overrun by a strong shock and undergoing conductively-driven thermal evaporation in the classical regime have been calculated by Orlando et al. (2005). Conduction inhibits the growth of Rayleigh-Taylor and Kelvin-Helmholtz instabilities, and the fragmentation of the cloud, but heats the evaporated material so that it quickly becomes part of the ambient flow. Simulations in the saturated regime confirm that the conduction front initially drives a shock into the cloud (Ferrara & Shchekinov 1993; Marcolini et al. 2005), substantially increasing its density. The resulting evolution then appears sensitive to assumptions concerning the cooling, with the cloud and the evaporation either settling into a quasi-steady-state (Ferrara & Shchekinov 1993), or displaying oscillatory behaviour (Marcolini et al. 2005). The mass-loss rate during the initial phase is comparable to that inferred from Eq. 5, but it is substantially reduced as the cloud is compressed and decreases in size (Marcolini et al. 2005). Somewhat different models including self-gravity have been calculated by Vieser & Hensler (2000) and Hensler & Vieser (2002).

Conductively-driven evaporation is perhaps the best-studied of the three processes highlighted in this section, but there are uncertainties in many physical processes whose influence on $\dot{M}_{\text{con}}$ remains poorly quantified. For instance, when $\sigma_0 \gtrsim 100$, viscous stresses have the potential to increase $\dot{M}_{\text{con}}$ significantly, but the exact enhancement depends on the uncertain degree to which they might also saturate (Draine & Giuliani 1984). On the other hand, if the electron mean free path is reduced, the heat flux will be inhibited. Bandiera & Chen (1994a,b) have emphasized that when $\sigma_0 \gtrsim 100$, the requirement of

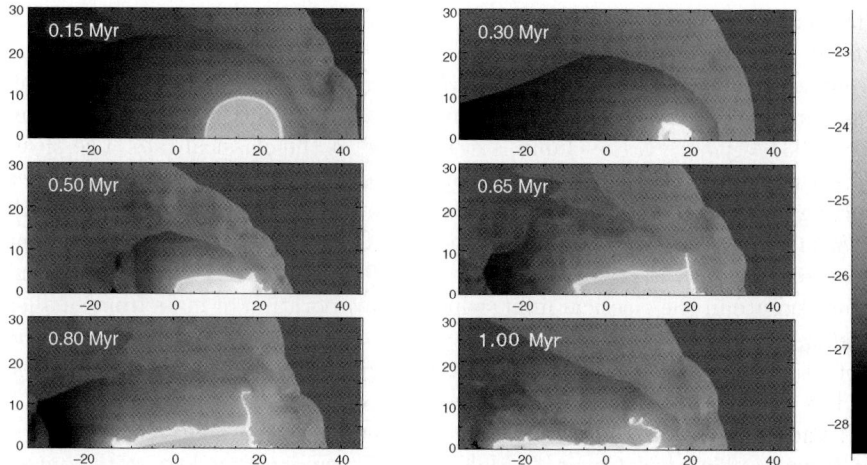

Fig. 3. As Fig. 2, but with heat conduction in the saturated regime ($\sigma_0 = 16$). A strong shock is driven into the cloud ahead of the conduction front and compresses it from all sides, so that its radius has decreased by a factor of 3 by $t = 0.3$ Myr. The subsequent evolution is also markedly different from the purely hydrodynamical case. (From Marcolini et al. 2005).

a zero net current means that hot electrons are stopped by electrostatic effects within a thin surface layer. The resulting heat flux is then considerably lower than that obtained if the electrons penetrate deep into the cloud and directly heat it through Coulomb collisions (Balbus & McKee 1982). The electron mean free path can also be reduced as a result of scattering by plasma instabilities, such as the ion-acoustic instability (Galeev & Natanzon 1984), and by whistler waves (Gary & Feldman 1977; Levinson & Eichler 1992). However, it is difficult to obtain a self-consistent model of these processes, and the presence of a strong magnetic field may suppress such instabilities. Partial ionization and non-equilibrium cooling are other possibilities for suppressing the conductivity (Böhringer & Hartquist 1987). Magnetic fields may not reduce the conductivity as much as previously thought (e.g., Balbus 1986; Rosner & Tucker 1989; Cho et al. 2003) unless the cloud is magnetically disconnected from its surroundings. An example of a numerical calculation of the time evolution of a conductively-evaporating cold cloud interacting with a hot supersonic wind is shown in Fig. 3.

2.3 Photoevaporation

The fate of a neutral clump exposed to a strong ionizing radiation field has been extensively studied over the years. Small, low-mass clumps are instantly ionized and rapidly dissipate i.e. are 'zapped'. In contrast, clouds which are sufficiently dense and large trap the ionization front, which becomes D-critical

(Dyson 1968; Bertoldi 1989) and moves into the cloud driving a shock front ahead of it. The ionized gas streams away perpendicular to the ionization front and expands supersonically into the interclump medium, reaching an asymptotic Mach number of ≈ 2 (e.g., Kahn 1969). This flow may absorb a large part of the incident ionizing flux and appear as a bright rim to the clump. The shock driven into the clump is focussed onto the clump axis, and can substantially increase the initial clump density (Sandford, Whitaker & Klein 1982). This 'radiation-driven implosion' may make the clump gravitationally unstable and lead to new star formation. Otherwise, the pressure overshoot causes the clump to reexpand, and it undergoes several radial oscillations (Lefloch & Lazareff 1994) before obtaining an equilibrium, cometary-shaped structure (Bertoldi & McKee 1990). Recombination may occur in the shadow of the clump, but this is prevented if there is a diffuse component to the radiation field (Pavlakis et al. 2001). The pressure in the evaporating flow declines rapidly, and eventually a termination shock forms. If the surrounding medium is supersonic a bow-shock is also formed. Photoevaporated flows also occur from the neutral disks which surround pre-main-sequence stars, known as 'proplyds' (O'Dell, Wen & Hu 1993; Bally et al. 1998).

Analytical equations for the mass injection rate of the photoevaporated flow as the clump is destroyed are presented in Bertoldi (1989) and Mellema et al. (1998). Good agreement with results from numerical simulations is obtained. However, it is possible to obtain a simple estimate by setting $\dot{M}_{\rm ph} = mFA$, where m is the mass per particle of the neutral material, and F and A are respectively the rate per unit time per unit area at which hydrogen ionizing photons reach the ionization front and its area. F is reduced by absorptions in the photoevaporating flow, and is approximately given by (Mellema et al. 1998)

$$F \approx \frac{F_0}{\left(1 + \alpha_{\rm B} F_0 r_{\rm if}/3c_{\rm i}^2\right)^{1/2}}, \quad (10)$$

where $\alpha_{\rm B} = 2.6 \times 10^{-13}\,{\rm cm}^3\,{\rm s}^{-1}$ is the case B recombination rate for H, $r_{\rm if}$ is the radius of curvature of the ionization front, $c_{\rm i}$ is the isothermal sound speed of the ionized gas, and $F_0 = \dot{S}/4\pi d^2$ is the flux delivered by the ionizing sources at the position of the clump. Typical assumptions are $r_{\rm if} \approx r$ and $A \approx \pi r^2$. $\dot{M}_{\rm ph}$ declines with time as the clump is destroyed and r decreases.

Photoevaporation may be suppressed if the ram or thermal pressure of the surrounding medium is greater than the pressure of the evaporating flow (Dyson 1994). Density inhomogeneities within the clump may affect the process of photoevaporation, as may the instability of inclined ionization fronts (Williams 2002). The role of magnetic fields on the structure of ionization fronts (Williams et al. 2000) may also affect $\dot{M}_{\rm ph}$.

2.4 Comparison of Mass-Injection Rates

While the exact rates of mass-loss by the three processes described in Sect. 2.1-2.3 remain somewhat uncertain, an appreciation of their relative importance can be obtained by considering several different objects.

The Helix Nebula (NGC 7293)

The Helix Nebula is famous for containing many extended cometary tail-like structures, which emanate from dense, neutral globules, and point away from the central star. The clumps have on average the following properties: $r_c \approx 10^{-3}$ pc, $T_c \approx 10$ K, $n_c \approx 10^6$ cm^{-3} (Dyson et al. 1989). A typical distance of a clump from the central ionizing star is $d \approx 0.1$ pc. The star has an ionizing photon flux $\dot{S}_{49} = \dot{S}/10^{49} \approx 3.5 \times 10^{-4}$. The clumps appear to be overrun by [OIII] gas with $T \approx 10^4$ K, $n \approx 10^3$ cm^{-3}, and a relative velocity, $v \approx 17$ km s^{-1} (Meaburn et al. 2005). Hence, the Mach number of the flow relative to the clumps is about 1.5. Using the analytical equations in Secs. 2.1 through to 2.3, we determine that $\dot{M}_{\rm ab} \approx 1.6 \times 10^{16}$ g s^{-1} and $\dot{M}_{\rm ph} \approx 2.4 \times 10^{17}$ g s^{-1}. The global saturation parameter, $\sigma_0 \approx 4 \times 10^{-7}$ for $\phi = 1$, and gas would like to condense onto the clumps at a rate $\dot{M} \approx 2 \times 10^{16}$ g s^{-1} (Cowie & McKee 1977), though this is likely prevented by the mass-loss that occurs through photoevaporation and ablation. The estimated lifetime of such clumps is in excess of 4×10^4 yr, compatible with the estimated age of the nebula. The origin of the knots is discussed in Dyson et al. (1989).

The Wolf-Rayet Nebula RCW 58

RCW 58 is a nebula surrounding the Wolf-Rayet star WR 40, and which has been formed by the current stellar wind sweeping up wind material from previous evolutionary stages, some of which is clumpy (Chu 1982; Smith et al. 1988). The clumps are ionized by the central star and typically have the following properties: $r_c \approx 0.1$ pc, $T_c \approx 10^4$ K, $n_c \approx 10^3$ cm^{-3} (Arthur, Henney & Dyson 1996). They are embedded in shocked stellar wind material, with $n \approx 1$ cm^{-3}, $v \approx 200$ km s^{-1}. The flow around the clumps has $\mathcal{M} \approx 0.6$. The H-ionizing flux from the star is $\dot{S}_{49} \approx 2.5$ (P. Crowther, private communication). For clumps at a distance of 1 pc from the star, $\sigma_0 \approx 4$, and the conductively-driven mass-loss is mildly saturated with $\dot{M}_{\rm con} \approx 3 \times 10^{20}$ g s^{-1}. A higher mass-loss rate is obtained for ablation: $\dot{M}_{\rm ab} \approx 10^{21}$ g s^{-1}. Since the cloud is already ionized it does not make sense to calculate a photoevaporation rate, but this may once have been the dominant process if the cloud was formerly neutral. Therefore, ablation would appear to control the rate at which mass is currently stripped from the clump.

Within a WBB

Of course, the mass-loss rate from each process varies within a bubble as the density, velocity, temperature, and distance from the ionizing source change. This is demonstrated in Fig. 4, where the top panel shows an example of the density and temperature structure within a bubble, while the mass-loss rates and lifetime of a specific clump are shown in the bottom panel.

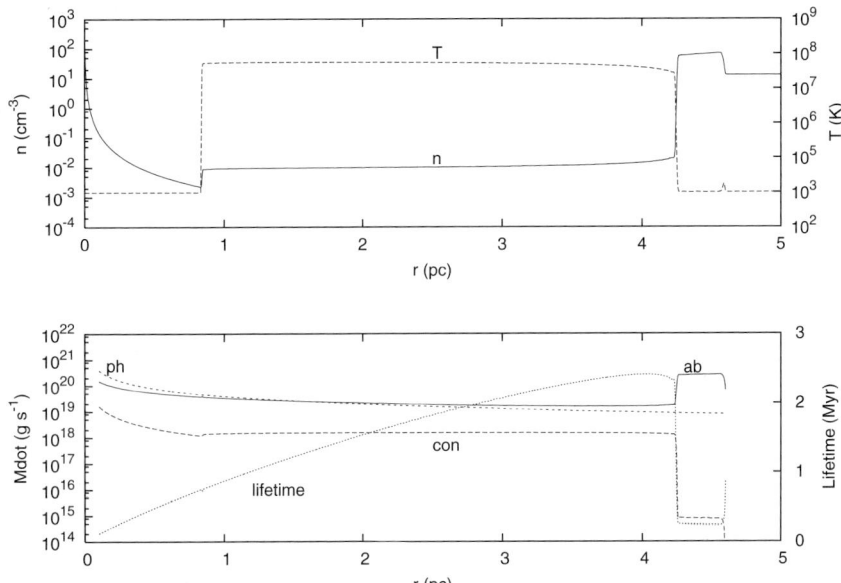

Fig. 4. Top: The internal structure (total number density and temperature) of a WBB of age 0.1 Myr expanding into a stationary medium with $n_H = 10$ cm^{-3}. The central star has $\dot{M} = 10^{-6}$ M$_\odot$ yr^{-1}, $v_\infty = 2000$ km s^{-1}, and $\dot{S}_{49} = 1.2$. Bottom: The mass-loss rate from clumps of radius 0.01 pc, $n = 10^7$ cm^{-3}, $T = 100$ K, due to hydrodynamic ablation (ab), photoevaporation (ph), and conductively-driven evaporation (con). No clumps are assumed to reside within the central 0.1 pc of the bubble. Clumps located within the hypersonic stellar wind are assumed to be surrounded by a sheath of hot gas bounded by a bowshock on their windward surface. The conductively-driven evaporation is highly saturated for clumps within the hot gas of the bubble given the paramaters chosen, but is in the condensational regime within the shell of swept-up ambient material. Also shown is an estimate for the lifetime of the clumps: $t = (\dot{M}_{ab} + \dot{M}_{con} + \dot{M}_{ph})/M_c$.

3 Intermediate-Scale Structure

While the stripping of mass from clumps has been extensively, though not definitively studied, to date there has been very little work on how intermediate-scale structures disperse/merge into the background flow. This problem has received some attention in studies of the interaction between multiple clouds and a flow (e.g., Jun, Jones & Norman 1996; Poludnenko, Frank & Blackman 2002; Steffen & López 2004), but our understanding of such interactions is still developing. A limitation of these works is that the clumps have been modelled as single-phase entities, and the density contrast between the clumps and their surroundings has, for numerical reasons, typically been $\sim 10^2$. The simulated clumps then have such short lifetimes that they are unable to significantly mass-load the flow. In reality, in many astrophysical flows the density contrasts are much larger. A different numerical approach which accounts for the much longer lifetimes of clumps is to set up sources which continuously inject mass into the flow (Falle et al. 2002). Multiple sources act as an efficient barrier to the flow if they are sufficiently close together that their combined mass injection rate is comparable to or exceeds the mass flux of the incident flow into the volume that they occupy (Pittard et al. 2005, see also Fig. 5). In such cases, the thermal pressure of the flow is greatly enhanced (at the expense of its ram pressure), and crucially becomes relatively uniform - these conditions are exactly those required to increase the probability of cloud collapse and new star formation.

Perhaps one of the best examples of intermediate scale structure is the comet-like tail extending from the Galactic Centre source IRS 7, a red supergiant (Yusef-Zadeh & Morris 1991; Serabyn, Lacy & Achtermann 1991). Although originally interpreted in terms of a global wind-wind collision between the slow dense wind from IRS 7 and a Galactic wind, this model had difficulty in explaining the length of the tail (Yusef-Zadeh & Melia 1992). However, it may be overcome if the wind of IRS 7 is clumpy enough that it becomes semi-porous to the Galactic wind, with much smaller bow-shocks forming around each clump (Dyson & Hartquist 1994).

4 The Global Effect of Mass-Loading on a Flow

The effect of mass-loading on a global flow can be studied in a one-fluid approximation if the following two assumptions are made: i) the clumps are sufficiently numerous that they can be considered to be continuously distributed; ii) the characteristic scale length of injection and mixing is much smaller than the dimensions of the global flow (so that the injected mass reaches the general flow velocity and temperature essentially instantaneously). For the steady injection of mass from an ensemble of clumps into the interior of a spherically expanding flow, the time-independent continuity and momentum equations are then

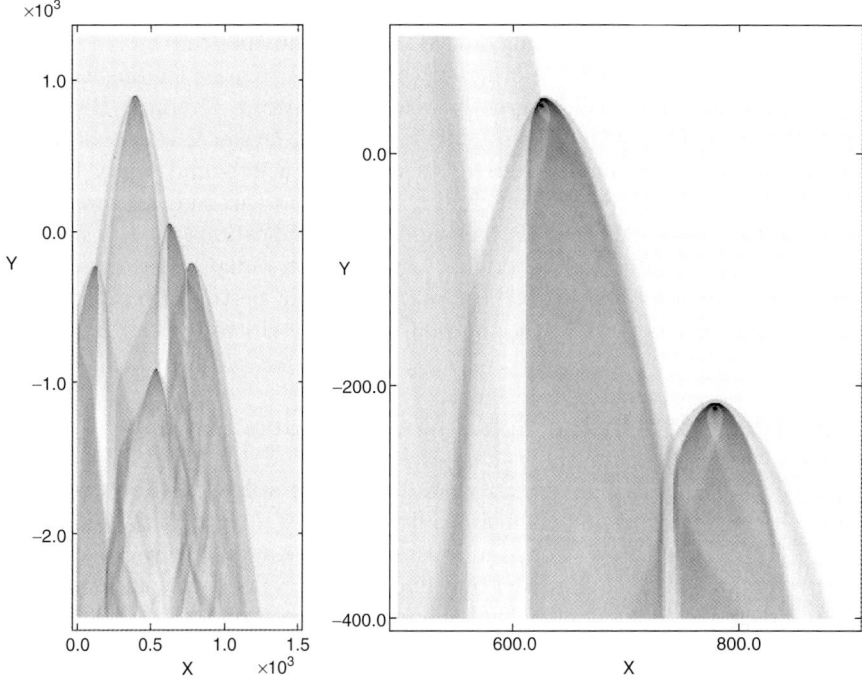

Fig. 5. The interaction of a hypersonic flow with multiple cylindrical mass sources when the combined mass injection rate is comparable to the mass flux of the incident flow through the volume that the mass sources occupy (from Pittard et al. 2005). The right panel shows an enlargement of the flow around two of the mass sources.

$$\frac{1}{r^2}\frac{\partial(r^2\rho u^2)}{\partial r} = Q \tag{11}$$

and

$$\rho v \frac{\partial v}{\partial r} + \frac{a^2}{\gamma}\frac{\partial \rho}{\partial r} + \frac{2\rho a}{\gamma}\frac{\partial a}{\partial r} = -Qv, \tag{12}$$

where r is the distance from the star, and a is the isothermal sound speed of the flow. Q is the mass loaded into the flow per unit time per unit volume. The clumps are assumed, on average, to be stationary with respect to the global flow, so that there is no net rate of momentum injection.

The continuity and momentum equation may be combined to give

$$(v^2 - a^2)\frac{\partial v}{\partial r} = -\frac{Q}{\rho}(v^2 + a^2) + \frac{2a^2 v}{r}. \tag{13}$$

If r is large (i.e. the flow is plane-parallel) and if $v > a$ (i.e. $M > 1$), then $v^2 > a^2$, $\partial v/\partial r < 0$, and the flow decelerates. On the other hand, if r is large and $v < a$ (i.e. $M < 1$), then $v^2 < a^2$, $\partial v/\partial r > 0$, and the flow accelerates. A key feature of mass-loading is that it tends to drive the ambient flow to Mach number unity (Hartquist et al. 1986). For an expanding, time-dependent flow, mass-loading may drive the Mach number to 0.6-0.7 (Arthur, Dyson & Hartquist 1993). Another feature is that if mass-loading is large, a global reverse shock will be greatly weakened (Arthur, Dyson & Hartquist 1994; Williams, Hartquist & Dyson 1995; Williams, Dyson & Hartquist 1999; Pittard, Hartquist & Dyson 2001b). In addition, a two-fluid study has revealed that shocks which overrun clumpy media are weakened and broadened (Williams & Dyson 2002). This has significant implications for the survival of the clumps. Finally, the injection of mass into a radiative medium has a stabilizing influence against isobaric and isentropic perturbations, and can suppress the development of the thermal instability (Pittard et al. 2003).

5 Sources with Large-Scale Mass-Loaded Flows

Theoretical studies of the effect of mass-loading on a flow fall into two categories. If the flow is treated as a single fluid and has a very simple geometry, it is often possible to obtain a similarity solution. Alternatively, one can use a hydrodynamical code to model and find a wider variety of solutions. This allows much greater flexibility, and the ability to incorporate a larger number of physical processes. In principle, a two-fluid approach can be used, in which a time-dependent spectrum of clumps is separately tracked. Of course, an unavoidable drawback of such computations is the loss of generality, as, for instance, the clump spectrum inside a starburst superwind is likely to be somewhat different to that in our local ISM.

In the following subsections I discuss mass-loading in a variety of astrophysical settings. The properties of winds mass-loaded by material from clumps (or stellar sources) has also received theoretical attention in the context of ultracompact H II regions (Dyson, Williams & Redman 1995), Herbig-Haro objects (see discussion in Hartquist & Dyson 1988, 1993), and globular cluster winds (Durisen & Burns 1981). The mass-loading of accretion flows in stellar clusters and AGNs has also been studied by David & Durisen (1989) and Toniazzo, Hartquist & Durisen (2001).

5.1 WBBs, PNe, and Superbubbles

The dramatic and beautiful impact of stellar winds on their environment has long been recognized (e.g., see the chapter by Jane Arthur). The first similarity solutions of this interaction, which assumed a smooth wind and environment, were obtained by Dyson & de Vries (1972) and Dyson (1973). The shocked ambient gas was found to cool very rapidly (as subsequently shown in a numerical simulation by Falle 1975), and the hot shocked wind was found to fill most of the bubble volume. The effect of mass-loading was first considered by Weaver et al. (1977), who obtained an approximate similarity solution for a WBB expanding into a constant density medium by assuming an isobaric shocked wind region. In this work, mass-loading of the interior of the WBB is assumed to occur through the conductively-driven evaporation of the cool swept-up shell. The mass transferred into the bubble interior may easily dominate the total mass within the bubble. Hanami & Sakashita (1987) used this same approach to generalize the Weaver et al. (1977) model to an ambient density with a power-law slope, as well as considering the conductive evaporation of embedded clumps. An analytical solution based on the Weaver et al. (1977) model for WBBs in a density gradient was obtained by García-Segura & Mac Low (1995). Similarity solutions of WBBs mass-loaded by embedded clumps and obtained without the assumption of a constant pressure shocked wind were described by Pittard, Dyson & Hartquist (2001a) and Pittard et al. (2001b). A similarity solution analogous to the Weaver et al. (1977) model, but taking into account the time-dependence of the stellar wind, was obtained by Zhekov & Perinotto (1996), while Zhekov & Myasnikov (1998) derived a similarity solution (based on the Weaver et al. (1977) model) with reduced thermal conduction.

Some of the best studied WBBs are around WR stars. Many of these display a clumpy morphology, and mass-loading is likely to be an important process. Evidence for mass-loading in WBBs was obtained by Smith et al. (1984), who observed a correlation between the velocity and ionization potential of ultraviolet absorption features towards the central star of the RCW 58 WBB. The slope of this relationship effectively measures the ratio of the mass flux to the pressure in the line-forming region, i.e. $\rho v/P$, and the value of this ratio implies that the rate of mass loading into the bubble is 40-50 times greater than the mass-loss rate of the WR star (Hartquist et al. 1986). A standard bubble without mass-loading is unable to explain the wide range of observed velocities, while the Weaver et al. (1977) model predicts a correlation of the opposite sense, irrespective of the assumed ambient density gradient (see Hanami & Sakashita 1987). In contrast, similarity solutions with mass loading from embedded clumps *are* able to reproduce this correlation (Hanami & Sakashita 1987; Pittard et al. 2001a,b), as can hydrodynamical simulations of the same process (Arthur et al. 1993, 1996). Further evidence for mass-loading in wind-blown bubbles is provided by spectroscopic data of core-halo PNe, which indicate halo temperatures in excess of that which can be obtained by photoionization alone (Meaburn et al. 1991). Dyson (1992) and Arthur et al. (1994) have shown that this is consistent with a transonic wind leaving a mass-loaded core region and shocking against clumps in the halo region.

Without some form of mass-loading, the hot gas in WBBs, PNe, and superbubbles may be too rarefied to produce observable X-ray emission. Irrespective of the entrainment process, mass-loading increases the interior density of the bubble while simultaneously reducing its temperature (the same thermal energy must now be shared between more particles), and as a consequence, the X-ray emission both softens and dramatically increases. We are therefore fortunate that the last three decades have been a golden age of X-ray astronomy, with the operation of successive facilities with vastly improved sensitivity, spatial- and energy-resolution. The observed temperatures in WBBs, PNe, and quiescent (i.e. those without recent SNe) superbubbles are all reasonably soft and therefore indicative of mass-loading (though other explanations are also possible). However, applications of the Weaver et al. (1977) and García-Segura & Mac Low (1995) models predict far too much X-ray emission (sometimes by a factor of 100), so at the very least the evaporation of mass from a dense swept-up shell must occur at a much reduced rate.

Only two WBBs have been detected to date at X-ray energies, NGC 6888 (Wrigge 1999) and S308 (Chu et al. 2003), both of which are limb-brightened. At first, this morphology seems to agree with the predictions of the Weaver et al. (1977) and García-Segura & Mac Low (1995) models since the emissivity is highest where the density is highest and the temperature lowest. However, the soft emission from the limb is easily absorbed by the ISM, with the result that such models predict a centre-filled appearance (Wrigge et al. 2005). Again, a reduction in the conductivity can improve the level of agreement. It remains to be seen whether bubbles with distributed mass-loading from embedded clumps can better reproduce the observed limb-brightening, though this process has some support from hydrodynamical simulations which often show instabilities breaking dense clumps off cold swept-up shells and subsequently becoming entrained in the hot bubble interior (e.g., see Fig. 6 in the chapter by Jane Arthur). An alternative explanation is that the X-ray emission from NGC 6888 and S308 arises from the shocked RSG wind, and is currently enhanced in these objects by a collision between the WR and RSG shells (Freyer, Hensler & Yorke 2006). However, this requires that the RSG wind is fast (~ 100 km s^{-1}), and speeds of this order are far from certain (see discussion in García-Segura, Langer & Mac Low 1996). Finally, there is a vast amount of evidence that the stellar winds themselves are clumpy, which has further implications for their interaction (see, e.g., Fig. 6).

Discerning the morphology of PNe is more difficult, as their angular size is smaller. Only Mz 3 and NGC 6543 are adequately resolved, and, like the WBBs around massive stars, are limb-brightened (Chu et al. 2001; Kastner et al. 2003). Unlike the WBBs around massive stars, internal absorption within the nebula is likely to be important, while the action of collimated outflows and heat conduction may also modify the observed X-ray morphology (Kastner et al. 2002). The duration for the presence of hot gas appears to be short, as only young PNe show diffuse X-ray emission.

On larger scales, the combined action of stellar winds and SNe from massive stars in OB associations sweep up the ambient ISM to form superbubbles. The physical structure of a superbubble is expected to be similar to that of a WBB formed around an isolated massive star. Superbubbles containing recent SNe have X-ray luminosities which exceed the predictions from the Weaver et al. (1977) model, but when in a quiescent state (i.e. without recent SN blasts) the X-ray luminosities are an order of magnitude lower than expected from the Weaver et al. (1977) model (Chu et al. 1995). In contrast to

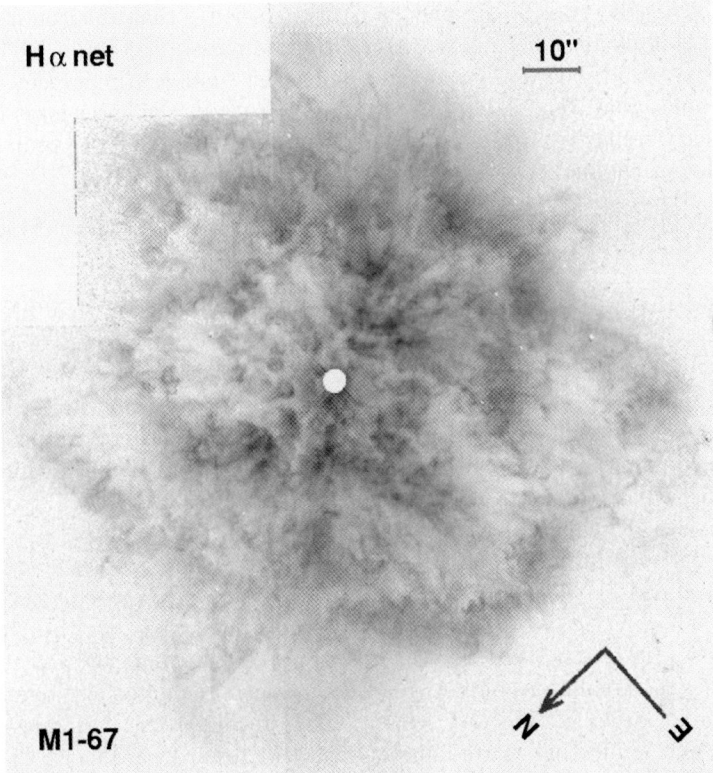

Fig. 6. The M1-67 nebula around WR 124. This is the youngest known WR nebula, and unlike older nebulae there is no obvious global shell surrounding a central cavity. Instead, the interaction between the current wind and its LBV progenitor starts close to the central star and extends to the outer boundary (reproduced from Grosdidier et al. 1998).

WBBs and PNe, quiescent superbubbles have a center-filled X-ray morphology, with brighter X-ray emission near the central star cluster, and hotter gas ($T \sim 10^7$ K). The diffuse X-ray emission from high-mass star-forming regions has also been studied on smaller scales (see, e.g., Townsley et al. 2003). In M 17, it appears that the hot plasma is mass-loaded by a factor of ~ 10 (Dunne et al. 2003). However, other processes, such as particle acceleration, and different electron and ion temperatures may also be important.

An interesting finding for clump-embedded, conductively-driven mass-loading is the occurence of a negative feed-back mechanism, which sets a maximum limit to the amount that a bubble can be mass-loaded (Pittard et al. 2001a). This arises as a consequence of the dependence of the evaporation rate on temperature and the lowering of temperature by mass loading. Since mass-loading to primary stellar wind mass ratios approaching those inferred for RCW 58 could not be obtained, a key conclusion is that ablation from embedded clumps must be the dominant driver of mass-injection in RCW 58 in particular, and perhaps also more generally (this finding is in agreement with the estimate that $\dot{M}_{\rm ab} > \dot{M}_{\rm con}$ in Sect. 2.4). However, the lower mass-loading to primary stellar wind mass ratios deduced for S 308 and M 17 probably do not rule out conductively-driven mass-loading in these sources.

5.2 SNRs and Starburst Superwinds

The fact that the ISM is known to be multi-phase means that SNRs are undoubtedly mass-loaded, and there is clear observational evidence of engulfed clumps within the SNR N63A (Chu et al. 1999; Warren, Hughes & Slane 2003). Young SNRs first interact with the circumstellar material ejected by their progenitor. The youngest SNR in the Galaxy, Cas A, contains bright, slow-moving knots of gas called quasi-stationary flocculi (QSFs), which have been suggested to arise from the circumstellar bubble of the WN progenitor (Chevalier & Kirshner 1978).

Similarity solutions for SNRs mass-loaded from embedded clouds have been obtained by McKee & Ostriker (1977), Chièze & Lazareff (1981), and White & Long (1991) for conductively-driven evaporation, and by Dyson & Hartquist (1987) for ablation-driven injection. A small number of papers based on numerical simulations of mass-loaded supernova remnants also exist in the literature. Cowie, McKee & Ostriker (1981) included the dynamics of the clumps and found that warm clumps are swept towards the shock front and are rapidly destroyed, while cold clumps are more evenly distributed and have longer lifetimes. Arthur & Henney (1996) studied the effects of mass loading by hydrodynamic ablation on supernova remnants evolving inside cavities evacuated by the stellar winds of the progenitor stars.

When SNRs overlap in regions with vigorous star formation, they may create highly pressurized superbubbles that burst out into intergalactic space. In the standard picture of such starburst superwinds, the wind remains subsonic until it reaches the boundary of the starburst region (Chevalier & Clegg 1985). However, the predicted X-ray luminosity of the thermalized SN and stellar wind ejecta is lower than observed unless the superwind is heavily mass-loaded (Suchkov et al. 1996), with the rate of mass-injection from the destruction of clouds several times larger than that due to SNe (see also Hartquist et al. 1997). Recently, high-spatial-resolution observations have provided a much clearer view of the mass-loading process in superwinds. Cool Hα emitting filaments and clumps are seen embedded within the superwind, and the soft X-ray emission appears to be associated with hot gas interacting with these structures (see Strickland et al. 2004, and references therein). Theoretical support comes from hydrodynamic models which show that the superwind sweeps up and incorporates large masses of material from within the galactic disk as it develops (Strickland & Stevens 2000).

Since superwinds are driven by overlapping SNRs, and the range and radiative energy losses of a remnant are affected by mass loading, it is desirable to have approximations which describe remnant evolution and range in clumpy media. The first steps towards this goal were taken by Dyson, Arthur & Hartquist (2002) and Pittard et al. (2003), where hydrodynamical simulations of SNRs undergoing mass-loading driven either by ablation or conduction were calculated. Significant differences between the evolution of the SNRs were discovered, due to the way in which conductive mass loading is extinguished at fairly early times, once the interior temperature of the remnant falls below $\sim 10^7$ K. At late times, remnants that ablatively mass load are dominated by loaded mass and thermal energy, while those that conductively mass load are dominated by swept-up mass and kinetic energy. These works may ultimately be used in superwind models which are akin to the McKee & Ostriker (1977) model of the interstellar medium.

5.3 AGN-SNR Interaction

Many theoretical explanations have been proposed for the origin of the broad emission line regions (BELR) in AGNs (see, e.g., Pittard et al. 2003b, and references therein), including the interaction of an AGN wind with supernovae and star clusters (Perry & Dyson 1985; Williams & Perry 1994). It is now clear that the bulk of the broad-line emission arises from an accretion-disk wind (Proga, Stone & Kallman 2000, see also the chapter by Stuart Lumsden), but interactions between an AGN wind and SNRs may make a non-negligible contribution to the emission of high ionization lines. These interactions will also mass-load the AGN wind (e.g. Smith 1996), and possibly could be used as a diagnostic of it. Hydrodynamical simulations of the early stages of this interaction have been presented by Pittard et al. (2001c) and Pittard et al. (2003b). The strong radiation field means that cool post-shock gas forms only for a very limited time, before being heated back up to the Compton temperature.

5.4 Intracluster Gas

The intracluster medium consists of hot, subsonic gas which is bound within the gravitational potential of the cluster. It was suggested almost 3 decades ago that the density of gas within the central regions is high enough to permit significant cooling within a Hubble time. This energy loss causes outer gas to flow subsonically in to the centre in order to maintain hydrostatic equilibrium, a process referred to as a 'cooling flow'. Recent X-ray observations appear to show a systematic deficit of low temperature emission in comparison to the standard isobaric cooling-flow model, which has led to the questioning of this model. Mass deposition rates significantly smaller than expected are also inferred (see, e.g., Peterson et al. 2003, and references therein). Attention has now focussed on the possibility that the gas is prevented from cooling

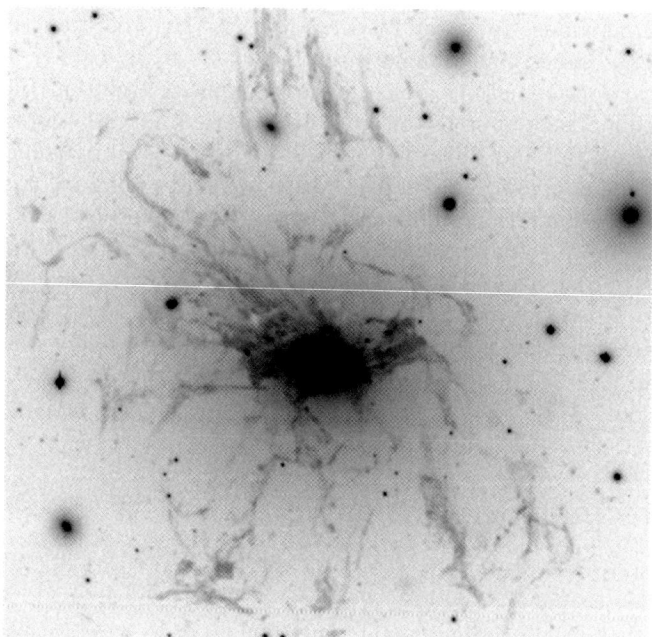

Fig. 7. An Hα image showing the filamentary structure around NGC 1275 (credit C. Conselice (Caltech), WIYN, AURA, NOAO, NSF).

by some compensating form of energy injection. For instance, the frequent occurence of X-ray cavities coincident with radio lobes around the central dominant galaxy demonstrates that AGNs can significantly influence the X-ray morphology of the hot gas (e.g., Fabian et al. 2002). However, the timescale for energy transfer from the relativistic plasma to the surrounding gas is poorly known, while the effectiveness of heat conduction at preventing cooling is enthusiastically debated.

An alternative interpretation is that the cooling plasma radiates its energy in the UV/optical bands rather than at X-ray wavelengths. This can occur if there is mixing between the cooling plasma and colder material (e.g., Fabian et al. 2002), and may be reprocessed into the IR if the gas is dusty. Luminous optical/UV nebulosity is common in cluster cooling flows (e.g., Heckman et al. 1989), and is particularly widespread around NGC 1275 in the Perseus cluster (Conselice, Gallagher & Wyse 2001) (see Fig. 7). While present information is rather sparse, and some of this emission will be powered by star formation, the total submillimetre to UV emission appears to be sufficient to account for the missing soft X-ray emission (Fabian et al. 2002). However, direct evidence for the UV emission resulting from this mixing is rather tentative in current FUSE data (Oegerle et al. 2001).

If cold clouds are embedded within the hot gas, they could be important for global mass injection into the hot surroundings. Indeed, since the intracluster gas is observed to be enriched with metals, material must either be injected by the host galaxies, through fountain flows or superwinds, or ablatively stripped from them. Mass-loading usually reduces the mean temperature of a flow, as the thermal energy is shared between more particles, but in clusters this situation is modified by the gravitational forces on the flow and the significant gravitational potential energy that the entrained material possesses. A preliminary study has shown that under such conditions it is possible to increase the radiated power at high temperatures (Pittard et al. 2004). Hence the differential luminosity distribution can have a positive slope with T, as required for agreement with the latest observations of clusters. In addition, ripples found in the X-ray emission map of the Perseus cluster (Fabian et al. 2003) resemble the structure of some mass-loaded accretion flows. The most likely explanation for their origin may be time variations in the source of outflowing material, but this interaction clearly needs further investigation.

6 Summary

The mass-loading of flows is now recognized as a fundamental process in astronomy, yet much work remains. We still do not have a good understanding of ablatively-driven evaporation or conductively-driven thermal evaporation and the ability of plasma instabilities to quench it. The rapidity with which cold material is mixed into a hotter flow is also unknown. The global properties of a variety of mass-loaded flows have now been studied, but in only a few instances have specific objects been modelled and a direct comparison with observations made.

Acknowledgements

It is a pleasure to thank John Dyson for his encouragement and friendship, and to my other collaborators, Tom Hartquist and Sam Falle, for theirs.

References

Arthur, S.J., Dyson, J.E., Hartquist, T.W. 1993 A&A **261**, 425
Arthur, S.J., Dyson, J.E., Hartquist, T.W. 1994 A&A **269**, 1117
Arthur, S.J., Henney, W.J. 1996 ApJ **457**, 752
Arthur, S.J., Henney, W.J., Dyson, J.E. 1996 A&A **313**, 897
Arthur, S.J., Lizano, S. 1997 ApJ **484**, 810
Balbus, S.A. 1986 ApJ **304**, 787
Balbus, S.A., McKee, C.F. 1982 ApJ **252**, 529

Bally, J., Sutherland, R.S., Devine, D., Johnstone, D. 1998 AJ **116**, 293
Bandiera, R., Chen, Y. 1994a A&A **284**, 629
Bandiera, R., Chen, Y. 1994b A&A **284**, 637
Bertoldi, F. 1989 ApJ **346**, 735
Bertoldi, F. 1990 ApJ **354**, 529
Böhringer, H., Hartquist, T.W. 1987 MNRAS **228**, 915
Boroson, B., McCray, R., Clark, C.O., Slavin, J., Mac Low, M.-M., Chu, Y., Van Buren, D. 1997 ApJ **478**, 638 [Erratum: 1997 ApJ **485**, 436]
Cantó, J., Raga, A.C. 1991 ApJ **372**, 646
Chevalier, R.A., Clegg, A.W. 1985 Nature **317**, 44
Chevalier, R.A., Kirshner, R.P. 1978 ApJ **219**, 931
Chièze, J.P., Lazareff, B. 1981 A&A **95**, 194
Cho, J., Lazarin, A., Honein, A., Knaepen, B., Kassinos, S., Moin, P. 2003 ApJ **589**, L77
Chu, Y.-H. 1982 ApJ **254**, 578
Chu, Y.-H., Chang, H., Su, Y., Mac Low, M.-M. 1995 ApJ **450**, 157
Chu, Y.-H., Guerrero, M.A., Gruendl, R.A., Williams, R.M., Kaler, J.B. 2001 ApJ **553**, L69 [Erratum: 2001 ApJ **554**, 233]
Chu, Y.-H., Guerrero, M.A., Gruendl, R.A., García-Segura, G., Wendker, H.J. 2003 ApJ **599**, 1189
Chu, Y.-H., et al. 1999 New Views of the Magellanic Clouds, Eds. Y.-H. Chu, N. Suntzeff, J. Hesser, D. Bohlender, IAU Symp. **190**, 143
Conselice, C.J., Gallagher, J.S., Wyse, R.F.G. 2001 AJ **122**, 2281
Cowie, L.L., McKee, C.F. 1977 ApJ **211**, 135
Cowie, L.L., McKee, C.F., Ostriker, J.P. 1981 ApJ **247**, 908
Cowie, L.L., Songalia, A. 1977 Nature **266**, 501
David, L.P., Durisen, R.H. 1989 ApJ **346**, 618
Draine, B.T., Giuliani, J.L. 1984 ApJ **281**, 690
Dunne, B.C., et al. 2003 ApJ **590**, 306
Durisen, R.H., Burns, J.O. 1981 MNRAS **195**, 535
Dyson, J.E. 1968 Ap&SS **1**, 388
Dyson, J.E. 1973 A&A **23**, 381
Dyson, J.E. 1992 MNRAS **255**, 460
Dyson, J.E. 1994 in Lecture Notes in Physics **431**, Star Formation Techniques in Infrared and mm-Wave Astronomy, Ed. T.P. Ray & S.V.W. Beckwith (Berlin:Springer), 93
Dyson, J.E., Arthur, S.J., Hartquist, T.W. 2002 A&A **390**, 1063
Dyson, J.E., de Vries, J. 1972 A&A **20**, 223
Dyson, J.E., Hartquist, T.W. 1987 MNRAS **228**, 453
Dyson, J.E., Hartquist, T.W. 1994 MNRAS **269**, 447
Dyson, J.E., Hartquist, T.W., Pettini, M., Smith, L.J. 1989 MNRAS **241**, 625
Dyson, J.E., Williams, R.J.R., Redman, M.P. 1995 MNRAS **237**, 700
Elmegreen, B.G., Lada, C.J. 1977 ApJ **214**, 725
Fabian, A.C., Allen, S.W., Crawford, C.S., Johnstone, R.W., Morris, R.G., Sanders, J.S., Schmidt, R.W. 2002 MNRAS **332**, L50
Fabian, A.C., et al. 2003 MNRAS **344**, L43
Falle, S.A.E.G. 1975 A&A **43**, 323
Falle, S.A.E.G., Coker, R.F., Pittard, J.M., Dyson, J.E., Hartquist, T.W. 2002 MNRAS **329**, 670

Ferrara, A., Shchekinov, Y. 1993 ApJ **417**, 595
Field, G.B. 1965 ApJ **142**, 531
Fragile, P.C., Anninos, P., Gustafson, K., Murray, S.D. 2005 ApJ **619**, 327
Fragile, P.C., Murray, S.D., Anninos, P., Breugel, W.V. 2004 ApJ **604**, 74
Freyer, T., Hensler, G., Yorke, H.W. 2006 ApJ **638**, 262
Galeev, A.A., Natanzon, A.M. 1984 Dokl Phys **275**, 6
Garcíia-Segura, G., Langer, N., Mac Low, M.M. 1996 A&A **316**, 133
Garcíia-Segura, G., Mac Low, M.M. 1995 ApJ **455**, 145
Gary, S.P., Feldman, W.C. 1977 J Geophs Res **82**, 1087
Giuliani, J.L. 1984 ApJ **277**, 605
Grosdidier, Y., Moffat, A.F.J., Joncas, G., Acker, A. 1998 ApJ **506**, L127
Hanami, H., Sakashita, S. 1987 A&A **181**, 343
Hartquist, T.W., Dyson, J.E. 1988 Ap&SS **144**, 615
Hartquist, T.W., Dyson, J.E., Pettini, M., Smith, L.J. 1986 MNRAS **221**, 715
Hartquist, T.W., Dyson, J.E. 1993 QJRAS **34**, 57
Hartquist, T.W., Dyson, J.E., Williams, R.J.R. 1997 ApJ **482**, 182
Heckman, T.M., Baum, S.A., van Breugel, W.J.M., McCarthy, P. 1989 ApJ **338**, 48
Hensler, G., Vieser, W. 2002 Ap&SS **281**, 275
Jun, B.-I., Jones, T.W., Norman, M.L. 1996 ApJ **468**, L59
Kahn, F.D. 1969 Physica **41**, 172
Kastner, J.H., Balick, B., Blackman, E.G., Frank, A., Soker, N., Vrtílek, S.D., Li, J. 2003 ApJ **591**, L37
Kastner, J.H., Li, J., Vrtílek, S.D., Gatley, I., Merrill, K.M., Soker, N. 2002 **581**, 1225
Klein, R.I., Budil, K.S., Perry, T.S., Bach, D.R. 2003 ApJ **583**, 245
Klein, R.I., McKee, C.F., Colella, P. 1994 ApJ **420**, 213
Landau, L.D., Lifshitz, E.M. 1959. Fluid Mechanics, Pergamon Press, Oxford
Lefloch, B., Lazareff, B. 1994 A&A **289**, 559
Levinson, A., Eichler, D. 1992 ApJ **387**, 212
McKee, C.F., Cowie, L.L. 1975 ApJ **195**, 715
McKee, C.F., Cowie, L.L. 1977 ApJ **215**, 213
McKee, C.F., Ostriker, J.P. 1977 ApJ **218**, 148
Marcolini, A., Strickland, D.K., D'Ercole, A., Heckman, T.M., Hoopes, C.G. 2005 MNRAS **362**, 626
Meaburn, J., Boumis, P., López, J.A., Harman, D.J., Bryce, M., Redman, M.P., Mavromatakis, F. 2005 MNRAS **360**, 963
Meaburn, J., Clayton, C.A., Bryce, M., Walsh, J.R. 1996 MNRAS **281**, L57
Meaburn, J., Nicholson, R.A., Bryce, M., Dyson, J.E., Walsh, J.R. 1991 MNRAS **252**, 535
Mellema, G., Kurk, J.D., Röttgering, H.J.A. 2002 A&A **395**, L13
Mellema, G., Raga, A.C., Cantó, J., Lundqvist, P., Balick, B., Steffen, W., Noriega-Crespo, A. 1998 A&A **331**, 335
O'Dell, C.R., Wen, Z., Hu, X. 1993 ApJ **410**, 696
Oegerle, W.R., et al. 2001 ApJ **560**, 187
Orlando, S., Peres, G., Reale, F., Bocchino, F., Rosner, R., Plewa, T., Siegel, A. 2005 A&A **444**, 505
Pavlakis, K.G., Williams, R.J.R., Dyson, J.E., Falle, S.A.E.G., Hartquist, T.W. 2001 A&A **369**, 263
Perry, J.J., Dyson, J.E. 1985 MNRAS **213**, 665

Peterson, J.R., et al. 2003 ApJ **590**, 207
Pittard, J.M., Arthur, S.J., Dyson, J.E., Falle, S.A.E.G., Hartquist, T.W., Knight M.I., Pexton M. 2003 A&A **401**, 1027
Pittard, J.M., Dyson, J.E., Hartquist, T.W. 2001a A&A **367**, 1000
Pittard, J.M., Dyson, J.E., Falle, S.A.E.G., Hartquist, T.W. 2001c A&A **375**, 827
Pittard, J.M., Dyson, J.E., Falle, S.A.E.G., Hartquist, T.W. 2003b A&A **408**, 79
Pittard, J.M., Dyson, J.E., Falle, S.A.E.G., Hartquist, T.W. 2005 MNRAS **361**, 1077
Pittard, J.M., Hartquist, T.W., Dyson, J.E. 2001b, A&A **373**, 1043
Pittard, J.M., Hartquist, T.W., Ashmore, I., Byfield, A., Dyson, J.E., Falle, S.A.E.G. 2004 A&A **414**, 399
Poludnenko, A.Y., Frank, A., Blackman, E.G. 2002 ApJ **576**, 832
Proga, D., Stone, J.M., Kallman, T.R. 2000 ApJ **543**, 686
Rosner, R., Tucker, W.H. 1989 ApJ **338**, 761
Sandford, M.T. II, Whitaker, R.W., Klein, R.I. 1982 ApJ **260**, 183
Serabyn, E., Lacy, J.H., Achtermann, J.M. 1991 ApJ **378**, 557
Smith, L.J., Pettini, M., Dyson, J.E., Hartquist, T.W. 1984 MNRAS **211**, 679
Smith, L.J., Pettini, M., Dyson, J.E., Hartquist, T.W. 1988 MNRAS **234**, 625
Smith, S.J. 1996 ApJ **473**, 773
Spitzer, L. 1962 Physics of Fully Ionized Gases, Interscience, New York
Spitzer, L., Härm, R. 1953 Phys Rev **89**, 977
Steffen, W., López, J.A. 2004 ApJ **612**, 319
Stone, J.M., Norman, M.L. 1992 ApJ **390**, L17
Strickland, D.K., Heckman, T.M., Colbert, E.J.M., Hoopes, C.G., Weaver, K.A. 2004 ApJSS **151**, 193
Strickland, D.K., Stevens, I.R. 2000 MNRAS **314**, 511
Suchkov, A.A., Berman, V.G., Heckman, T.M., Balsara, D.S. 1996 ApJ **463**, 528
Toniazzo, T., Hartquist, T.W., Durisen, R.H. 2001 MNRAS **322**, 149
Townsley, L.K., Feigelson, E.D., Montmerle, T., Broos, P.S., Chu, Y.-H., Garmire, G.P. 2003 ApJ **593**, 874
Vieser, W., Hensler, G. 2000 Ap&SS **272**, 189
Warren, J.S., Hughes, J.P., Slane, P.O. 2003 ApJ **583**, 260
Weaver, R., McCray, R., Castor, J., Shapiro, P., Moore, R. 1977 ApJ **218**, 377
White, R.L., Long, K.S. 1991 ApJ **373**, 543
Williams, R.J.R. 2002 MNRAS **331**, 693
Williams, R.J.R., Dyson, J.E. 2002 MNRAS **333**, 1
Williams, R.J.R., Dyson, J.E., Hartquist, T.W. 1999 A&A **344**, 675
Williams, R.J.R., Dyson, J.E., Hartquist, T.W. 2000 MNRAS **314**, 315
Williams, R.J.R., Hartquist, T.W., Dyson, J.E. 1995 ApJ **446**, 759
Williams, R.J.R., Perry, J.J. 1994 MNRAS **269**, 538
Wrigge, M. 1999 A&A **343**, 599
Wrigge, M., Chu, Y.-H., Magnier, E.A., Wendker, H.J. 2005 ApJ **633**, 248
Xu, J., Stone, J.M. 1995 ApJ **454**, 172
Yusef-Zadeh, F., Melia, F. 1992 ApJ **385**, L41
Yusef-Zadeh, F., Morris, M. 1991 ApJ **371**, L59
Zhekov, S.A., Myasnikov, A.V. 1998 New Ast **3**, 57
Zhekov, S.A., Perinotto, M. 1996 A&A **309**, 648

The Effects of Cosmic Rays on Interstellar Dynamics

T.W. Hartquist[1], A.Y. Wagner[2], and S.A.E.G. Falle[3]

[1] School of Physics and Astronomy, University of Leeds, Leeds LS2 9JT, UK
twh@ast.leeds.ac.uk
[2] School of Physics and Astronomy, University of Leeds, Leeds LS2 9JT, UK
ayw@ast.leeds.ac.uk
[3] Department of Applied Mathematics, University of Leeds, Leeds LS2 9JT, UK
sam@amsta.leeds.ac.uk

1 Introduction

Almost all models of interstellar dynamics have been of a hydrodynamic nature. However, the magnetic and cosmic ray pressures are comparable to the thermal pressure throughout a sizable fraction of the volume of the interstellar medium (e.g. Ferrière 1998). Since the early work by Ipavich (1975) and Jokipii (1976) on models of a cosmic-ray driven Galactic wind, some workers have investigated the effects of cosmic rays on interstellar shocks, supernova remnant evolution, and accretion flows on the scale of a cluster of galaxies as well as revisited the possible importance of cosmic rays for the distribution of gas in the Galactic halo. This chapter provides a brief summary of this sort of work.

Section 2 contains an introduction to the kinetic equation governing the evolution of the cosmic ray distribution function and a description of how that equation was initially applied to the problem of the acceleration of cosmic rays in shocks. Section 3 concerns the treatment of cosmic rays as a fluid and the application of the treatment to the construction of models of acceleration in adiabatic shocks in which the effects of cosmic rays on the thermal fluid flow are included. Section 4 includes brief summaries of work including the effects of cosmic rays on the thermal fluid in adiabatic shocks while retaining a kinetic treatment of the cosmic rays. The suppression of thermal instability in an initially uniform medium and of thermally induced overstability in radiative shocks by cosmic rays is the focus of section 5. The effects of cosmic rays on supernova remnant evolution is the subject of section 6. Cosmic ray support of gas in the Galactic halo, cosmic-ray driven winds, and the relevance of cosmic rays for the Parker instability are the topics of section 7. The possible role of cosmic rays in accretion flows in clusters of galaxies is addressed in section 8.

2 The Cosmic Ray Transport Equation and Its Application to Acceleration in Shocks

Blandford & Eichler (1987) have given one of the clearer derivations of the cosmic ray transport equation. In this article, we primarily consider plane-parallel problems in which the only spatial dependence is on z and the large scale velocity is $u\hat{z}$. We assume that the large-scale magnetic field is in the z direction and that the only waves present are small amplitude Alfvén waves propagating along the z axis. Then the transport equation governing the isotropic part of the distribution function, f, is

$$\frac{\partial f}{\partial t} + \frac{\partial}{\partial z}\left((u+v_A)f\right) - \frac{\partial}{\partial z}\left(\kappa\frac{\partial f}{\partial z}\right) = \frac{1}{3p^2}\left(\frac{\partial}{\partial z}(u+v_A)\right)\frac{\partial(p^3 f)}{\partial p}. \quad (1)$$

v_A is the Alfvén speed and equation (1) is appropriate only if backward propagating waves are absent. p is the magnitude of the particle momentum. κ is the coefficient for diffusion in the z direction. For relativistic cosmic rays,

$$\kappa \approx \frac{1}{3}\left(\frac{B_0}{\delta B}\right)^2 cr_g, \quad (2)$$

where B_0 is the strength of the large scale field, c is the speed of light, and r_g is the gyroradius.

$$\frac{(\delta B)^2}{8\pi} \approx k_z P(k_z), \quad (3)$$

where k_z is the wavenumber and $P(k_z)$ is the wave power spectrum.

The presence of waves moving in both the backwards and forwards direction would give rise to additional terms. The inclusion of waves traveling in both directions would be necessary for the study of second-order Fermi acceleration, a process which we ignore here.

First-order Fermi acceleration in strong adiabatic shocks was addressed independently by Krymsky (1977), Axford, Leer, & Skadron (1978), Bell (1978a,b) and Blandford & Ostriker (1978). Assume that cosmic rays do not modify the flow of thermal gas and that the flow is steady with

$$u = u_s \qquad z < 0 \quad (4a)$$
$$u = u_t \qquad z > 0, \quad (4b)$$

where u_s and u_t are constant. Take $u \gg v_A$. Thus, to a good approximation equation (1) becomes

$$\frac{\partial}{\partial z}\left(uf - \kappa\frac{\partial f}{\partial z}\right) = \frac{\partial u}{\partial z}\frac{1}{3p^2}\frac{\partial(p^3 f)}{\partial p}, \quad (5)$$

which can be rearranged to give

$$\frac{\partial}{\partial z}\left(\kappa\frac{\partial f}{\partial z} + \frac{up}{3}\frac{\partial f}{\partial p}\right) = u\frac{\partial}{\partial z}\frac{\partial(p^3 f)}{\partial p^3}. \tag{6}$$

Equation (5) must be solved for $z < 0$ and for $z > 0$. f must be continuous at $z = 0$. The continuity of f at $z = 0$ together with equation (6) implies that $\kappa(\partial f/\partial z) + \frac{up}{3}(\partial f/\partial p)$ is also continuous at $z = 0$. The solution is

$$f(z,p) = f_-(p) + [f_+(p) - f_-(p)]\exp\left\{\int_0^z \frac{u_s}{\kappa}dz\right\} \quad z < 0 \tag{7a}$$

$$= f_+ \quad z > 0, \tag{7b}$$

where

$$f_+(p) = qp^{-q}\int_0^p f_-(p')p'^{q-1}dp', \tag{8}$$

with

$$q = \frac{3u_s}{u_s - u_t}. \tag{9}$$

For a $\gamma = 5/3$ thermal gas, $q = 4$.

Thus, the theory based on a neglect of cosmic-ray modification of the shock predicts that the postshock spectrum of cosmic rays will be a power law in momentum.

3 Fluid Treatment of Cosmic-Ray Modified Adiabatic Shocks

Drury & Völk (1981) and Völk, Drury, & McKenzie (1984) used a fluid treatment of cosmic rays to study their effects on adiabatic shock structures.

Multiplication of equation (1) by $(4\pi p^4/3\gamma m)dp$ and integration over the range $p = 0$ to $p = \infty$ yields

$$\frac{\partial P_C}{\partial t} + \frac{\partial F_C}{\partial z} = (u - v_A)\frac{\partial P_C}{\partial z}. \tag{10}$$

γ is the relativistic factor $(1 - \beta^2)^{-\frac{1}{2}}$ where β is the particle speed divided by c, and m is the mass of a cosmic ray particle. We have assumed that the Alfvén waves are propagating in the $-\hat{z}$ direction in the frame moving with the bulk flow of the thermal fluid.

$$P_C = \frac{4\pi}{3}\int_0^\infty \frac{p^4}{\gamma m}f dp. \tag{11}$$

is the cosmic ray pressure, and

$$F_C = \frac{\gamma_C}{\gamma_C - 1}(u - v_A)P_C - \frac{\bar{\kappa}}{\gamma_C - 1}\frac{\partial P_C}{\partial z}. \tag{12}$$

$\gamma_C = 4/3$ for very relativistic particles.

$$\bar{\kappa} = \frac{\int_0^\infty \frac{p^4}{\gamma m} \kappa \frac{\partial f}{\partial z}}{\int_0^\infty \frac{p^4}{\gamma m} \frac{\partial f}{\partial z}} \tag{13}$$

κ ordinarily depends on p, but $\bar{\kappa}$ clearly does not.

$-u(\partial P_C/\partial z)$ is the rate per unit volume per unit time at which energy is transferred to the thermal fluid from the cosmic rays. $v_A(\partial P_C/\partial z)$ is the rate per unit volume per unit time at which energy is input into the waves from cosmic rays to cause their growth in amplitude or compensate for at least some of their dissipation. In principle, one can examine a model in which the wave energy density is generated by the above term and dissipated by a set of specified processes; a relationship between $\bar{\kappa}$ and the wave energy density might also be assumed. In practice, Drury & Völk (1981) neglected the $v_A(\partial P_C/\partial z)$ term, as they considered hyper-Alfvénic flow, and took $\bar{\kappa}$ to be constant. Völk et al. (1984) assumed $\bar{\kappa}$ to be constant and prompt dissipation of the waves and included the term as a heat source term in the energy equation of the thermal gas.

That is, under the assumptions of Völk et al. (1984), the energy equation for a $\gamma = 5/3$ thermal gas is

$$\frac{\partial}{\partial t}\left(\frac{1}{2}\rho u^2 + \frac{3}{2}P\right) + \frac{\partial}{\partial z}\left(\frac{1}{2}\rho u^3 + \frac{5}{2}uP\right) = -(u - v_A)\frac{\partial P_C}{\partial z}. \tag{14}$$

ρ and P are the mass density and pressure of the thermal gas. The continuity and momentum equations for the thermal gas are

$$\frac{\partial \rho}{\partial t} + \frac{\partial (\rho u)}{\partial z} = 0 \tag{15}$$

$$\frac{\partial \rho u}{\partial t} + \frac{\partial}{\partial z}\left(\rho u^2 + P + P_C\right) = 0 \tag{16}$$

Equations (10) through (16) constitute a closed set which can be solved to obtain insight into cosmic-ray modified flows parallel to the large-scale magnetic field.

Drury & Völk (1981) and Völk et al. (1984) used the two-fluid equations given above to investigate cosmic-ray modified, adiabatic shocks. As mentioned above, Drury & Völk (1981) neglected the $v_A(\partial P_C/\partial z)$ term appearing in equations (10) and (14). They found that for distant upstream values of $M = [\rho u^2/(\gamma_G P + \gamma_C P_C)]^{\frac{1}{2}}$ of less than about 6, the distant downstream cosmic ray pressure is a monotonic function of the assumed distant upstream cosmic ray pressure. For distant upstream values of M of about 6 and greater, multiple solutions of the steady shock equations exist for distant upstream values of the cosmic ray pressure that are small compared to the distant upstream ram pressure. For one of these solutions, the distant downstream cosmic ray pressure is a substantial fraction of the distant upstream ram pressure, no matter how small the distant upstream cosmic ray pressure is. The stability of the various solutions has been studied (e.g. Mond & O'C. Drury 1998). The general conclusion has been that for a distant upstream value $M \gtrsim 6$, the distant downstream cosmic ray pressure is at least comparable to and usually greatly exceeds the distant downstream thermal pressure in adiabatic shocks, independent of the value of the distant upstream cosmic ray pressure. Cosmic ray acceleration in existing strong steady adiabatic shocks models is almost certainly more efficient than in many real interstellar shocks.

In cases in which the distant downstream cosmic ray pressure is a sizable fraction of the distant upstream pressure, the contrast between distant downstream and distant upstream densities in the thermal fluid is closer to 7 than to 4, as would occur in a strong adiabatic shock with $\gamma = 5/3$ if cosmic rays did not modify it. Also in such cases, a cosmic ray precursor through which thermal gas is gradually warmed and decelerated is present. A subshock in the thermal fluid is sometimes present.

Völk et al. (1984) investigated the extent to which the inclusion of the $v_A(\partial P_C/\partial z)$ term would lower the efficiency of cosmic ray acceleration. Obviously the term's effect is greatest for weaker shocks.

4 Kinetic Treatment of Cosmic-Ray Modified Adiabatic Shock

A number of authors (e.g. Bell 1987; Falle & Giddings 1987; Jones & Kang 2005) have used finite difference methods to solve the coupled equations (1) (including a source term), (11), (14), (15), and (16) with $v_A = 0$. These authors have considered time-dependent, as well as steady, solutions. The main conclusions drawn by Falle & Giddings (1987) are: a) for very subrelativistic steady shocks and $M \geq 4$ in the distant upstream flow, the distant downstream pressure is independent of the cosmic ray injection rate provided that the injection rate is not too small; b) in some time-dependent solutions for parameters for which the steady state solutions would give a divergent cosmic ray pressure, the structures of the postshock thermal fluids' states are quasi-steady, even though the cosmic ray spectra are evolving; c) for high enough values of M, sufficiently small values of γ_C, and sufficient particle injection in the past, there are cosmic-ray dominated shocks even in the absence of current sources of cosmic rays.

Another approach to the kinetic treatment involves the use of Monte Carlo methods (e.g. Ellison & Eichler 1984; Ellison, Drury, & Meyer 1997). In the work of Ellison & Eichler (1984), a distribution of scattering centres, moving with the thermal fluid was assumed for particles of each momentum, and the trajectories of roughly 10^5 particles of different momenta were calculated. At each point in the shock the cosmic ray pressure was calculated from the momenta of the particles, and the cosmic ray pressure was employed in the fluid equations governing the flow of the thermal fluid. Iteration was continued until convergence to a steady solution was obtained. Comparisons between some results of Monte Carlo simulations and suitably chosen simulations involving the use of finite difference methods to solve the cosmic ray transport equation have shown good agreement (Ellison, Berezhko, & Baring 2000)

5 The Modification of the Thermal Instability and Radiative Shocks by Cosmic Rays

We take $\rho \mathcal{L}(\rho, T)$ to be the difference between the heating and cooling rates per unit volume per unit time due to radiative processes. This term is often included on the right hand side of (14), and in this section, we will assume that it is.

To study the thermal stability of a static uniform medium, we take the background values of ρ, P, P_C and T to be ρ_0, P_0, P_{C0} and T_0. A small amplitude perturbation with associated values of density, velocity, thermal pressure and cosmic ray pressure equal to $\rho_1, u\hat{z}$ and P_{C1} is superposed. If we take v_{A0} (the value of the Alfvén speed in the background medium) to be negligible, four of the equations governing the evolution of the small amplitude quantities are found from (10) through (16) to be

$$\frac{\partial P_{C1}}{\partial t} + \gamma_C P_{C0} \frac{\partial u}{\partial z} - \bar{\kappa} \frac{\partial^2 P_{C1}}{\partial z^2} = 0 \tag{17}$$

$$\frac{\partial}{\partial t}\left(\frac{3}{2}P_1\right) + \frac{5}{2}P_0 \frac{\partial u}{\partial z} = \rho_1 \mathcal{L}(\rho_0, T_0) + \rho_0 \left.\frac{\partial \mathcal{L}}{\partial \rho}\right|_{\rho=\rho_0} \rho_1 + \rho_0 \left.\frac{\partial \mathcal{L}}{\partial T}\right|_{T=T_0} T_1 \tag{18}$$

$$\frac{\partial \rho_1}{\partial t} + \rho_0 \frac{\partial u}{\partial z} = 0 \tag{19}$$

$$\rho_0 \frac{\partial u}{\partial t} + \frac{\partial}{\partial z}(P_1 + P_{C1}) = 0 . \tag{20}$$

The fifth is

$$P_1 = \frac{k_B}{\mu}\rho_0 T_1 + \frac{k_B}{\mu}T_0 \rho_1 , \tag{21}$$

where k_B is Boltzmann's constant and μ is the mean mass per particle.

Wagner et al. (2005) showed that the assumption that all small quantities vary as $\exp(i(\omega t+kz))$ together with the above equations leads to the following dispersion relation

$$\zeta^4 - i\zeta^3\left(\frac{k}{k_C} + \frac{k_T}{k}\right) - \zeta^2\left(\frac{k_T}{k_C} + \phi + 1\right)$$
$$+ i\zeta\left(\frac{k_T}{k} + \frac{k}{k_C} + \frac{3(k_T - k_\rho)}{5}\right) + \frac{3(k_T - k_\rho)}{5k_C} = 0 , \tag{22}$$

where

$$\zeta \equiv \frac{\omega}{ak} \quad (23)$$

$$a \equiv \left(\frac{5}{3}\frac{P_0}{\rho_0}\right)^{\frac{1}{2}} \quad (24)$$

$$k_T \equiv \frac{2\mu}{3k_B a}\left.\frac{\partial \mathcal{L}}{\partial T}\right|_{T=T_0} \quad (25)$$

$$k_\rho \equiv \frac{2\mu\rho_0}{3k_B a T_0}\left.\frac{\partial \mathcal{L}}{\partial T}\right|_{T=T_0} \quad (26)$$

$$k_C \equiv \frac{a}{\kappa} \quad (27)$$

$$\phi \equiv \frac{3\gamma_C P_{C0}}{5P_0}. \quad (28)$$

The dispersion relation in the absence of cosmic rays is a cubic having two roots corresponding to radiatively modified sound waves and one corresponding to an isobaric perturbation. Obviously, the inclusion of cosmic rays gives rise to the existence of an additional mode associated with compression of the cosmic rays. Wagner et al. (2005) concluded that stability obtains when any of the following conditions is met.

$$\frac{k_T - k_\rho}{\gamma_g k_C} > 0 \quad (29)$$

$$\frac{k_T \phi + \frac{3}{5}(k_T - k_\rho) + \frac{k^2}{k_C}}{k_T + \frac{k^2}{k_C}} > \frac{1}{2}\left(\phi + \frac{k_T}{k_C} + 1\right)$$
$$- \sqrt{\frac{1}{4}\left(\phi + \frac{k_T}{k_C} + 1\right)^2 - \frac{3k_T - k_\rho}{5k_C}} \quad (30)$$

$$\frac{k_T \phi + \frac{3}{5}(k_T - k_\rho) + \frac{k^2}{k_C}}{k_T + \frac{k^2}{k_C}} < \frac{1}{2}\left(\phi + \frac{k_T}{k_C} + 1\right)$$
$$+ \sqrt{\frac{1}{4}\left(\phi + \frac{k_T}{k_C} + 1\right)^2 - \frac{3k_T - k_\rho}{5k_C}}. \quad (31)$$

For the case in which cosmic rays are not included, stability obtains if

$$k_T - k_\rho > 0 \quad (32)$$

(Field 1965).

Wagner et al. (2005) examined the stability criteria (29), (30), and (31) in various limiting cases. For $\phi \gg 1$, $\left|\frac{k_T}{k_C}\right|$, $\left|\frac{k_\rho}{k_C}\right|$, $\left|\frac{k}{k_C}\right|$, conditions (30) and (31) give

$$0 \lesssim \frac{k_T}{k_T + \frac{k^2}{k_c}} \lesssim 1 \quad (33)$$

for stability. k_C is always positive, and, consequently, (33) implies that when the cosmic ray pressure is very high and the perturbation is on a scale such that $k^2 << |k_C k_T|$, sound waves will never grow no matter what $\mathcal{L}$ is.

For certain cooling laws and Mach numbers, radiative shocks that are not modified by cosmic rays are overstable (e.g. Langer, Chanmugam, & Shaviv 1981; Pittard, Dobson, Durisen, Dyson, Hartquist, & O'Brien 2005). An overstable shock's speed oscillates. Wagner et al. (2006) investigated the effects of the cosmic ray modification of radiative shocks in the anticipation that the overstability of some radiative shocks is damped. They numerically solved equations (10), (12), (14) with the $\rho \mathcal{L}(\rho, T)$ term added to the right hand side, (15), and (16) and

$$P = \frac{k_B}{\mu} \rho T \qquad (34)$$

They took $v_A = 0$ in (10), (12) and (14) and, thus, focused on flows that are hyper-Alfvénic everywhere.

Wagner et al. (2006) found that the overstability was damped for all power law cooling functions for all values of the distant upstream Mach number (defined in this case as $(3\rho u^2/5P)^{1/2}$) greater than 3. The damping is a consequence of the distant downstream cosmic ray pressure being larger than the distant downstream thermal pressure no matter what the value of the distant upstream cosmic ray pressure is in these radiative shocks. As we mentioned in section 3, the distant downstream cosmic ray pressure in steady adiabatic cosmic-ray modified shocks is large compared to the distant downstream pressure in all cases in which the shock Mach number exceeds about 6. Very efficient cosmic ray acceleration is found in the radiative shock models at even lower Mach numbers.

The efficiency of the cosmic ray acceleration is probably limited by a mechanism or mechanisms not included in the current models. For instance, Drury & Falle (1986) have identified an instability that will certainly lead to the transfer of energy from the cosmic rays to the thermal fluid and, thus, limit the efficiency of cosmic ray acceleration. We are currently investigating models that incorporate this.

6 The Modification of Supernova Remnants by Cosmic Rays

In many circumstances, a plane-parallel model provides an adequate description of a radiative shock, because the cooling and cosmic ray diffusion lengthscales are small compared to the shock's radius of curvature. Thus, often plane-parallel models are used to describe radiative shocks driven by supernova remnants (see the chapter by John Raymond). They are also frequently used in the study of thin regions of adiabatic shocks driven by supernova remnants, such as regions in which the collisionally induced production of particular ions occurs (again, see the chapter by John Raymond). However, plane-parallel models are not appropriate for the study of the global evolution of remnants.

Chevalier (1983) found similarity solutions for a somewhat restricted type of two-fluid model of cosmic ray modified adiabatic spherical blast wave. He assumed $\gamma_C = 4/3$ and that $\bar{\kappa} = 0$, and that the ratio of the cosmic ray pressure to the total pressure at the shock front, w, is independent of time. The similarity solutions for different values of w differ from one another. For $w = 0$ and $w = 1$, the solutions correspond to Sedov (1959) solutions for adiabatic indices of 5/3 and 4/3 respectively.

Other authors (e.g. Dorfi 1994, and references therein) have numerically integrated the coupled partial differential equations of a two-fluid model for cosmic ray modified spherical blast waves for less restrictive assumptions than Chevalier (1983). Dorfi (1994) presented results for two cases. In each, the supernova energy, ejected mass, ambient number density of thermal particles, and ambient temperature are 10^{51}ergs, $5\,\mathrm{M}_\odot$, $0.3\mathrm{cm}^{-3}$, and 8000 K, respectively. In one case, the fraction of the kinetic energy of the material approaching the shock (as measured in the shock frame) converted to cosmic ray energy at the shock was assumed to be small, while in the other it was taken to be 10^{-3}. The heating of gas due to the dissipation of Alfvén waves is included in the latter of the models but not in the former. Radiative cooling became important in the second model before it did in the first. In both models, cosmic rays limit the importance of "catastrophic cooling" and the associated development of secondary shocks like those found by Falle (1975) in his investigation of radiative spherical remnants that are not modified by cosmic rays. At an age of 6×10^5 years, only about one percent of the remnant energy is in cosmic rays for the first model; for the second model the relevant number is about thirty percent.

Drury, Markiewicz, & Völk (1989) and Markiewicz, Drury, & Völk (1990) derived a set of ordinary differential equations from the full two-fluid system of partial differential equations. The ordinary differential equations govern various features of a spherical blast wave modified by cosmic rays including the outer radius of the shock precursor (arising from a finite value of $\bar{\kappa}$) and the radius of the subshock in the thermal fluid. The use of these equations allows the exploration of many cases for a much more modest computational investment than the use of partial differential equations. Kang & Drury (1992) have resolved the discrepancies between results obtained with the simplified method and those obtained by Jones & Kang (1990) who numerically solved the partial differential equations.

7 Cosmic Rays and Gas in the Galactic Halo

As mentioned in section 1, Ipavich (1975) and Jokipii (1976) did early work on cosmic-ray driven galactic winds. Chevalier & Fransson (1984) and Hartquist,

Pettini, & Tallant (1984) were the first to invoke cosmic ray support of gas in order to explain observations of the thermal gas in the halo of the Milky Way. Ultraviolet absorption data for C VI and Si VI obtained against bright halo stars seemed to indicate that gas in which such features formed is not very hot ($T \leq 10^5$K) and dynamically rather quiescent (Pettini & West 1982). Consequently, Chevalier & Fransson (1984) and Hartquist & Morfill (1986) developed two-fluid static plane-parallel models of thermal gas supported by cosmic ray streaming in the direction opposite to that of the gravitational field. Ultraviolet absorption data obtained subsequently indicated the presence of hot gas in the halo and show that the regions in which the C IV and Si IV features form are generally not as dynamically quiescent (e.g. Indebetouw & Shull 2004, references therein) as the Pettini & West (1982) work led theorists to believe.

More recent work on cosmic-ray driven winds includes that of Breitschwerdt, McKenzie, & Völk (1991, 1993) and Zirakashvili et al. (1996). Breitschwerdt et al. (1991, 1993) neglected the rotation of the Galaxy but adopted a realistic dependence for the gravitational field as a function of distance from the plane of the Galaxy. Zirakashvili et al. (1996) included Galactic rotation in a Weber & Davis (1967) type model.

Breitschwerdt et al. (1991) set $\bar{\kappa} = 0$ and assumed that the thermal gas behaves adiabatically in the construction of their "reference model". The wave pressure was included in the equation of motion for the thermal gas, and an equation governing the wave pressure was included. The wave pressure equation is a flux conservation equation with a source term due to cosmic ray streaming and, in principle, a sink term arising from wave damping, but for their "reference model", Breitschwerdt et al. (1991) took the sink term to be zero. They assumed the flow to diverge as $(1 + (z/z_0)^2)^{-1}$ where $z_0 = 15$ kpc, and z is the height above a point in the disc 10 kpc from the Galactic centre. In the reference model, which is of a steady outflow, the cosmic ray and thermal pressures are comparable at $z = 10$ kpc, where the wave pressure is a factor of a few lower than each of the other pressures and is about 2×10^{-14} erg cm^{-3}. At $z = 200$ kpc, each of the wave and cosmic ray pressures is about 2×10^{-17} erg cm^{-3}, but the thermal pressure is two orders of magnitude lower. The terminal speed is about 300 km s^{-1}.

Breitschwerdt et al. (1993) focused on $z < 1$ kpc, where cosmic ray diffusion cannot be neglected and mass loading of the wind by supernova remnants occurs. They also examined the stability of static cosmic-ray supported halo gas models and found appropriate criteria.

The inclusion of rotation in their dynamical models of the Galactic wind allowed Zirakashvili et al. (1996) to estimate the rate at which the wind and the magnetic field carry away angular momentum. They concluded that a moderate fraction of the Galaxy's angular momentum, including that associated with its stellar component, may have been transported away by the wind and magnetic field.

Alan Watson has noted in section 6 of his chapter that an upper bound to the energy to which a cosmic ray can be accelerated in a shock is proportional to the shock's radius of curvature and to the magnetic field strength. This conclusion is based on a comparison of the acceleration time, of order κ/u_s^2, to the age of the object, of the order R/u_s, where R is the size, and the assumption that κ is of order cr_g. It implies that supernova remnants are too small to produce the highest energy cosmic rays. Jokipii & Morfill (1985) noted that cosmic rays can be accelerated to much higher energies at the termination shock of the Galactic wind than in supernovae, but as is clear from Alan's discussion the origin of the highest energy cosmic rays is still unknown.

As mentioned by Hartquist (1990) the growth of the Parker (1966) instability might be affected if the source of cosmic rays were included. Parker (1966) took a plane stratified medium supported in part by a magnetic field that is perpendicular to the gravitational field as an initial state. The pressures of the thermal gas and cosmic rays contribute to the support. He included no source term for cosmic rays; i.e. their number is constant in his model. The initial state is unstable, resulting in gas flowing down magnetic field lines as they increasingly bow out of the disc. Sources of cosmic rays would clearly hinder the infall of the gas.

8 Accretion in Clusters of Galaxies

The intracluster gas in clusters of galaxies emits X-rays. In the 'cooling flow'model of such gas, plasma from the hotter regions flows subsonically, giving rise to a quasi-hydrostatic density distribution, to cooler regions (e.g. Sarazin 1988). High resolution spectroscopic observations show less emission from lower temperature plasma than expected from the "cooling flow" model (e.g. Peterson et al. 2003). A heating mechanism not included in the cooling flow model may be operating.

In a static atmosphere supported by cosmic rays the dissipation of waves excited by cosmic ray streaming provides a heating rate per unit volume per unit time of $\rho v_A g$ where g is the strength of the gravitational field (Hartquist & Morfill 1986). Böhringer & Morfill (1988) suggested that the cosmic ray modification of cluster cooling flows could be significant. However, Colafrancesco, Dar, & De Rújula (2004) have argued that the "traditional" theory of cosmic ray acceleration in supernova remnants implies that too little cosmic ray energy is produced for the cosmic rays to affect cluster flows substantially. They have also claimed that if cosmic ray acceleration takes place close to the central regions of galaxies, as would be expected for the "traditional" theory, cosmic rays would not deposit energy as uniformly throughout a cluster as is required.

Colafrancesco et al. (2004) invoked cosmic ray acceleration throughout intracluster space due to the ejection of cannon balls from supernovae. These authors claimed that such cannon balls, which have relativistic γs of the order of 10^3, are responsible for the large recoil velocities of some neutron stars and will appear as gamma ray bursts if travelling nearly parallel to the line of sight towards the observer. They also sugested that they carry an order of magnitude more energy than is input into cosmic rays by the "traditional" method of cosmic ray acceleration in supernova remants. They have argued, somewhat unconvincingly, that the energy deposition rate per unit volume per unit time by cosmic rays accelerated by the passage of cannon balls should vary as ρ^2, just as the cooling rate per unit time per unit volume does.

Mass, momentum, and energy exchange between outflowing material from the central galaxy of a cluster and infalling matter may give rise to a multiple shock structure in the intracluster medium even if the flow is steady (Pittard et al. 2004, see also section 5.4 of the chapter by Julian Pittard). Each shock will be the source of cosmic ray acceleration. Consequently, it is possible for dynamicaly and thermally important cosmic rays to be produced in the intracluster medium, even if cannon balls are not relevant.

Thus, cosmic rays may play roles in the largest scale flows of the current epoch usually addressed by astrophysicists.

References

Axford, W.I., Leer, E., Skadron, G. 1978 15th International Cosmic Ray Conference, Plovdiv, 1977, **11**, 132–137
Bell, A.R. 1978a MNRAS **182**, 147
Bell, A.R. 1978b MNRAS **182**, 443
Bell, A.R. 1987, MNRAS **225**, 615
Blandford, R., Eichler, D. 1987 PhR **154**, 1
Blandford, R.D., Ostriker, J.P. 1978 ApJL **221**, L29
Böhringer, H., Morfill, G.E. 1988 ApJ **330**, 609
Breitschwerdt, D., McKenzie, J.F., Völk, H.J. 1991, A&A **245**, 79
Breitschwerdt, D., McKenzie, J.F., Völk, H.J. 1993 A&A **269**, 54
Chevalier, R.A. 1983 ApJ **272**, 765
Chevalier, R.A., Fransson, C. 1984 ApJL **279**, L43
Colafrancesco, S., Dar, A., De Rújula, A. 2004 A&A **413**, 441
Dorfi, E.A. 1994 ApJS **90**, 841
Drury, L.O'C., Falle, S.A.E.G. 1986 MNRAS **223**, 353
Drury, L.O'C., Markiewicz, W.J., Völk, H.J. 1989 A&A **225**, 179
Drury, L.O'C., Völk, J.H. 1981 ApJ **248**, 344
Ellison, D.C., Berezhko, E.G., Baring, M.G. 2000 ApJ **540**, 292
Ellison, D.C., Drury, L.O'C., Meyer, J.-P. 1997 ApJ **487**, 197
Ellison, D.C., Eichler, D. 1984 ApJ **286**, 691
Falle, S.A.E.G. 1975 MNRAS **172**, 55
Falle, S.A.E.G., Giddings, J.R. 1987 MNRAS **225**, 399
Ferrière, K. 1998 ApJ **497**, 759
Field, G.B. 1965 ApJ **142**, 531

Hartquist, T.W. 1990 in The Evolution of the Interstellar Medium, ed. L. Blitz, ASP Conf. Series **12**, 99
Hartquist, T.W., Morfill, G.E. 1986 ApJ **311**, 518
Hartquist, T.W., Pettini, M., Tallant, A. 1984 ApJ **276**, 519
Indebetouw, R., Shull, J.M. 2004 ApJ **607**, 309
Ipavich, F.M. 1975 ApJ **196**, 107
Jokipii, J.R. 1976 ApJ **208**, 900
Jokipii, J.R., Morfill, G.E. 1985 ApJL **290**, L1
Jones, T.W., Kang, H. 1990, ApJ **363**, 499
Jones, T.W., Kang, H. 2005 Astroparticle Physics **24**, 75
Kang, H., Drury, L.O'C. 1992 ApJ **399**, 182
Krymsky, G.F. 1977 Akademiia Nauk SSSR Doklady **234**, 1306
Langer, S.H., Chanmugam, G., Shaviv, G. 1981 ApJL **245**, L23
Markiewicz, W.J., Drury, L.O'C, Völk, H.J. 1990 A&A **236**, 487
Mond, M., Drury, L.O'C. 1998 A&A **332**, 385
Parker, E.N. 1966 ApJ **145**, 811
Peterson, J.R., Kahn, S.M., Paerels, F.B.S., et al. 2003 ApJ **590**, 207
Pettini, M., West, K.A. 1982 ApJ **260**, 561
Pittard, J.M., Dobson, M.S., Durisen, R.H., et al. 2005 A&A **438**, 11
Pittard, J.M., Hartquist, T.W., Ashmore, I., et al. 2004 A&A **414**, 399
Sarazin, C.L. 1988 X-ray Emission from Clusters of Galaxies (Cambridge: Cambridge University Press, 1988)
Sedov, L.I. 1959 Similarity and Dimensional Methods in Mechanics (New York: Academic Press, 1959)
Völk, H.J., Drury, L.O'C., McKenzie, J.F. 1984 A&A **130**, 19
Wagner, A.Y., Falle, S.A.E.G., Hartquist, T.W., Pittard, J.M. 2005 A&A **430**, 567
Wagner, A.Y., Falle, S.A.E.G., Hartquist, T.W., Pittard, J.M. 2006, A&A, in press
Weber, E.J., Davis, L.J. 1967 ApJ **148**, 217
Zirakashvili, V.N., Breitschwerdt, D., Ptuskin, V.S., Völk, H.J. 1996 A&A **311**, 113

The Status of Observations and Speculations Concerning Ultra High-Energy Cosmic Rays

A.A. Watson

School of Physics and Astronomy, University of Leeds, Leeds LS2 9JT, UK
a.a.watson@leeds.ac.uk

1 Introduction

For the purposes of this chapter I have defined Ultra High-Energy Cosmic Rays (UHECRs) as those cosmic particles or photons that have energies above 3×10^{18} eV. This limit is chosen as there has been much speculation (and even much belief) that at higher energies the majority of cosmic rays have their origins outside our galaxy. Although there is a lack of firm observational evidence to support such a hypothesis, it is perfectly possible that either of the exotic regions highlighted in the title of this book, 'Star Forming Regions and Active Galaxies', could be the sources. Below, after a brief historical review, I will describe the present status of observations of particles above 3×10^{18} eV and discuss some of the ideas that have been advanced to explain the observations.

There is great uncertainty about the properties of the highest energy cosmic rays. The major questions concerning possible anisotropies in the distribution of arrival directions, the mass composition and the long-sought steepening in the energy spectrum near 10^{20} eV (the Greisen-Zatsepin-Kuz'min effect) remain unanswered: but there has been significant progress and there is promise of a rapid increase in our knowledge over the next 10 years. In this article I will outline how UHECRs are detected and review the observations of the energy spectrum, the mass composition and distribution of arrival directions, all three of which needed to be interpreted in a consistent manner. I will also discuss some of the ideas that have been advanced to explain the origin of the highest energy cosmic rays.

2 Interest in the Highest Energy Cosmic Rays and Methods of Detection

2.1 Some Historical Background

Above about 10^{14}eV the flux of cosmic rays is so low that it is barely practical to detect them directly using instruments on balloons or in space. As the energy rises one must rely on the extensive air-showers that the particles create when they hit the Earth's atmosphere. Even at 10^{14} eV, where the maximum number of $\sim 10^5$ particles is reached at ~ 6 km above sea-level, some particles survive so that remnants produced by the primary are detectable. Because of scattering, electrons and photons can be found at large distances from the axis of such showers although about 50% lie within the Molière radius, about 70 m at sea level. Thus, a sampling technique can be used to study the phenomenon.

The discovery of extensive air-showers is usually credited to Pierre Auger who, in 1938, observed an unexpectedly high rate of coincidences between Geiger counters separated by a few metres. Further investigations by his team showed that even when the counters were as far as 300 m apart, the rate of coincidences was significantly in excess of the chance expectation. Speculating that the primaries were photons, and using the newly developed ideas of quantum electrodynamics, Auger demonstrated that the incoming entities had energies as high as 10^{15} eV (Auger et al. 1938, 1939). Earlier Rossi (1934) had reported experimental evidence for extensive groups of particles ('sciami molto estesi di corpscoli') which produced coincidences between counters rather distant from each other. Kolhörster and colleagues made very similar observations to those of Auger and his group (Kolhörster et al. 1938). However, it was Auger who was in a position to explore this new phenomenon and, with his inferences about the primaries, extend the range of known energies by nearly 6 orders of magnitude, a remarkable achievement with such simple equipment. More than 65 years later cosmic rays remain the most extreme example of the departure of matter from a state of thermal equilibrium.

Most early studies of air-showers were made by spreading particle counters of various kinds over large areas. By the late 1950s, the MIT group, under the leadership of Rossi, had established that cosmic rays as energetic as 10^{19} eV exist and techniques to measure the direction of the incoming primary to within about 2° had been developed (Bassi, Clark & Rossi 1953; Linsley, Scarsi & Rossi 1961). At Volcano Ranch, Linsley set up an array of scintillators covering 8 km^2 and in 1963 presented the first evidence for a cosmic ray of 10^{20} eV (Linsley 1963). The array was the first of the 'giant arrays' and many of the ideas incorporated in its design were used in subsequent larger installations.

The significance of the claim for a primary of 10^{20} eV was not immediately appreciated but soon after the discovery of the 2.7 K cosmic microwave radiation in 1965, Greisen (1966) and Zatsepin & Kuz'min (1966) independently pointed out that if the highest energy particles were protons and if their sources were uniformly distributed throughout the Universe, then there would be interactions between the cosmic rays and the microwave background that would modulate the spectrum of the highest energy particles in the region of 10^{20} eV.

A particularly important reaction is:

$$p + \gamma_{2.7\,\text{K}} \to \Delta + (1232) \to p + \pi^0 \quad \text{or} \quad n + \pi^+. \tag{1}$$

In the rest-frame of the proton (the cosmic ray), the microwave background photon will look like a very high energy γ-ray because of the relativistic Doppler effect. When the relativistic Lorentz factor, Γ, of the proton is $\sim 10^{11}$ the $\Delta+$ resonance will be excited. In each reaction (1) the proton loses about 15% of its energy. Over cosmological distances sufficient reactions take place that the observed spectrum would become significantly depleted of UHECRs compared with what might have been present at the time of acceleration. It follows that if protons of $\sim 10^{20}$ eV are observed then they must have originated from nearby. For example, a cosmic ray of 5×10^{19} eV has only a 50% chance of having come from beyond 100 Mpc. This opens the prospect of seeing point sources of cosmic rays at the very highest energies as inter-galactic magnetic fields are not expected to bend the trajectories of the particles by more than a few degrees over 50 Mpc. Conversely, if such high-energy particles are seen and established as protons, a lack of anisotropy might herald exotic physics, such as the breakdown of Lorentz invariance. Measurement of the energy spectrum, the mass composition and the distribution of arrival directions of the highest energy particles is thus of paramount importance.

If high-energy particles can escape from the acceleration region as nuclei then both the cosmic microwave radiation and the diffuse infra-red radiation field are important factors. The key resonance is now the giant dipole resonance (typical energy ~ 10 MeV). In each interaction a neutron or a proton is removed from the nucleus so that a mixture of species arrives at Earth. The composition of the mixture depends upon the paths travelled through the radiation fields and what is produced at the source.

Protons and nuclei also lose energy by pair production, with a threshold corresponding to $\Gamma \sim 10^9$. Here the energy losses are small but nearly continuous and may be important if protons of energies $1 - 10 \times 10^{18}$ eV are of extragalactic origin. The reactions of equation 1 are important too in the context of γ - and ν - astronomy: the neutrons from photodisintegration are also a source of neutrinos, as has been discussed by Ave et al. (2005) and Hooper, Taylor & Sarkar (2005).

By the early 1960s, before the discovery of the cosmic microwave background, it was suspected that the energy spectrum might extend to 10^{21} eV and alternative methods of detection were explored. Of particular note are the pioneering discussions of Suga (1962) and Chudakov (1962) who pointed out the possibility of using the Earth's atmosphere as a massive volume of scintillator. Fluorescence light in the 350 – 450 nm band would be produced by the excitation of atmospheric nitrogen by the electrons of an air-shower at an intensity sufficient to allow the light to be detectable above the night sky background. Experiments with alpha particles and 57 MeV protons established the necessary design data. The light is emitted isotropically and it was predicted that showers of $\sim 10^{19}$ eV should be detectable at distances of $\sim$20 km. The idea was further advanced by Greisen (1965) who subsequently built a prototype instrument at Cornell. In 1968 Tanahashi and collaborators first succeeded in detecting fluorescence light from showers of $\sim 10^{19}$ eV at Mt Dodaira, near Tokyo (Hara et al. 1970). However, neither the Cornell nor the Japanese sites were climatically suitable for exploiting the promise of this method and the challenge was taken up by Keuffel's group at the University of Utah in what became known as the Fly's Eye project (Baltrusaitas et al. 1985). In principle the fluorescence technique has the potential of allowing a near-calorimetric estimate of the primary energy as so much of it is transferred to electrons and positrons that can be tracked with the fluorescence light. The light output, $\sim$4 photons per metre of electron track, is proportional to the energy loss. The output is comparable to that produced by the Cherenkov process in air but, although the latter builds up in the forward direction, the relatively small lateral spread has made it of limited use in shower detection.

A further technique that appeared to offer promise was intensively studied in the 1960s, following the pioneering work of Jelley and his collaborators (Jelley et al. 1995), and early 1970s (see Allan (1971) for an authoritative review). This was the detection of showers through low frequency (<100 MHz) radio emission that arises primarily from the dipole radiation associated with the charge separation of electrons and positrons in the shower. Other mechanisms are Cherenkov emission (because of the small negative charge excess) and emission from charge separation in the geo-electric field. The latter emission process was difficult to assess and its apparent dependence on thunderstorm activity led to work on radio emission being terminated in $\sim$1975. Very recently interest has revived (Falcke et al. 2005) and it may be that with modern techniques of waveform analysis and an enhanced ability to monitor geo-electric effects that the phenomenon will have a role to play in the development of very large areas to study the highest energy particles.

The low flux of events is a significant problem: above 10^{19} eV the rate is only about 1 per km^2 per year but even this figure is not well-known. Above 10^{20} eV it falls to less than 1 per km^2 per century. Following the pioneering efforts at Volcano Ranch, a surface array of 12 km^2 was built at Haverah Park, UK using water-Cherenkov detectors, while arrays of scintillators were constructed at Yakutsk, Siberia (25 km^2) and at Narribri, Australia and at Akeno in Japan (both 100 km^2). The Yakutsk array, now covering a smaller area, is the only one of these still in operation. Until very recently, only the Utah Group had used the fluorescence technique, first as Fly's Eye (Baltrusaitis et al. 1985) and then as the HiRes instrument (Sokolsky 1998) which ceased operation in March 2006.

The exposures achieved by the various instruments are summarised in Table 1 where estimates of the number of events and the integral rates above 10^{19} eV are given.

Table 1. Exposure and Event Numbers from various instruments

	km^{2}sr yr	$>3 \times 10^{18}$eV	$>10^{19}$ eV	rate $> 10^{19}$ eV km^{-2} sr^{-1} yr^{-1}
AGASA	1600	7000	827	0.52
HiresI mono	~5000	1616	403	0.08
HiResII mono		670	95	
HiRes Stereo	~2500	~3000	~500	~0.20
Yakutsk	~900	1303	171	0.19
Auger	1750	3525	444	0.25

2.2 The Pierre Auger Observatory

One of the instruments listed in Table 1 is the Pierre Auger Observatory from which the first results were reported in 2005. This device was designed to combine the best features of surface detectors (SD) and fluorescence detectors (FD) in a 'hybrid' detector and there is promise of definitive measurements from it within the next few years. It is therefore appropriate to describe it briefly and to show examples of the quality of the information that is becoming available. An example of a hybrid event is shown in Figure 1.

Construction of a prototype for the Auger Observatory began shortly after a ground-breaking ceremony in March 1999. The performance of this prototype has been described (Pierre Auger Collaboration 2004) and only relatively minor changes were made to the design of the sub-systems for the instrument now under construction. The first data-taking run began in January 2004 and continued into June 2005. At the start ~150 water-Cherenkov detectors, each of 10 m^2 and 1500 m from each other, were operational, together with two fluorescence detectors, one of which was only partially complete. At the end of the run, the number of operational tanks had increased to ~780 and three of the eventual 4 fluorescence detectors were fully-functional. It proved to be a relatively straight-forward matter to take high-quality data as the collection area grew. A large number of 'stereo events' has been recorded in which two fluorescence detectors recorded signals in coincidence with information from the surface detectors. In August 2005 the first tri-ocular event was detected. The status of the Observatory was described at the 2005 International Cosmic Ray Conference held in Pune, India (see Mantsch (2005) for a short summary of the results). Completion of the construction is expected in early 2007.

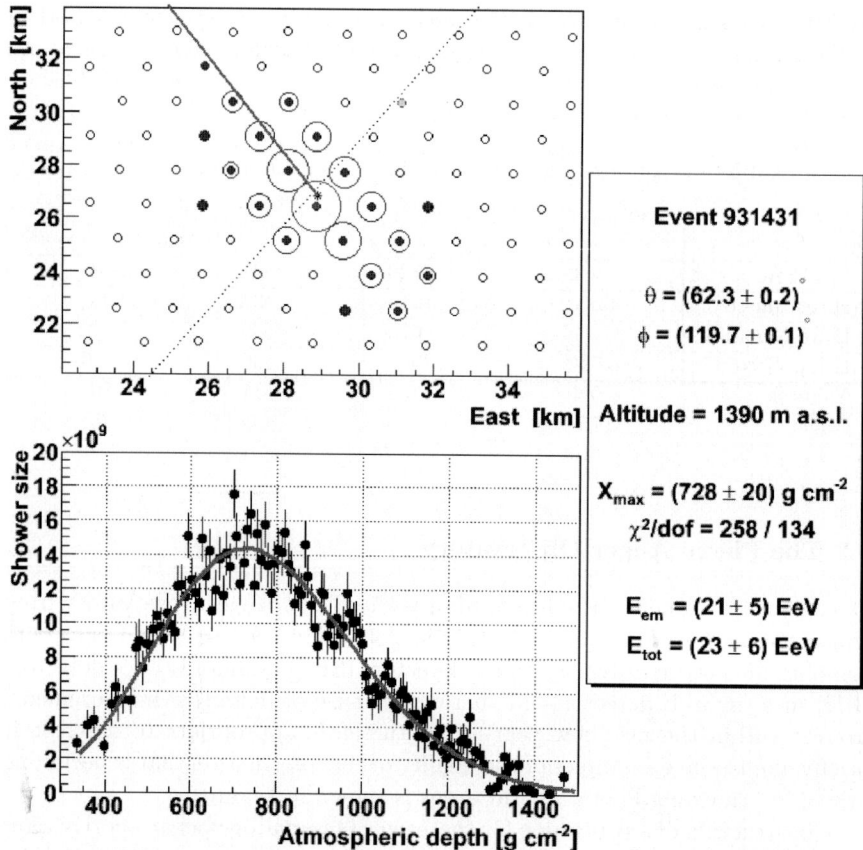

Fig. 1. This is a typical hybrid event inclined at a relatively large angle and at an energy where a large number of tanks (21) have been triggered. The complete profile of the shower is seen. In the upper part of the figure, triggered tanks are shown as filled discs, with the size of the surrounding circles indicating the strongest signals. The solid line represents the projection of the shower axis on the ground. The dotted line shows the intersection between the plane containing the shower axis and the horizontal plane at the altitude of the fluorescence detector. The core position is indicated with a star lose to the red tank. The lower diagram shows a fit of the shower profile to a particular model. Copyright: the Pierre Auger Collaboration.

The high-level of understanding that is derived from simultaneous observations of the fluorescence signals and the tank signals is well-illustrated by results from the detection of the scattered light from the Central Laser Facility (Pierre Auger Collaboration 2005a). This facility, located close to the centre of the array, hosts a 355 nm frequency-tripled YAG laser that generates pulses of up to ∼7 mJ. The scattered light seen from such a pulse by the fluorescence detectors ∼30 km away is comparable to that expected from a shower initiated by a primary of 10^{20} eV: the laser can be pointed in any direction. Some of the light from it is fed into an adjacent tank via an optical fibre so that correlated timing signals can be registered. In this way it has been established that the angular resolution of the surface detectors is $\sim 1.7°$ for $3 \times 10^{18} < E < 10^{19}$ eV and $\sim 0.6°$ for hybrid events. It has been shown (Pierre Auger Collaboration 2005b) that the accuracy of reconstruction of the position of the laser, using the hybrid technique, is better than 60 m. The corresponding figure for the root mean square spread, if a monocular reconstruction is made, is ~ 570 m. There is always at least one tank response for each shower detection at a fluorescence station; these data give a preliminary indication of the geometrical power of the hybrid method. A sample set of results illustrating the principle is shown in Figure 2.

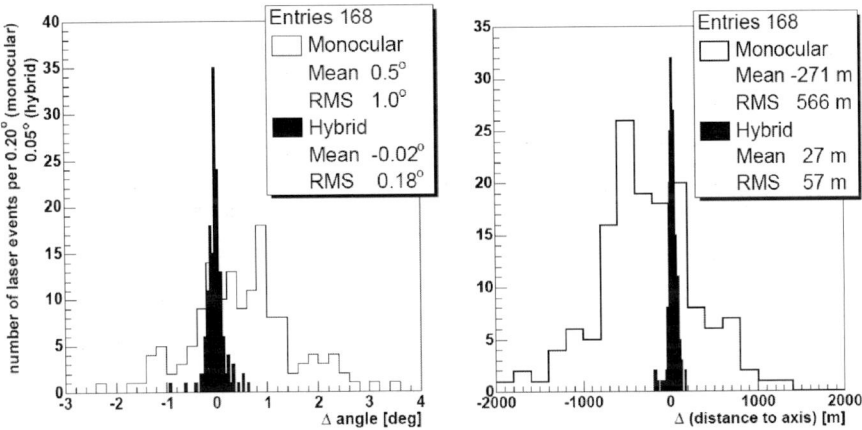

Fig. 2. Determination of the hybrid angular and core position resolutions using laser simulated hybrid events. The space angle of the laser beam determined by the hybrid reconstruction is shown in the left panel. The broader plot in the background is the angle determined with the fluorescence detector alone. The position of the laser "core" determined by hybrid reconstruction and by the fluorescence detector alone are shown in the panel on the right. Both plots are illustrative of the power of fluorescence detector – surface array combination. Copyright: the Pierre Auger Collaboration.

In Figure 3 details of an event recorded by the SD system are displayed. Each photomultiplier output is connected to a Flash Analogue to Digital Converter (FADC) with 25 ns resolution. The FADC signal for one of the 18 detectors that triggered is shown. Signals from detectors far from the axis are more spread in time than those close to the shower axis because of geometrical path differences. In Figure 4 a comparison of the FADC signals from the water-tanks in a near-vertical event and an inclined event is made. Even far from the shower-axis the signal in the inclined event is very narrow with

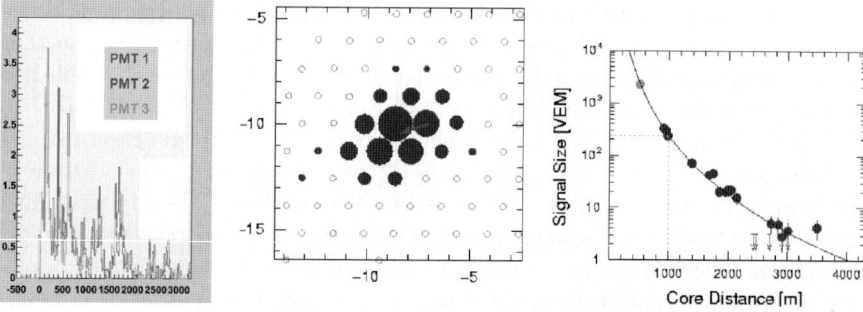

Fig. 3. An example of a surface array event with an energy of ∼70 EeV and zenith angle of about 48°. On the left is a typical FADC trace of one of the more distant stations; the vertical scale is in VEM and the horizontal scale is in nanoseconds. In the center is the hit pattern on the surface array with the axes in kilometers. On the right is the fall-off of signal size with distance. The arrows correspond to stations that did not trigger. Note that the signal at 1000 m (S(1000)) is well-defined. Copyright: the Pierre Auger Collaboration.

most of the energy arriving within 250 ns. The Auger Observatory has the capability to record air-showers arriving at all angles from the zenith up to 90°. This is one of the advantages of the deep-water (1.2 m) Cherenkov detectors that form the SD array (Pierre Auger Collaboration 2004) and opens the possibility of discovering neutrinos of very high energy as the observation of a group of signals such as in Figure 4a, but at a large angle (>70°), would be hard to interpret in other ways.

A study of the 20 highest energy events has demonstrated that the properties of these showers are in no way unexpected when compared with the more numerous lower energy events that have been recorded so far (Pierre Auger Collaboration 2005c): a large event (E > 1.4×10^{20} eV) was detected but fell with its core outside the fiducial area as then defined and is not included in the spectrum determination described below.

3 The Mass Composition of Cosmic Rays and Hadronic Interactions

A cursory comparison of the rate of events above 10^{19} eV makes clear that the differences between the integral rates are much larger (by more than a factor of 2) than can be accounted for by Poissonian variations. In deducing the primary energy of an incoming cosmic ray from observations made with an array of surface detectors alone (as at AGASA or at Yakutsk) one must make assumptions about the hadronic interactions that occur in the shower and also about the mass of the primary particle. For example, for the case of an array of water-Cherenkov detectors, the Santiago group (Zas et al. 2005)

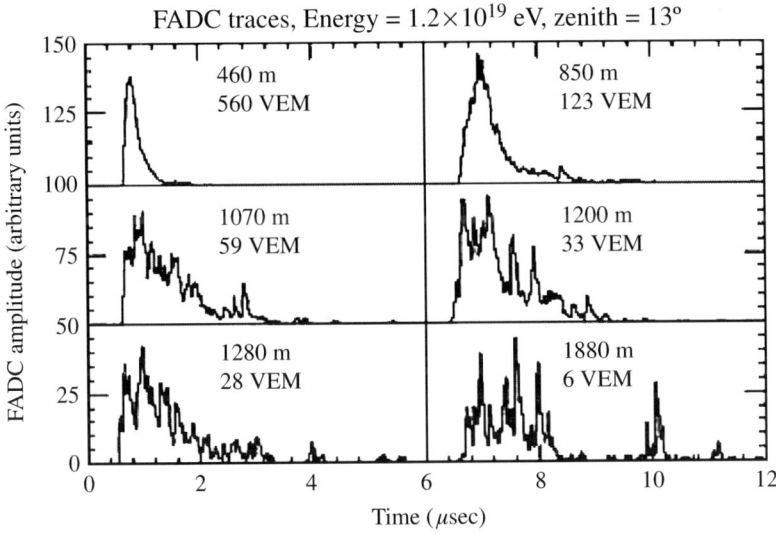

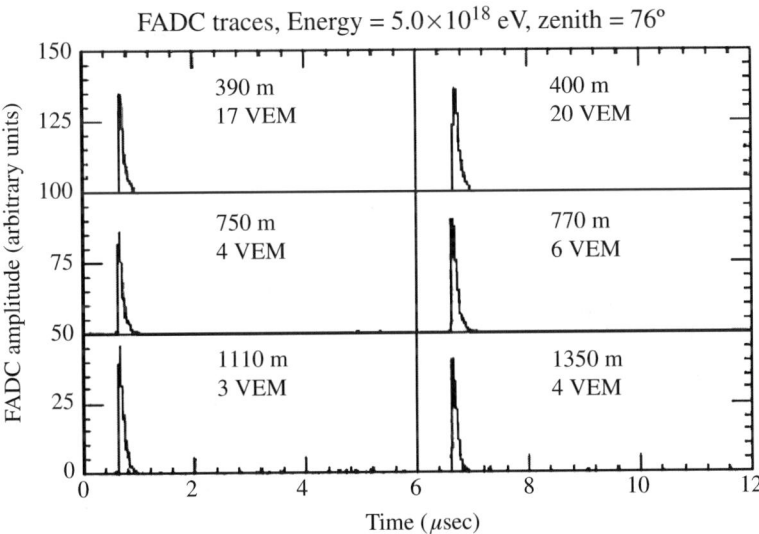

Fig. 4. The upper plot shows FADC traces from a 13° shower: the radius of curvature was 4 km, and the density ratio 134 over a distance ratio of 3.7. By contrast the lower figure shows a shower of 76° which has a radius of curvature of 27 km and a density ratio of 7.5 for a distance ratio of 3.5. Copyright: the Pierre Auger Collaboration.

have made calculations of the signal at 1000 m from the shower axis (S(1000)) produced by 10^{19} eV primaries of p, Fe and photons. S(1000), as discussed below, turns out to be a convenient measure of the size of the shower. With the QGSJET01 model of hadronic interactions, the corresponding values of S(1000) in the vertical direction are 40, 51 and 9.4 VEM (vertical equivalent muons respectively) while for the SIBYLL model the proton value is 34.5 VEM. Knowledge of both the model and the mass is thus essential if the primary energy is to be derived using any surface array. Knowledge of the mass and of the hadronic interactions are also important for the fluorescence method, but to a lesser extent, as some of the energy in the shower is transferred to high-energy muons and neutrinos that travel into the Earth. The estimate of the energy due to ionisation loss in the air must be supplemented by this missing energy; the magnitude of the effect is discussed below.

The discovery of even the mean mass of the primary particles is extremely difficult and several techniques have been used in the attempts to make it. It is easy to understand that the lower energy per nucleon of an iron primary implies that the depth of maximum of the resultant shower will be higher in the atmosphere than for a shower produced by a proton of the same energy. Similarly the lower energy per nucleon means that iron showers will contain relatively more muons. The depth of shower maximum, X_{max}, quantity can be measured directly with fluorescence detectors, as is clear from Figure 1. The muon numbers are harder to obtain as they are not very numerous in showers and can only be measured with detectors surrounded by heavy shielding to cut out the more numerous electrons, an expensive endeavour. It is usually assumed that Fe defines the heaviest mass component at these energies because of its place in the abundance of the elements, but there is no a priori reason to exclude even heavier nuclei.

Important developments in our understanding of hadronic interaction models at the highest energies have been recently reported. In particular a revision of the QGSJET series of models (QGSJETII) has been described by Ostapchenko and Heck (2005). Their new analysis of high-energy interactions, resulting from a more exact treatment of diffractive processes, has led to new parton distribution functions that are consistent with cross-section measurements. There is a reduction in the number of secondary particles produced in hadron-nucleus and nucleus-nucleus collisions compared with what had been estimated before, leading to an increase in the number of electrons observed at ground level. Of particular importance, for mass inferences at the highest energy, are the results on the depth of shower maximum, X_{max}, which is now predicted to be some 10% deeper in the atmosphere for both proton and iron primaries, and the demonstration that there is a substantial fall in the number of muons expected. The predictions for X_{max} and the muon number with energy for proton and iron primaries are shown in Figures 5 and 6. Note that the muon signal now calculated is 83% of that predicted by the model, QGSJET01, for proton primaries, and 90% of what had been previously been predicted for iron nuclei at 10^{19} eV. The QGSJET01 model had previously

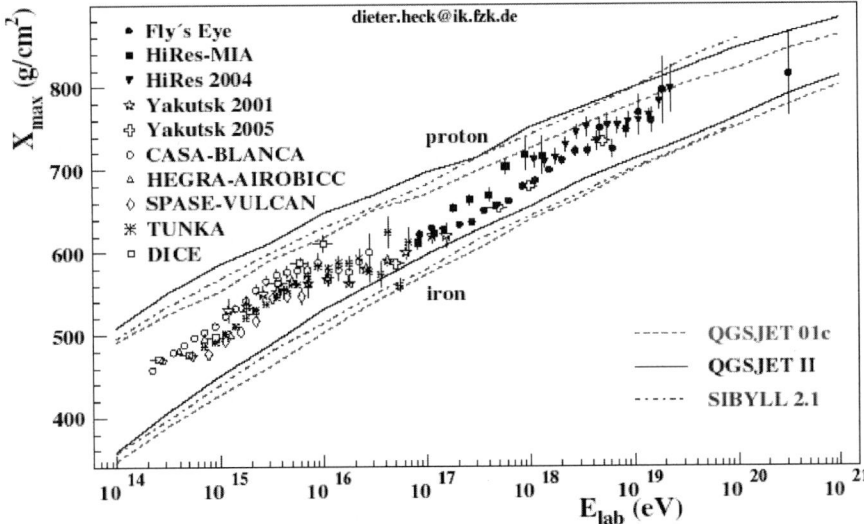

Fig. 5. Position of the shower maximum calculated from three hadronic interaction models compared with cosmic ray data from a variety of experiments. From Ostapchenko & Heck (2005).

been widely regarded as the best description of hadronic interactions at extreme energies.

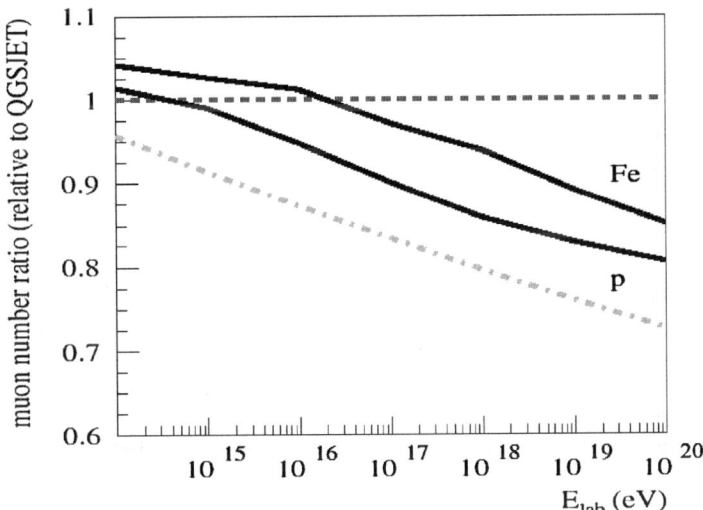

Fig. 6. The ratio of the number of muons in showers as a function of energy calculated for protons and Fe nuclei using the QGSJETII model and the SIBYLL model. From Ostapchenko & Heck (2005).

Qualitatively, the behaviour of X_{max} and the muon number with energy calculated with the new model is similar to what has been found previously using the Sibyll 2.1 model as reported by Alvarez-Muñiz et al. (2002) in a paper particularly worth studying for the clear discussion of the differences between the QGSJET and SIBYLL families of models. If the QGSJETII model is correct it is clear that significant reinterpretations of existing data on the muon component and on X_{max} are called for. It has been pointed out previously (Watson 2003, 2004 and Dova et al. 2004, 2005) that the composition above 10^{19} eV might not be as proton-rich as has often been assumed, and the new model results appear to reinforce this idea. While it is almost certain that there will be more surprises as accelerator measurements are made at higher energies, as with the LHC, the possibility of a higher mean mass than has been assumed hitherto for the UHECRs needs to be kept in mind when interpreting data on the energy spectrum and on cosmic ray propagation and origin.

4 The Cosmic Ray Energy Spectrum Above 3×10^{18} eV

Recent measurements of the high-energy cosmic ray spectrum reported by the AGASA (Takeda et al. 2003) and HiRes (Abbasi et al. 2004) collaborations have yielded conflicting results. Factors leading to the difference between the fluxes at high energy (see Table 1) are the serious limitations that come from the use of only a surface detector (SD) array or only a fluorescence detector (FD) system when measuring the primary spectrum. In the former case, observations at ground level are used to infer the energy of the primary particle and, as discussed above, assumptions about the nature of the primary and of features of hadronic interactions at energies well-above accelerator measurements are unavoidable. In the latter case, as with any fluorescence instrument, the monitoring of the atmosphere is a major problem. Additionally, by contrast with SD arrays, there are significant uncertainties in determination of the aperture of a fluorescence detector as assumptions about the mass of the primary and of the energy spectrum are necessary. The methods developed by the Auger Collaboration largely circumvent these difficulties and should lead to a reliable estimate of the primary energy spectrum once the systematic uncertainties are fully understood and sufficient events have been recorded. The Collaboration has reported the first precision measurement of the high-energy cosmic ray spectrum from the Southern Hemisphere (Pierre Auger Collaboration 2005d).

For this first analysis attention was restricted to events with zenith angle, $\theta < 60°$. The strategy was to reconstruct the arrival direction for each event recorded by the SD, then to estimate the magnitude of the signal at 1 km from the shower axis, S(1000), as a measure of the size of the shower. S(1000) is chosen as the ground-parameter as it can be measured to better than 10%, independent of knowledge about how the signal falls off with distance from the shower axis. In addition, as shown in the pioneering studies of Hillas (1970, 1971), the magnitude of this particular ground-parameter is at least three times less susceptible to stochastic fluctuations and variations in primary mass than are measurements made close to the shower axis. These two *unrelated* properties of S(1000) make it a valuable parameter to use when estimating the primary energy.

Two cosmic rays of the same energy, but incident at different zenith angles, will yield different values of S(1000) with S(1000) expected to be smaller at larger zenith angles because of attenuation of the shower by the additional atmosphere. Thus, a necessary step is to find the relation between the ground-parameter measured at one zenith angle and that measured at another. The approach adopted here is to use the well-established technique of the constant intensity cut (CIC) method which has been recently reappraised (Alvarez-Muñiz et al. 2003). The principle of this method is that the high level of isotropy of cosmic rays supports the proposition that showers created by primaries of the same mass and energy will be detected at the observation level at the same rate. Here the rates of events above different values of S(1000) are found for different zenith angles and all azimuth angles so that events come from a broad band of sky. This method is used to establish the relationship between $S(1000)_{38°}$ and $S(1000)_\theta$, where the subscripts refer to the reference angle chosen as $38°$ and θ is the angle of incidence conventionally measured from the vertical. The average thickness of the atmosphere above the Auger Observatory is 875.5 g cm^{-2}.

The link between $S(1000)_{38°}$ and the primary energy is best established using data from the fluorescence detectors. On clear, moonless nights fluorescence signals can be observed simultaneously with SD events. For every FD event for which the shower core falls within the fiducial SD area, at least one tank is struck so that the time at which the tank was triggered can be used to enhance to the reconstruction accuracy of the FD geometry, as illustrated in Figure 2. Further, as the FD instruments are used primarily as calibration devices in this application, the selection of events can be made in a very selective manner. This approach was adopted for the first spectrum described by the Pierre Auger Collaboration (2005d). The selection criteria required that the FD tracks had to be longer than 350 g cm^{-2}, that the contribution of the Cherenkov light to the signals collected be less than 10% and that there were contemporaneous measurements of the aerosol content of the atmosphere.

When estimating the energy of an event from the fluorescence yield, a correction must be made for the missing energy that is carried by high-energy muons and neutrinos that penetrate into the ground. A study of this conver-

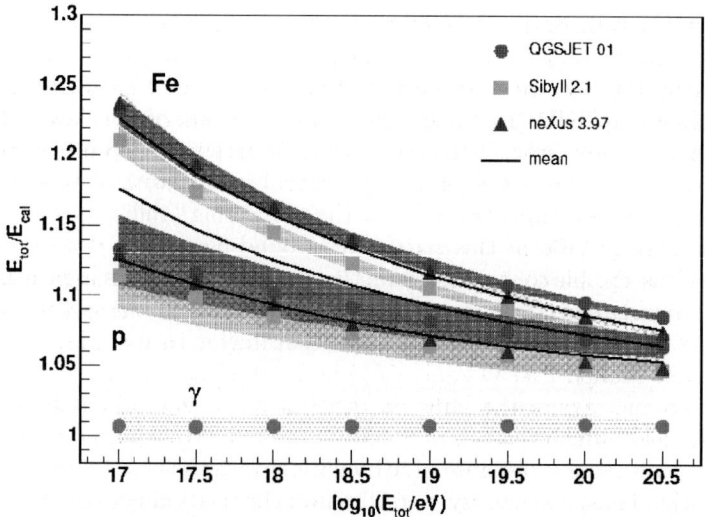

Fig. 7. The mean and root mean square values of the ratio of the primary energy (E_{tot}) to the calorimetric energy (E_{cal}) for different models and primary particles. The calorimetric energy is deduced in practice from the FD measurements. From Pierog et al. (2005).

sion factor has recently been made for nucleonic primaries (Figure 7, Pierog et al. 2005). At 10^{19} eV it is evident that the correction for missing energy is ~10% with a systematic uncertainty, due to our lack of knowledge of the nuclear mass and the hadronic interactions, estimated as ~7%. The corrections and the associated systematic uncertainties are, of course, dependent on hadronic interaction models and ideas may have to be revised when LHC data are available.

In the derivation of these figures, a mean nuclear mass between those of proton and iron has been assumed for the primaries. If the high-energy beam contained a mixture of photons and nuclei, use of this average energy calibration relation would lead to an under-estimate of the energy of any photons in the primary beam and an over-estimate of the energy of any nuclei that are present. This is because a photon shower has an S(1000) that is ~4 to 5 times smaller than a proton or Fe-nucleus of the same energy (Zas et al. 2005). The magnitudes of the under- or over- estimates would depend on the fraction of photons present. The upper limits to the fraction of photons that have so far been reported are, at the 95% confidence level, 16% above 10^{19} eV (Pierre Auger Collaboration 2006), 50% above 4×10^{19} eV (Ave et al. 2000, 2002) and 70% at 10^{20} eV (Risse et al. 2005).

For a surface array alone, the primary energy is derived under certain assumptions about the mass of the primary particle and about the hadronic interactions. A suitable conversion relation based on QGSJETII models has yet to be derived for the AGASA array (Takeda et al. 2003), and the mass is clearly uncertain as can be seen from the comparisons with the X_{max} measurements made in Figure 5. For example, if the energies of the AGASA events near 10^{20} eV are estimated for protons and the SIBYLL1.6 model or iron with the QGSJET98 model, the ratio of the energies is 1.33. However, for SIBYLL1.6 and iron compared with QGSJET98 and protons, the ratio is 1.02. This illustrates the sensitivity of energy estimates to the mass and hadronic model assumptions.

The exposure of the Auger Observatory has had to be determined with some care as the number of operating tanks of the SD increased from ~150 to ~780 over the run. It was computed as 1750 km^2 sr yr and is independent of energy above 3×10^{18} eV. This exposure is slightly above that achieved by the AGASA group (Takeda et al. 2003) and a factor of about 3 lower than the exposure of the monocular HiRes I (Abbassi et al. 2004). The HiRes stereo exposure, from which the most reliable 'fluorescence-only' energy spectrum is expected to come, is not yet available. The average array size during the time of the Auger exposure was 22% of what will be available when the southern site has been completed.

For all of the HiRes detector modes of operation the aperture increases with energy and is never energy independent. For example, the stereo aperture rises from 6×10^2 km^2 sr at 3×10^{18} eV to 1.5×10^3 km^2 sr at 10^{19} eV. The problems encountered in determining the apertures outlined above are discussed in recent reports (Zech et al. 2005; Abu-Zayyad et al. 2005). While understanding is clearly growing, it is evident that the position is not yet finalised. Furthermore, the aperture depends on a detailed understanding of the atmospheric conditions, which need to be determined from shower-to-shower. It is possible for example, that the variation of aperture with energy is less rapid than is claimed at present, in which case the spectral slope would be slightly flatter and any claims for the GZK-steepening would be less firm. Furthermore, for the spectra so far reported the HiRes team has assumed an average atmosphere using data recorded on a monthly basis, rather than measurements on a much shorter timescale. No stereo-spectrum has yet been reported. This situation contrasts sharply with the approach taken in determining the spectrum with the Auger Observatory.

A comparison of spectra measured by Auger, AGASA, HiResI and Yakutsk is made in Figure 8. In Figure 9 (Watson 2005) the ratio of the values from Auger, AGASA and HiResI are compared with an E^{-3} line passing through the first Auger point at 3.55×10^{18} eV which contains 1216 events. It is clear that further statistics are needed and that a better understanding of the conversion to energy in the case of SD arrays is required. The differences between the fluorescence measurements by Auger and HiResI are relatively small except at the highest energies where the number of events recorded with Auger is presently too low to define the flux above 10^{20} eV. The difference between the AGASA result and the fluorescence measurements probably arises, at least in part, because of the mass and hadronic model assumptions made by the

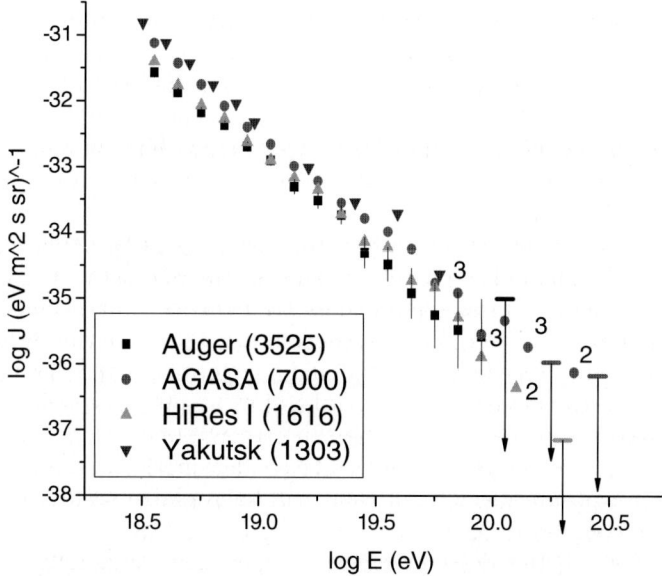

Fig. 8. The differential spectra from Auger, AGASA, HiResI and Yakutsk are compared on a plot of log J vs. log E. The numbers shown in the legend correspond to the events reported above 3 EeV. The numbers (3, 2) by some points refer to the last bin of each data set in which > 0 events were recorded. From Watson (2005).

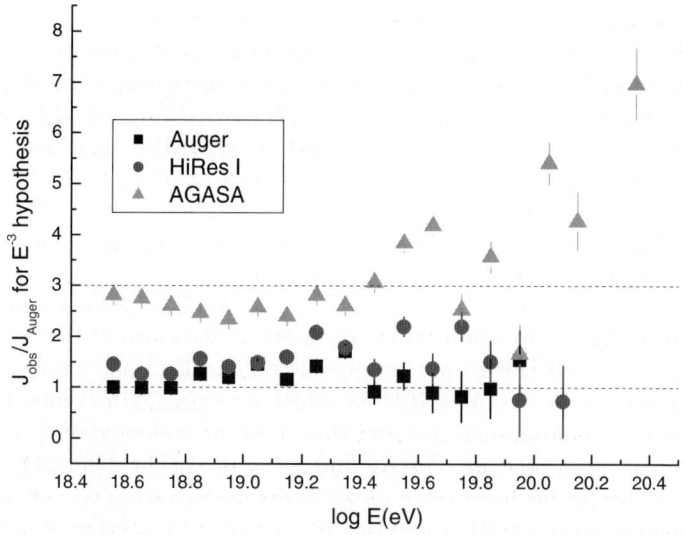

Fig. 9. The ratio of the values of each point with respect to a fit of E^{-3} to the first point of the Auger spectrum at 3.55 EeV which contains 1216 events. The purpose of the plot is to illustrate the differences between the different measurements in a straight-forward manner. Yakutsk data are not included in this plot as they are so discordant (see Figure 8). From Watson (2005).

AGASA group. If these differences become established, a promising route to extracting model and mass information will open.

5 Studies of Arrival Directions at High Energies: Is There Evidence for Clustering?

For some time the AGASA group (Teshima et al. 2003) have insisted upon clusters of cosmic rays above 4×10^{19}eV on an angular scale of 2.5° from a data set of 59 events. Such clusters are claimed to occur much more frequently than expected by chance with 10^{-4} given for the chance probability. A search of HiRes data (Abbassi et al. 2004) did not reveal clusters with the same frequency. Recently, Finley and Westerhoff (2004) have presented an analysis using the directions of the 72 events most recently released by the AGASA group. They have taken the 30 events originally described by Hayashida et al.(1999) as the trial data set and used the additional 42 events to search for pairs, adopting the criteria established by the AGASA group. Two pairs were found but such a result is estimated as having a probability of 19% of occurring by chance.

A further search for clusters has been made by the HiRes group using 27 events from their own data and 57 from AGASA above 4×10^{19} eV (Finley & Westerhoff 2005). Using a novel likelihood method, the authors state that "no statistically significant clustering of events consistent with a point source is found": the most significant signal seen is the triplet cluster of events already known from the AGASA data. If the energy scales of the two instruments are then normalised by reducing the HiRes threshold to 3×10^{19} eV (as seems reasonable from Figure 8), the HiRes sample is increased to 40 events. The sample now includes an event close to the AGASA triplet in the enhanced sample but, as the HiRes group make very clear, it is impossible to evaluate a chance probability that is valid for this observation because of the *ad hoc* nature of the energy shift. Accordingly, they have identified a direction (near $\alpha = 169°$ and $\delta = 57°$) and stated that if 2 events from the next 40 observed by HiRes with the same energy selection fall in a bin of 1° around this direction the chance probability will be 10^{-5}. The target of 40 events was expected to be reached by the time the HiRes instrument was decommissioned in March 2006 and a final statement about the reality of clustering is expected soon.

A feature of the HiRes instrument is the excellent angular resolution achieved with stereo events: a resolution of ∼0.5° has been obtained above 10^{19} eV. This is significantly better than was reached at AGASA and surpasses what has so far been achieved with the surface detectors of the Auger Observatory. Such resolution is particularly valuable for small-scale anisotropy searches. Recently, the HiRes group reported (Finley et al. 2005) the results of a search for correlations with BL Lac active galaxies. Claims of associations have been made previously based on data from AGASA and Yakutsk (Tinyakov and Tkachev 2001, 2002) but are regarded as controversial because of the way in which a postiori selections were made. An analysis of correlations of BL Lacs from the Veron 10th catalogue with m < 18 has also been reported (Gorbunov et al. 2004) using 271 HiRes events above 10^{19} eV. The HiRes group have confirmed the earlier claims and, as it seems reasonable to suppose that the correlations are indicative of neutral particles, have extended their search to lower energies. For all 4495 HiRes stereo events, they find a significance $F = 2 \times 10^{-4}$, where F is the fraction of sets of simulated events that have a likelihood of clustering by chance greater than is found in the data. The group has broadened the search to objects in the BL Lac population that have high polarisation and to those BL Lacs that have been identified as sources of γ-rays at 1 TeV. Small values of F, corresponding to probabilities less than 0.5%, are found in several cases and these selections are now regarded as hypotheses to be tested with additional HiRes data.

6 Speculations About the Origin of the Highest Energy Cosmic Rays

It seems likely that all of the UHECR observatories have recorded at least one event above 10^{20} eV. What these particles are and in particular what fraction may be photons is presently in doubt but if a large fraction does turn out to be protons then their very existence implies that their origin must be within ~50 Mpc. Thus, if the inter-galactic magnetic field strength were $\sim 10^{-9}$G, we would expect to see evidence of point sources whereas the cosmic ray sky looks rather isotropic at $\sim 4 \times 10^{19}$ eV where there are a reasonable number of events.

There is considerable interest in the idea that cosmic rays of all energies are accelerated in shocks by diffusive processes and currently it is popularly believed (e.g. Drury 1994) that cosmic rays with energies up to about 10^{15} eV are energised by 'diffusive shock acceleration' with possible ways of extending this limit to greater energies under debate (Hillas 2005). Supernova explosions are identified as the likely sites, although so far there is no direct evidence of acceleration of protons or nuclei by supernova remnants at any energy. The diffusive shock process, which has its roots in the seminal ideas of Fermi (1949), has been extensively studied since its conception in the late 1970s. In a review article that remains timely, Hillas (1984) has shown that the maximum energy attainable is given by $E = kZeBR\beta c$, where B is the magnetic field in the region of the shock, R is the size of the shock region and βc is the shock speed, with k being a constant less than 1. The same result has been obtained by a number of people and most who are expert in this field accept Hillas's conclusion. However, some claim that the diffusive shock acceleration process can be modified to give much higher energies than indicated by the equation and that the lobes associated with radio galaxy, in particular, are possible acceleration sites. A more complete discussion of cosmic ray acceleration in shocks is given in the chapter by Tom Hartquist, Alex Wagner and Sam Falle.

It is difficult to see how an energy of 3×10^{20} eV, as has been claimed for a Fly's Eye event (Bird et al. 1995), can be accounted for if the size of the shock region is 10 kpc and the magnetic field is 10 µG (values thought typical for the lobes of radio galaxies), as even the optimum estimate of the energy is lower by a factor of 3 than the observational upper limit. It could be that the magnetic fields are less well-known than is usually supposed, a line of argument that also comes from the arrival direction work just discussed. Stronger fields ease the problems of acceleration and would help to make the directions of arrival of the cosmic rays more isotropic.

Because of the problem with accelerating protons to $> 10^{20}$ eV, a number of proposals have been made that dispense with the need for electromagnetic acceleration. In general, attention has been focused on the very highest energy events ($>10^{20}$ eV). However, it is my view that proposers of some of the more exotic 'top down' processes often overlook one or more important points. Any mechanism able to explain the very highest energy events must also explain those above about 3×10^{18} eV, where the Galactic component may disappear. The spectrum above this point is probably too smooth to imagine that there are two or more radically different components – although this might be seen by some as an almost philosophical argument. In addition, the solutions proposed must produce particles at the top of the atmosphere that can generate showers of the type we see and now understand rather well: Dirac monopoles, for example, cannot produce the kind of shower profile (see Figure 1) that is repeatedly observed. Finally, source energetics cannot be ignored: there seems little point in inventing a mechanism to 'solve' the GZK problem that requires a source generating unrealistically energies. For example, the decay of the Z^o- resonance, postulated as being excited by the interaction of neutrinos of $10^{21} - 10^{22}$ eV with relic neutrinos, (Weiler 1982), has attracted considerable attention. However, the source will surely need to be able to accelerate protons to $\sim 1000 \times 10^{20}$ eV to create the neutrinos and this does not seem easy to achieve.

An overview of the various mechanisms proposed has been given relatively recently (Nagano and Watson 2000) and I will only discuss one of these here as it has attracted much attention and is also the proposal which is most likely to be subjected to critical tests fairly soon. It has been suggested that UHECR arise from the decay of super-heavy relic particles. In this picture, the cold dark matter is imagined to contain a small admixture of super-heavy particles with a mass $>10^{12}$ GeV per particle and a mean lifetime greater than the age of the Universe (Berezinsky, Kachelreiss & Vilenkin 1997). It is argued that such particles can be produced during the reheating supposed to follow inflation or through the decay of hybrid topological defects such as monopoles connected by strings. I find it hard to judge how realistic these ideas are but the decay cascade from a particular super-heavy candidate (Benalli, Ellis & Nanopoulis 1999) has been studied in some detail (Birkel & Sarkar 1998; Rubin 1999; Sarkar & Toldra 2002). A feature of the decay cascade is that a flux of photons and neutrinos is predicted which may be detectable with a large enough installation. Neutrinos and photons dominate over protons as quark-anti-quark pairs are more likely to be produced than quark-triplets. The anisotropy question has been examined by several authors and specific predictions for the anisotropy that would be seen by a Southern Hemisphere observatory have been made. The observation of the predicted anisotropy, plus the identification of appropriate numbers of neutrinos and photons, would be suggestive of a super-heavy relic origin. The super-heavy relic idea has been one of the motivations for the search for photon primaries at the highest energies. The present limits (Ave et al. 2002, 2003; Risse et al. 2005) are coming close to predictions so that the model of super-heavy relic particles may need to be revised or abandoned fairly soon. It may be, of course, that radio-photon density is sufficient to suppress the UHE photon flux through pair production but this quantity is very poorly known.

Perhaps the most remarkable resolution of the problem of the origin of ultra-high energy cosmic rays would be through the identification of a large fraction of heavy nuclei in the cosmic ray beam. This seems to be an increasingly reasonable speculation understanding of hadronic interaction models grows (see section 3). A large fraction of heavy nuclei would ease acceleration difficulties, go some way to explaining the high degree of isotropy and resolve some of the discrepancies between the energies inferred by the ground arrays and the fluorescence detectors. Perhaps, even a galactic source of the UHECR could be a possibility. The issue of mass composition is one that the Pierre Auger Observatory is expected to be used to resolve.

7 Summary

The answers to the key questions about the pattern of arrival directions, the energy spectrum and the mass of the highest energy cosmic rays may come from the full HiRes stereo data set and from the current data run of the Pierre Auger Observatory. When an energy spectrum based on fluorescence detectors has been finalised and compared with one from surface detectors in which mass and hadronic model assumptions are of key importance (such as that from AGASA) insight into the issues of mass and models should be forthcoming. The solution to the long-standing problem of the origin of the highest-energy cosmic rays may be in sight.

Acknowledgements

It is a delight to acknowledge the pleasure that my friendship with John Dyson has given me over many years and also his long-standing interest in cosmic rays, both of which predate his return to his native city. It has been fantastic to have his physical insight and ready wit so close at hand for over a decade and I wish him a long, active and very happy retirement.

Research on Ultra High Energy Cosmic Rays at the University of Leeds is supported by PPARC, UK.

References

Abbasi, R.U. et al. 2004 ApJ **610**, L73 (astro-ph/0404137)
Abbasi, R.U. et al. 2004 Phys Rev Lett **92**, 151101
Abu-Zayyad T., Bergmann, D.R. 2005 for the HiRes Collaboration Proc. 29th ICRC (Pune): usa-matthews-John-abs1-he14-poster
Allan, H.R. 1971 Progress in Elementary Particle and Cosmic Ray Physics **10**, 169
Alvarez-Muñiz J. et al. 2002 Phys Rev **D66**, 033011 2002 (astro-ph/0205302)
Alvarez-Muñiz J. et al. 2003 Phys Rev **D66**, 123004 (astro-ph/0209117)
Auger, P. et al. 1938 Comptes Rendue **206**, 721
Auger, P. et al. 1939 Rev Mod Phys **11**, 288
Ave, M. et al. 2000 Phys Rev Lett **85**, 2244
Ave, M. et al. 2002 Phys Rev D **65**, 063007
Ave, M. et al. 2005 Astropart. Physics **23**, 19
Baltrusaitis, R.M. et al. 1985 NIM A **240**, 410
Bassi, P., Clark, G., Rossi, B. 1953 Phys Rev **92**, 441
Benakli, K., Ellis, J., Nanopolous, D.V. 1999 Phys Rev D **59**, 047301
Berezinsky, V., Kachelreiss, M., Vilenkin, A. 1997 Phys Rev Lett **22**, 4302
Bird, D. et al. 1995 ApJ **441**, 144
Birkel, M., Sarkar, S. 1998 Astroparticle Physics **9**, 297
Chudakov, A. 1962 Proc 5th Interamerican Seminar on Cosmic Rays, La Paz, Bolivia **2**, p XLIV
Dova, M.T. et al. 2004 Astroparticle Physics **21**, 597 (astro-ph/0312463)
Dova, M.T. et al. 2005 Proc. 29th ICRC 2005 (Pune): uk-analisa-M-abs1-HE14-oral
Drury L. O'C. 1994 Contemporary Physics **35**, 232
Falcke, H. et al. 2005 Nature **435**, 313
Fermi, E. 1949 Phys Rev **75**, 1169
Finley, C.B., Westerhoff, S. for the HiRes Collaboration: usa-finley-C-abs1-HE14-oral (ICRC2005)
Finley, C.B., Westerhoff, S. 2004 Astroparticle Physics **21**,359 (astro-ph/0309159)
Greisen, K. 1964 Proc. 9th ICRC 1965 (London) **2**, 609
Greisen, K. 1966 Phys Rev Letters **16**, 748
Gurbunov, D.S. et al. 2004 JETP Letters **80**, 145
Hara, T. et al. 1970 Acta Phys Acad Sci Hung **29**, Suppl 3 361
Hayashida, N. et al. 1999 ApJ **522**, 225
Hillas, A.M. 1969 Acta Phys Acad Sci Hung **29**, Suppl 3 355
Hillas, A.M. 1984 Ann Rev Astronomy & Astrophysics **22**, 425
Hillas, A.M. 2005 J Phys G **31**, R95
Hillas, A.M. et al. 1971 Proc. 12th ICRC (Hobart) **3**, 1001
Hooper, D., Taylor, A., Sarkar, S. 2005 Astroparticle Physics **23**, 11
Jelley, J.V. et al. 1965 Nature **205**, 327
Kolhörster, W. et al. 1938 Naturwiss. **26**, 576
Linsley, J. 1963 Phys Rev Letters **10**, 146

Linsley, J., Scarsi, L., Rossi, B. 1961 Phys Rev Letters **6**, 458
Mantsch, P. 2005 Highlight talk: Proc. 29th ICRC 2005 (Pune)edited by Tonwar, S.
Nagano, M., Watson, A.A. 2000 Rev Mod Phys **27**, 689
Ostapchenko, S., Heck, D. 2005 29th ICRC, Pune, Eds.Acharya, B.S. et al. **7**, p135.
Pierog, T. et al. 2005 Proc. 29th ICRC, Pune, Eds. Acharya, B.S. et al. **7**, p103.
Pierre Auger Collaboration 2005a Proc. 29th ICRC 2005(Pune): usa-malek-abs-he15-poster
Pierre Auger Collaboration 2005b Proc. 29th ICRC 2005 (Pune): usa-mostafa-M-abs-he14-oral
Pierre Auger Collaboration 2005c Proc. 29th ICRC 2005 (Pune): usa-matthews-James-abs2-he14-oral
Pierre Auger Collaboration 2005d Proc. 29th ICRC 2005 (Pune): usa-sommers-P-abs1-he14-oral
Pierre Auger Collaboration 2006 Astroparticle Physics, submitted
Pierre Auger Collaboration: Abraham, J. et al. 2004 NIM A **523**, 50
Risse, M. et al. 2005 Phys Rev Lett **95**, 171102 (astro-ph/0502418)
Rossi, B. 1934 Supplemento a la Ricerca Scientifica **1**, 579
Rubin, N.A. 1999 M Phil Thesis, University of Cambridge
Sarkar, S., Toldra, R. 2002 Nuclear Physics B **621**, 495
Sokolsky, P. 1998 Proc. Workshop on Observing Giant Cosmic Ray Showers from Space: AIP Conference Series No 433 p 65
Suga, K. 1962 Proc 5th Interamerican Seminar on Cosmic Rays, La Paz, Bolivia **2**, p XLIX
Takeda, M. et al. 2003 Astroparticle Physics **18**, 135
Teshima, M. et al. 2003 Proc. 28th ICRC (Tsukuba) 1 437
Tinyakov, P.G., Tkachev, I.I. 2001 JETP Letters **74**, 445
Tinyakov, P.G., Tkachev, I.I. 2002 Astroparticle Physics **18**, 165
Watson, A.A. 2003 Proc. Int. Conf. 'Thinking, Observing and Mining the Universe', Sorrento (World Scientific) p 277 (astro-ph/0312475)
Watson, A.A. 2004 Nucl Phys B (Proc. Suppl.) **136**, 290 (astro-ph/0408110 and astro-ph/0410514)
Watson, A.A. 2005 Procedings of the TAUP Conference in Zaragoza, September 2005
Weiler, T.H. 1982 Phys Rev Lett **49**, 234
Zas, E. et al. 2005 Santiago de Compostela Group, private communication
Zatsepin, G.T., Kuz'min, V.A. 1966 ZH Eksp Teor Fiz Pis'ma Red **4**, 144
Zech, A. for the HiRes Collaboration 2005 Proc. 29th ICRC (Pune): fra-zech-A-abs1-he14-poster

Part IV

Starburst Galaxies and Active Galactic Nuclei

The Messier 82 Starburst Galaxy

A. Pedlar[1] and K.A. Wills[2]

[1] JBO, University of Manchester, Jodrell Bank, Nr Macclesfield, Cheshire. SK11 9DL
   email: ap@jb.man.ac.uk
[2] Department of Physics & Astronomy, University of Sheffield, The Hicks Building, Sheffield, S3 7RH
   email: K.Wills@sheffield.ac.uk

1 Introduction

All galaxies show some degree of star formation, although the star formation rate differs greatly between different classes of galaxies. Early types, such as ellipticals, clearly show significantly less star formation than late types, such as spirals and irregulars. Much of the current star formation in spiral galaxies is associated with spiral arms, and appears to be due to the interaction of stellar density waves which drive shocks into neutral gas. In normal spiral galaxies this yields star formation rates of the order of a solar mass ($M_\odot$) per year. Hence, given a mass of gas in a typical late type galaxy of $\sim 10^9$ $M_\odot$ this rate of star formation can be sustained at a relatively constant rate for approximately the lifetime of the galaxy.

However, in many galaxies, particularly irregulars in interacting systems, the star formation rates (SFRs) are orders of magnitude higher than in normal galaxies. A number of methods to measure the SFR can be used (see section 1.2), and although there is significant scatter in the estimates, there seems little doubt that the derived rates would exhaust all the neutral gas in such a galaxy on timescales of order 10^8 years or less. Hence, the star formation that we are currently seeing in such galaxies cannot be sustained over their lifetimes, and, therefore, on galactic timescales they must be exhibiting a burst of star formation– hence the term 'starburst'. This term was first used by Weedman et al. (1981) in the context of a multiwavelength study of NGC 7714.

1.1 Why Study Starbursts?

Most galaxies probably pass through a period of intense star and heavy element formation as part of their evolution (Eggen et al. 1962), which is, of course, consistent with the increase in the star formation rate with increasing redshift (Cram et al. 1998). This results in starburst galaxies being relatively common in the Early Universe. For example, most of the radio sources in the Hubble Deep Field appear to be starbursts rather than Active Galactic Nuclei (Muxlow et al. 2005).

However, these high redshift objects cannot be investigated in detail, and it is necessary to study relatively nearby starbursts to understand their physics. Most nearby starbursts appear to be due to interactions and mergers, in principle similar to processes thought to be taking place in the Early Universe, although clearly the differences in metallicity need to be taken into account before reliable extrapolations can be carried out, such as, for example, the relation between star formation rate and radio luminosity.

In addition to being objects of interest in their own right, nearby starbursts can provide us with samples of HII regions, supernova remnants, molecular clouds etc, all observed at essentially the same distance and often with the same instrumental parameters and sensitivity. Supernova remnants (SNRs) have proved particularly useful for constraining the global supernova, and hence star formation, rate. At radio wavelengths the distribution of the SNRs is unaffected by extinction and traces the structure of the starburst approximately 10^7 years ago. This can be compared with the compact HII region distribution which traces the current starburst. The study of supernovae and supernova remnants in starburst galaxies provides unique information on their properties and evolution. The starburst remnants appear to represent a population of remnants at least an order of magnitude younger than those in our own galaxy. In addition, starburst galaxies allow investigation of the supernova remnants and HII regions in environments with considerably higher pressures and densities than our own galaxy.

1.2 Estimating Star Formation Rates

The most widely used diagnostic of the star formation rate in galaxies is the far infrared (FIR) luminosity (Rieke et al. 1980). The FIR emission originates from dust grains which have been heated by photons from young stars created in the starburst. It is assumed that most of the energy in photons emitted by young stars eventually results in heating the dust, which re-radiates it as black-body thermal continuum in FIR photons. Hence, the star formation rate for stars more massive than $5M_\odot$ can be estimated (Cram et al. 1998) from the FIR luminosity ($L_{60\mu}$) as

$$\mathrm{SFR}(M \geq 5\mathrm{M}_\odot) = \frac{L_{60\mu}}{5.1 \times 10^{23}} \ \mathrm{M}_\odot \ \mathrm{yr}^{-1}. \tag{1}$$

The star formation rate can also be estimated from measurements of ultraviolet photons from O and B stars although this method is often severely compromised because of the extinction intrinsic to the starburst. An indirect measure of these photons can, in principle, be made by determining the parameters of the HII regions ionised by them. This can be carried out via

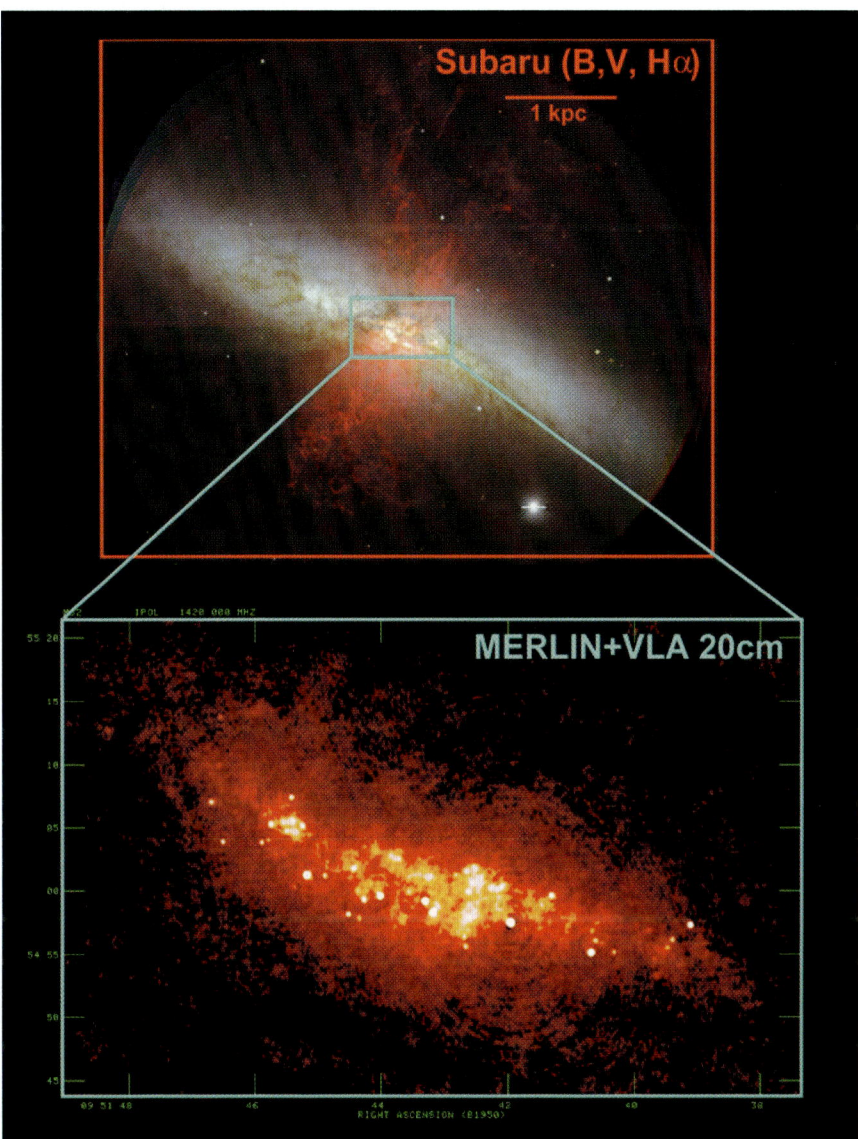

Fig. 1. The Subaru Hα and continuum image (Copyright Subaru Telescope, National Astronomical Observatory of Japan, all rights reserved) with the combined MERLIN/VLA 20 cm image inset (produced by T. W. B. Muxlow, see Wills et al. (1999)).

optical (e.g. Hα) emission lines (Kennicutt 1983) although quantitative measurements of individual starbursts are also compromised by dust extinction. Another method of deriving the HII region parameters is via thermal free-free emission at cm wavelengths, which is unaffected by dust extinction. Hence, if the thermal free-free luminosity, L_T, of a starburst is known, the flux of Lyman continuum photons s^{-1} (N_{UV}) can be estimated (Condon 1992) from

$$N_{UV} \geq 6.3 \times 10^{52} T_e^{-0.45} \nu^{0.1} L_T, \qquad (2)$$

where T_e is the electron temperature in units of 10^4 K, ν is the frequency in GHz and L_T is the thermal free-free luminosity in units of 10^{20} Watt Hz^{-1}. N_{UV} can then be converted into a star formation rate via the relation

$$\mathrm{SFR}(M \geq 5\mathrm{M}_\odot) = \frac{N_{UV}}{3.5 \times 10^{53}} \ \mathrm{M}_\odot \ \mathrm{yr}^{-1}. \qquad (3)$$

The main difficulty with this method is the problem of separating the the free-free emission from the non-thermal synchrotron emission which will dominate the radio luminosity, particularly at the longer wavelengths.

The diffuse non-thermal radio continuum luminosity can also be used to infer (Cram et al. 1998) the star formation rate in galaxies from the relation

$$\mathrm{SFR}(M \geq 5\mathrm{M}_\odot) = \frac{L_{1.4\ \mathrm{GHz}}}{4.0 \times 10^{21}} \ \mathrm{M}_\odot \ \mathrm{yr}^{-1}, \qquad (4)$$

which uses the 1.4 GHz luminosity which is dominated by synchrotron emission. As the most likely sources of relativistic electrons responsible for the diffuse non-thermal emission are supernova explosions, it is generally assumed that the non-thermal radio luminosity is determined by the supernova rate in a galaxy and, hence, the star formation rate. Thus, a supernova rate of ν_{sn} per year will imply a star formation rate of

$$\mathrm{SFR}(M \geq 5\mathrm{M}_\odot) = 25\nu_{sn} \ \mathrm{M}_\odot \ \mathrm{yr}^{-1}. \qquad (5)$$

Hence, this can account for the tight correlation between the far infrared and radio continuum luminosity as both measurements are linked via a common star formation rate (Cram et al. 1998). Note that, as pointed out by Condon (1992), the star formation rate inferred from supernova rates and UV photon fluxes are sensitive to the initial mass function in different ways. Hence, in principle, accurate measurements of star formation rate using these methods could be used to, at least, constrain the initial mass function in starburst galaxies.

2 Messier 82

"Feb. 9th, 1781. Nebula without a star, near the preceding (M81) both appearing in the same field of the telescope. This one is less distinct than the

preceding: the light is faint and elongated with a telescopic star in its extremity. (first) Seen by J.E. Bode at Berlin Obs. on Dec. 31st, 1774, and by P. Mechain in August 1779."

This entry in Charles Messier's 1784 catalogue of faint nebulae is obviously the first reference to M82! M82 is associated with M81 and NGC 3077, forming a small group at a distance of $\sim$ 3 Mpc. Classification is difficult on account of its almost edge-on orientation, but it appears to be a dwarf irregular with a size several times smaller than the Milky Way.

Particular interest in the galaxy began following the discovery by Lynds & Sandage (1963) of a system of Hα filaments extending along the galactic minor axis out to 3 kpc from the plane of the galaxy. They found that these filaments appeared to be expanding from the centre of the galaxy and suggested a time of $\sim 1.5 \times 10^6$ years since all matter in the filaments was near to the centre. This was considered to be as a result of some sort of explosion within M82 and for a while it was known as *"The Exploding Galaxy"* although more recent studies show that the filaments represent the outflow associated with a starburst-driven wind.

The starburst itself remained largely unobserved, as dust in the associated molecular clouds results in large visual extinctions. However, both FIR and radio observations revealed the existence, and relatively small extent ($\sim$700pc), of a central starburst. Estimates of the star formation rate using several methods (Pedlar 2001) indicated that SFR(M>5M$_\odot$) was $\sim$ 2M$_\odot$ per year, which implies a total rate, assuming a standard Initial Mass Function (IMF), of $\sim$ 10M$_\odot$ per year. Hence, this central 300-400 pc of M82 has an SFR approximately 3 times the SFR of our entire Galaxy.

2.1 Multi-Wavelength Studies

Optical Studies

The central few hundred parsecs of M82 have been studied optically by O'Connell & Mangano (1978) and found to contain a number of star clusters, giant HII regions and dust lanes. In contrast, outside the nuclear regions, star formation rates were found to be abnormally low, and deficient in clusters, HII regions and individual supergiant stars. They suggested that active star formation had occurred in parts of the disk but not in the last $\sim 2 \times 10^7$ years. HST observations (O'Connell et al. 1995) revealed $\sim$200 'super star clusters' in the line of sight to the central starburst. These objects have sizes ranging between 3 and 9 parsecs and masses between 10^4 to $10^6 M_\odot$. Melo et al. (2005) estimated the density of super starclusters to be $\sim$ 620 per kpc^2 which is much higher, for example, than in NGC 253. Studies of these clusters have led to the conclusion that in addition to the current burst of star formation, an earlier burst occurred $\sim$600 Myr ago (de Grijs et al. 2001). However, it is important to note that much of the starburst is obscured by at least 20 magnitudes of visual extinction (Mattila & Meikle 2001) and, hence, the clusters which can be studied at optical wavelengths only represent a small proportion of the total population. Note that a number of compact HII regions have been detected at 15GHz (McDonald et al. 2002) and 23GHz (Rodriguez-Rico et al. 2004). They appear to be ionised by optically obscured super star clusters in the current starburst.

The most dramatic optical image of M82 is from the Subaru telescope shown in Fig. 1 (Ohyama et al. 2002). This image shows blue light from the disk of the galaxy against which a series of dusty filaments can be seen in absorption. However, the most spectacular features are the red Hα filaments which can be seen extending $\sim$ 3 kpc from the plane of the galaxy. A plethora of models for these filaments have been developed over the years, including some which interpreted the filaments as evidence of infall. However, Axon & Taylor (1978) found a significant increase in the splitting of the Hα and [NII] lines with distance from the core of the galaxy, which they claimed was due to expansion and suggested an expulsion velocity of $\sim$ 400 km s^{-1} for the filamentary material. Hence, rather than a single explosion, the filaments appear to be due to a 'starburst driven superwind', an interpretation which has been confirmed by X-ray studies (Strickland et al. 1997).

Infrared Studies

The spectrum of M82 is dominated by infrared (IR) emission (see Fig. 2). The IR emission can be split into three regions of the spectrum where radiation from different parts of the starburst dominate. The predominant cause of the near-IR emission is light from late giant stars (Rieke et al. 1980). The mid-IR emission is most likely due to thermal radiation from warm dust, heated by young stars (Telesco & Gezari 1992) and has a steep positive spectral index. The FIR contribution to the spectrum comes from warm dust and also a cooler 'cirrus' component.

Rieke et al. (1980) found the nucleus of M82 to be prominent at 2.2 μm and suggested that this represents a smooth disk of stars centred on the nucleus. Near-IR maps (1.2-2.2 μm) produced ten years later by Telesco et al. (1991) show a strong central peak which was identified with the nucleus, a secondary peak $\sim$ 10$''$ to the west and lower level emission stretching along the major axis of the visible galaxy. They also presented maps at 10.8, 19.2 and 30 μm, all of which show peaks separated by 10$''$ roughly along the major axis of the galaxy. These peaks appear to straddle the 2 μm nucleus. It appears that the 10.8 μm map and those at longer wavelengths illustrate a nearly edge-on ring of warm dust centred on the nucleus whereas the images in the near-IR $\sim$ 2 μm) indicate regions of high star formation, dominated by the flux of late-type stars.

The observational evidence for the correlation between the non-thermal radio and FIR emission is very widespread (Helou et al. 1985). However, with the improvement in resolution, it was realised that the correlation breaks down on sub-galactic scales. For example, Telesco & Gizani (1992) showed that in

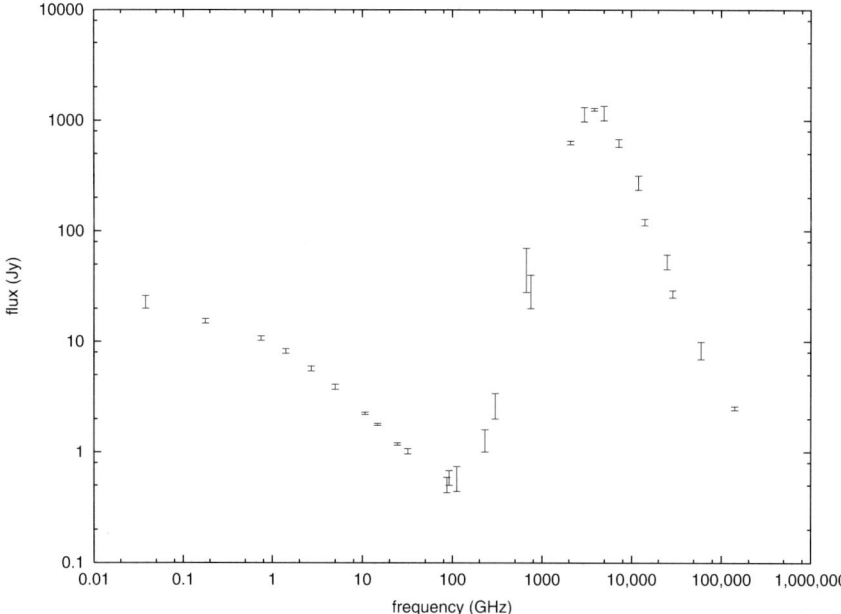

Fig. 2. The spectrum of the integrated emission of M82. Data have been obtained from Thronson et al. (1989), Carlstrom (1988), Carlstrom & Kronberg (1991), Smith et al. (1990) and tabulated flux densities from various references contained within Klein et al. (1988).

M82, the radio emission is equally bright near the nucleus and at the western IR lobe, whereas the IR emission at the nucleus is less than half as bright as the lobe. Although the presence of thermal (free-free) radio emission may account for some of the differences between the IR and radio distributions (Carlstrom 1987), the 'excess' diffuse radio emission near the 2 μm nucleus must have another explanation. Since the radio-IR correlation reflects the coexistence of SNRs and OB stars, the most obvious conclusion would be that the difference in the IR/radio ratios at the nucleus and the lobe implies a relative deficiency of OB stars near the nucleus. However, Telesco & Gizani (1992) suggested that the relative numbers of SNRs and OB stars may be comparable at the two locations. Instead, they proposed that the non-thermal radio emission detected near the nucleus is generated by SNRs expanding into the void left by the energetic outflow of gas from the starburst, powered by supernovae and OB winds. They suggested that the IR emission at the nucleus is relatively low because the dust was also swept away during the outflow.

Telesco et al. (1991) presented 2.2 μm images of the centre of M82 which indicate a wing on either side of the nucleus extending to ∼ ± 30″ along the major axis. They, therefore, proposed that the centre of M82 contains a bar with a projected length of nearly a kiloparsec. They also noted that the axis defined by the orientation of the bar is tilted ∼ 4° with respect to the major axis of the more extended emission, suggesting that the major axis of the bar is tilted to our line of sight. The inclination of the disk of M82 to our line of sight is highly uncertain but has been calculated by a number of authors to be somewhere between 8° and 29° from edge-on. Notni & Bronkalla (1983) suggested that the southern side of M82 is the more distant side since they observed that the light from the southern half is more heavily reddened than the light from the northern side. Assuming that the disk is inclined at 10° to the line of sight and using the angle of 4° observed between the bar and M82's major axis, Telesco et al. (1991) inferred that the bar must actually be inclined $\sim 22°$ to the plane of the sky, with the eastern, receding end being more distant from us than the western end.

X-Ray Studies

The X-ray emission of M82 has a very complex structure, comprising extended soft and medium temperature plasma and also point-like hard X-rays near to the centre of the galaxy.

X-ray observations using the High Resolution Imager (HRI) of the Einstein Observatory showed the soft X-ray emission to be in the form of a halo of patchy diffuse emission which extends over several kiloparsecs along the minor axis, both north and south of the nuclear region (Watson et al. 1984). This diffuse emission is thermal in origin, probably associated with shock heated clouds from a galactic wind (Strickland et al. 1997) and showed substantial spatial correlation with the Hα filamentary structure within the optical halo. More recent observations by XMM-Newton (Stevens et al. 2003) show the extraplanar X-ray emission extends out to a height of at least 14 kpc in the north and 7.5 kpc in the south. In the north the superwind appears to have blown a hole in the HI gas, which may explain why the X-ray halo is more extended in this direction.

HRI observations of the hard X-rays from the central $0.5 - 1$ kpc of M82 (Watson et al. 1984) show signs of time variability and spectral change. Several discrete unresolved sources were also detected in the central region and possible identifications were made with the positions of optical features and compact radio features. High resolution observations by Chandra have revealed at least 20 discrete X-ray sources in the central kiloparsec, few of which appear to be related to the radio SNR discussed below. In fact, the majority of the X-ray sources have properties consistent with X-ray binaries. One of these sources may be associated with a ~ 500 M$_\odot$ black hole, representing a possible new class of object with a mass intermediate between stellar-mass and supermassive black holes (Kaaret et al. 2001).

Radio Continuum Studies

One of the more spectacular radio images of M82 is shown in Fig. 1 and uses combined MERLIN and VLA data at 20 cm (Wills et al. 1999). We will now briefly review the history of radio observations of this source.

Messier 82 is a relatively strong radio continuum source and is included in the 3C catalogue as 3C 231. One of the first detailed radio studies of the galaxy was by Hargrave (1974) who used the 5 km array to image M82 at 5 GHz with $\sim 2''$ angular resolution. The radio source is extended over a $50'' \times 15''$ region which he proposed is largely powered by a compact radio source (41.95+575) embedded in the diffuse structure. This compact object was first observed by Bash (1968) and studied further by Wilkinson (1971). Kronberg & Wilkinson (1975) used the Greenbank interferometer to image M82 at 8 GHz and reported the possible existence of 9 compact sources within the extended source. However, it was not until the mid 1980s that MERLIN and the VLA revealed the extensive number of compact sources present. Both Unger et al. (1984), using MERLIN, and Kronberg et al. (1985), using the VLA, reported the presence of ~ 30 compact objects over the central 750 pc. At the time, most extragalactic radio sources were thought to be powered by single supermassive black holes, and, hence, these radio observations of M82 suggested a very different mechanism. Unger et al. (1984) found the brightness temperatures of the sources to be $> 10^5$ K, which ruled out the idea of HII regions and suggested the sources were either radio supernovae or young supernova remnants[3].

More recent observations have indicated that most of the more luminous compact sources in M82 are, in fact, SNRs. Ten year monitoring of the compact radio sources from 1981 to 1991 (Kronberg & Sramek 1992) has shown that, although they were time variable in this period, they have not shown the rapid variations expected of radio supernovae. They also showed that no new sources had appeared since 1981 which suggested that previous estimates of the star formation rates were probably too high (eg Rieke et al. 1980). The identification of the compact sources as SNRs was finally established with MERLIN observations at 5 GHz (Muxlow et al. 1994). They detected in excess of 40 discrete sources, all of which were resolved by their 50 mas (0.75 pc) beam. Of these, several show complete shells and many more show structure consistent with partial shells. Muxlow et al., by assuming a typical SNR expansion velocity of 5000 km s^{-1}, estimated an average age for the remnants of ~ 200 years (compared with ~ 2000 and 3000 years, respectively, for Galactic and LMC remnants). In Fig. 4 we show a plot of the flux density against size of the remnants in M82 compared with samples in the LMC and Arp220 (both scaled to 3.2 Mpc). From this plot it can be seen that most of the M82 remnants are more luminous and more compact than the most luminous Galactic supernova remnant, Cassiopeia A. One question which needed to be addressed was whether the sizes of the remnants are functions of age,

---

[3] A radio supernova (RSn) is associated with the initial supernova explosion interacting with circumstellar gas and is consequently very compact ($\ll 1$ pc) and shows large flux changes over periods of a few years, whereas a supernova remnant (SNR) represents the interaction of the supernova with the interstellar medium and has a size typically of $1 - 100$ pc and shows gradual changes in flux density.

or whether the sizes are determined by the ambient density. One method of distinguishing between these possibilities is to measure the expansion velocities of the remnants directly using VLBI. Initially EVN measurements were carried out at two epochs and an expansion velocity of one of the brighter remnants, 43.31+592 was measured to be $\sim 10,000$ km s^{-1} (Pedlar et al. 1999). If the remnant is in free expansion this implies an age of ~ 35 years, or shorter if significant deceleration has occurred. Recent Global VLBI measurements (McDonald et al. 2001) of these remnants (Fig. 3) have angular resolutions of a few milliarcseconds corresponding to ~ 0.06 pc and, hence, direct measurements of deceleration are feasible. Given that 43.31+592 must have originated before it appeared on the early images in 1972 (Kronberg & Wilkinson 1975), it is already possible to rule out Sedov expansion suggesting that at least this remnant is still expanding close to free expansion (Fig. 5).

It, therefore, appears that the sizes of at least some of the remnants give estimates of their ages, and, hence, assuming a common expansion velocity of $10,000$ km s^{-1}, or simply assuming all the remnants more compact than Cassiopeia A are less than 330 years old, gives a supernova rate of ~ 0.07 yr^{-1}. This can be used to calculate a star formation rate ($\geq 5M_\odot$) of ~ 1.8 M$_\odot$ yr^{-1}, which is consistent with the ($\geq 5M_\odot$) star formation rates derived from the FIR emission, non-thermal radio continuum and the thermal continuum – all of which give values close to 2 M$_\odot$ yr^{-1} (Pedlar 2001).

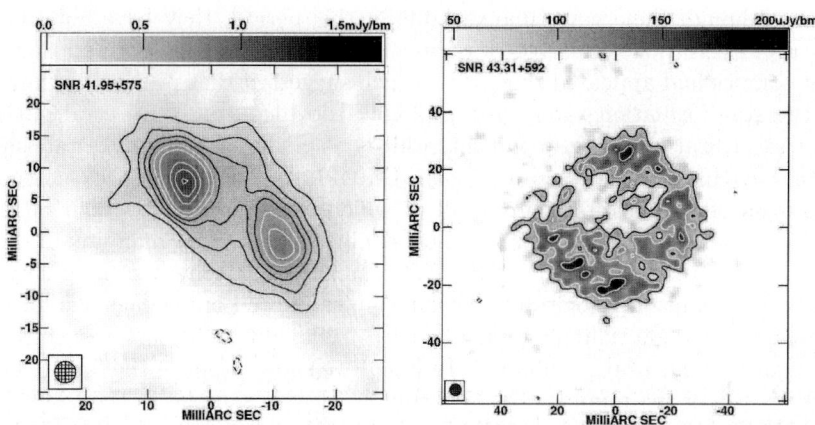

Fig. 3. 18 cm Global VLBI images of the compact sources 41.95+575 and 43.31+592 at angular resolutions of 3.3 and 4.0 mas respectively. The images were produced from epoch 2001.2 data by R. Beswick (Private Communication).

The most luminous compact source in M82 (41.95+575) appears to be anomalous when compared to the other sources. Not only is it more than an order of magnitude more luminous than typical remnants in M82, but its luminosity has been decreasing rapidly (8.5% per annum) since its discovery in

Flux Density vs Diameter plot for Extragalactic SNR

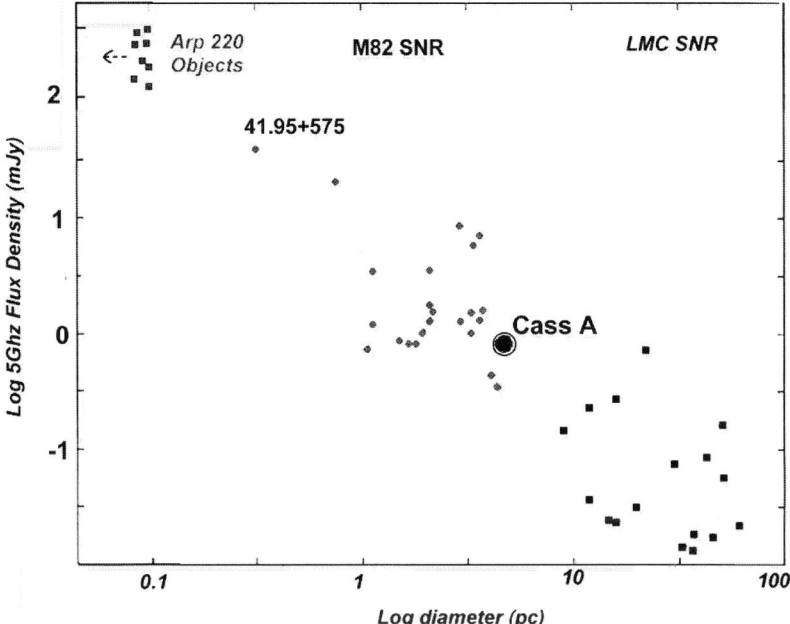

Fig. 4. A plot of size versus 5 GHz flux density for the compact sources seen in M82 (data from Muxlow et al. 1994) compared with Cassiopeia A. The flux densities for supernova remnants in the Large Magellanic Cloud scaled to be at the distance of M82 are shown, together with the approximate parameters of the Arp220 estimated from Smith et al. (1998).

the mid 1960s. Furthermore, recent global VLBI measurements (McDonald et al. 2001) show an elongated structure (Fig. 3, left) which could be consistent with collimated ejection rather than a shell. Unlike the shell supernova discussed above, the object is not expanding at $\sim 10,000$ km s^{-1} and limits of < 2500 km s^{-1} have been determined. One explanation for this object is that it is a supernova remnant confined by a high density medium, although other possibilities (e.g. a gamma-ray burst radio afterglow) require investigation.

As can be seen in Fig. 4, the current radio luminosity and size of 41.95+575 appears to be intermediate between the typical remnants in M82 and the compact objects recently discovered in Arp220 (Smith et al. 1998) and Mkn273 (Carilli et al. 2000). Note that in the 1960s 41.95+575 was of comparable luminosity to the Arp220/Mkn273 objects. Thus, it is possible that the compact objects seen in Arp220 and Mkn273 may be similar to 41.95+575.

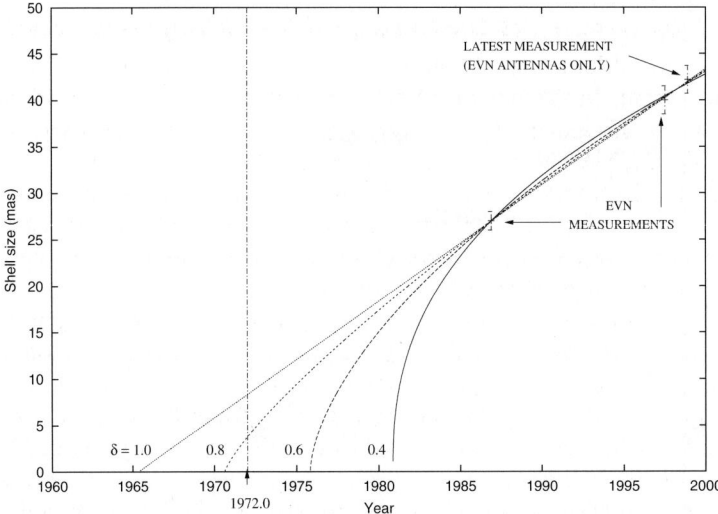

Fig. 5. A plot of size versus time for the compact supernova remnant 43.31+592 in M82 from McDonald et al. (2001). The curves show four possible models with deceleration parameters from 0.4 to 1.0 (Free-Expansion). Note that 43.31+592 was present in radio images taken in 1972, which puts a lower limit on its age and rules out deceleration parameters <0.7.

At low radio frequencies, although the continuum is dominated by synchrotron emission (Fig. 2), the ionised gas is detectable via free-free absorption (Wills et al. 1997). Many of the spectra of the supernova remnants show low frequency (<1 GHz) turnovers consistent with free-free absorption by ionised gas with emission measures of $\sim 10^5$ pc cm^{-6}. Also, localised regions of free-free absorption can be seen against the diffuse synchrotron emission at 0.4 GHz. Interestingly, there is a region of deep absorption, ~ 100 pc in extent, centred on the most luminous compact radio source 41.95+575.

At ~ 100 GHz the continuum appears to be largely free-free emission consistent with ionised gas excited by 5×10^8 O8 stars (Carlstrom 1991). High resolution observations at 15 GHz (McDonald et al. 2002) have detected a number of compact HII regions with emission measures as high as 10^8 pc cm^{-6}, which require ~ 500 O5 stars to provide the necessary excitation.

2.2 Gas Dynamics in M82

Neutral hydrogen is one of the best tracers of gas dynamics in galaxies and is particularly valuable in an optically obscured starburst galaxy such as M82. Neutral hydrogen observations of the region surrounding M82 (Gottesman & Weliachew 1977; Cottrell 1977) provided direct evidence that some kind of interaction is occurring. There appeared to be an asymmetric cloud surrounding

M82 with a bridge extending towards its neighbour, M81. These features are almost certainly the product of a tidal encounter with M81 which possibly occurred several hundred million years ago. VLA D-array observations of the M81-M82 group by Yun et al. (1993) beautifully illustrate the "filamentary tidal features threading all three galaxies". Their observations suggested that the M81-M82 HI bridge was tidally drawn out of M82, leading to the starburst activity. An alternative mechanism involves the interaction with a neighbour inducing non-circular motions in the central regions which may lead to the formation of a bar (Noguchi 1988). Gas flows are focussed inwards via the bar and interact to produce starburst activity.

Neutral hydrogen emission studies of external galaxies are limited by instrumental sensitivity to angular resolutions greater than 5 arcseconds. However, by using neutral hydrogen *in absorption* it is possible to observe neutral gas on scales limited only by the angular resolution of the instrument. The work of Weliachew et al. (1984) on neutral hydrogen absorption in M82 used the VLA in the B-array configuration to yield a beam of $6'' \times 3.8''$. This revealed significant neutral hydrogen absorption over the M82 continuum. They also found that the position of maximum optical depth shifts from west to east with increasing velocity, indicative of rotation. The average HI optical depth shows two concentrations connected by a faint bridge and this was interpreted as a rotating ring of gas with a radius of 250 pc, maximum rotational velocity ~ 140 km s^{-1} and systemic velocity 225 km s^{-1}.

More recently, a high resolution study of neutral hydrogen gas in M82 was undertaken with MERLIN (Wills et al. 1998). With high resolution, the large numbers of compact sources in M82 make it possible to probe the ISM of the starburst along many lines of sight with sub-arcsecond resolution. Since the compact sources are typically $0.1 - 0.2''$ in size (Muxlow et al. 1994), the detection of absorption against individual compact sources allows a probe of the neutral hydrogen distribution on scales $\sim 1.5 - 3$ pc. Of the 33 supernova remnants detected in this study, 26 showed absorption suggesting that the majority of sources are embedded within or located behind neutral gas clouds.

Neutral hydrogen absorption observations by Wills et al. (2002), using the VLA in the A-array configuration (angular resolution $\sim 1.3''$), detected four HI shells in the central kiloparsec of M82. These shells are $\sim 30 - 50$ pc in diameter with expansion velocities ~ 30 km s^{-1} – implying ages of $\sim 10^6$ years. Similar shells have been detected, for example, in the LMC, albeit with larger diameters, smaller expansion velocities and a lower density per unit area (Kim et al. 1999). The typical kinetic energies of the M82 shells are $10^{51} - 10^{52}$ erg, which exceeds the energy input to the compact LMC shells by at least an order of magnitude. This difference could result from the higher star formation rate in the M82 starburst (e.g. M82 0.05 SNR yr^{-1}, Muxlow et al. 1994; LMC 0.004 SNR yr^{-1}, Meaburn 1991). There appears to be no association between the shells and the overall distribution of the SNRs in the central starburst of M82, which is expected as the shells are younger than the typical SNR (representing star formation regions 10^7 years ago). However, there is an association between the current regions of star formation (as given by the ionised gas distribution; Achtermann & Lacy 1995) and the HI shells – in each case the ionised gas peaks are typically 60 pc north of the HI shells, which could suggest that the starburst has moved inwards by this distance during the lifetime of the shells. However, as well as being influenced by the sources driving the expanding shells, the gas dynamics in M82 are also affected by a bar potential (Wills et al. 2000).

The molecular dynamics of M82 have been studied in a number of molecular lines. Single dish observations of the CO 1-0 transition (Nakai et al. 1987) detected a double structure in the centre of M82 which was interpreted as evidence for a molecular ring with a radius of 200 pc. One of the first high angular

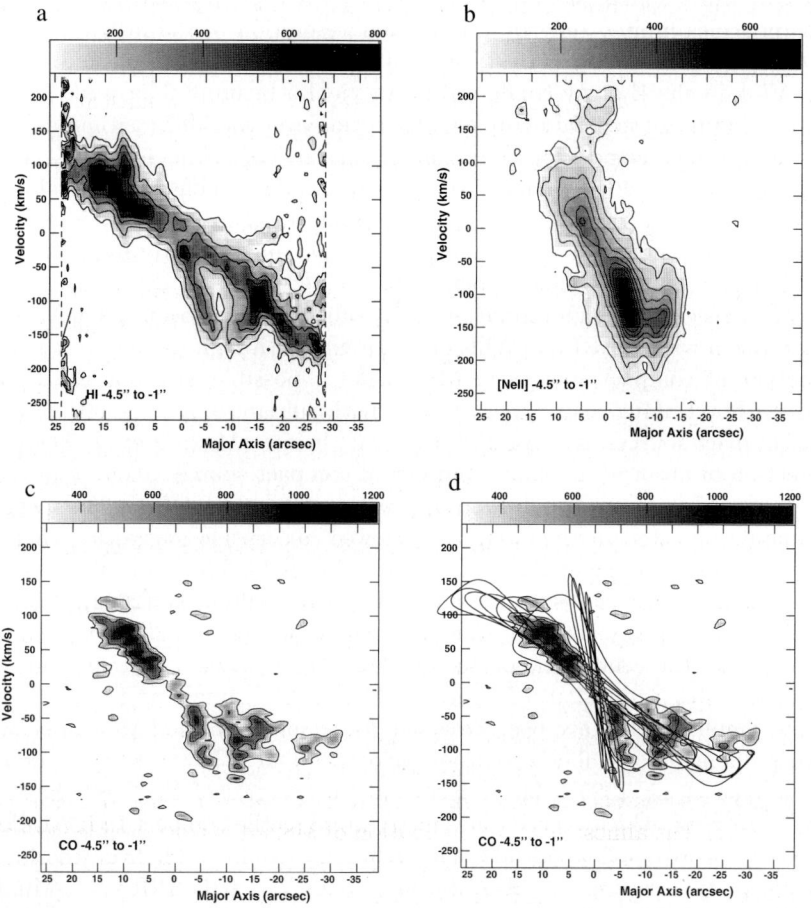

Fig. 6. Position-velocity diagrams of HI (a), [NeII] (b) and CO (c) taken parallel to the major axis of M82. In (d) we show the CO with possible bar orbits superimposed. From Wills et al. (2000).

resolution interferometric studies of molecular lines was by Brouillet & Schilke (1993). In it the south-west part of M82 was studied on arcsecond scales in the HCN (1-0) line. Shen & Lo (1995) studied the CO emission at similar angular resolution and suggested a model in which the CO was concentrated in molecular spiral arms at 125 and 390 pc from the nucleus. More recently, a mosaic of high sensitivity CO observations over a 4 kpc region (Walter et al. 2002) has resulted in the detection of $\sim 10^9$ $M_\odot$ of molecular gas outside the central kpc. Much of the gas appears to be associated with molecular streamers and appears to be part of a ~ 200 km s^{-1} molecular outflow associated with with the starburst driven wind.

The dynamics of the ionised gas can be studied both in the near infrared via [NeII] lines (Achtermann & Lacy 1995) and the radio/mm band via recombination lines (Rodriguez-Rico 2004). The basic dynamics of the ionised gas in the disk follows a velocity gradient which is significantly steeper than the main gradients observed in the neutral and molecular gas (Wills et al. 2000). The [NeII] lines suggest a structure dominated by a nuclear ring and two ionised 'lanes'. Such components are found in other barred galaxies, supporting the suggestion that M82 may contain a nuclear bar. The ionized ring is approximately circular with a radius of 86 pc and an inclination of 73°. The centre of the ring closely corresponds to M82's kinematic center.

The bar model was investigated more fully (Wills et al. 2000) by comparing the molecular, atomic hydrogen and ionised gas dynamics, and several dynamical features which could be interpreted as the signature of gas moving in a bar potential were identified. The CO, HI and [NeII] position-velocity diagrams (Fig. 6) showed two main velocity gradients, which can be interpreted as due to gas moving on x1- and x2-orbits within a bar potential. This is consistent with a flat rotation curve velocity of 140 km s^{-1}. A fit to the velocity data implied a bar potential of total length 1 kpc and a non-axisymmetry parameter 0.9. The angular velocity of the bar is 217 km s^{-1} arcsec^{-1} with an inclination angle of 80° and a projected angle between the bar and the major axis of the galaxy of 4°.

In an alternative explanation for the 'bar signature' velocity features they are considered to be due to an expanding molecular superbubble. This hypothesis was first suggested by Weiss et al. (1999) on the basis of position-velocity plots and possible evidence for a CO shell was presented by Matsushita et al. (2000). Wills, Pedlar & Muxlow (2002) noted that the superbubble was in the vicinity of one of the neutral hydrogen shells (see above) and the ~ 100 pc HII region was associated with the brightest remnant, 41.95+575. Unfortunately, given the almost edge-on inclination of M82, it is not possible to decide unambiguously whether the non-circular gas dynamics observed in this region are due to an expanding shell or motions in a bar potential.

3 Future Prospects

It is unfortunate that M82 will not be observable with ALMA, as subarcsecond, high sensitivity studies of both molecular gas and mm continuum emission of this object would have undoubtably produced spectacular results. Nevertheless, further high sensitivity studies of the type reported by Walter et al. (2002) can be carried out by both IRAM and CARMA.

In the radio, the large increase in sensitivity of e-MERLIN, EVLA and e-VLBI will result in significant advances in the studies of the more evolved supernova remnants as well as HII regions. Eventually, the SKA will enable studies of neutral hydrogen emission to be carried out on sub-arcsecond scales. However, the main impact of these radio instruments will be to allow detailed studies, comparable to those undertaken in M82, in more distant, or less active, starburst galaxies.

4 Acknowledgements

It is a pleasure to thank Tom Muxlow and Rob Beswick for their help in the production of the article. We also thank the Subaru telescope staff for permission to use their beautiful image of M82 in Fig. 1.

References

Achtermann, J.M., Lacy, J.H. 1995 ApJ **439**, 163
Axon, D.J., Taylor, K. 1978 Nature **274**, 37
Bash, F.N. 1968 ApJ **152**, 375
Brouillet, N., Schilke, P. 1993 A&A **277**, 381
Carilli, C.L., Taylor, G.B. 2000 ApJ **532**, 95 Carlstrom, J.E. In: *Galactic and Extragalactic Star Formation*, ed by R.E Pudritz and M. Fich (Dordrecht: Kluwer 1987) p571
Carlstrom, J.E. 1988 PhD Thesis, University of California, Berkeley
Carlstrom, J.E., Kronberg, P. 1991 ApJ **366**, 422
Condon, J.J. 1992 ARA&A **30**, 575
Cottrell, G.A. 1977 MNRAS **178**, 577
Cram, L., Hopkins, A., Mobasher, B., Rowan-Robinson, M. 1998 ApJ **507**, 155
Eggen, O.J., Lynden-Bell, D., Sandage, A.R. 1962 ApJ **136**, 748
Gottesman, S.T., Weliachew, L. 1977 ApJ **211**, 47
Grijs, R. de, O'Connell, R.W., Gallagher, J.S. 2001 AJ **121**, 768
Hargrave, P.J. 1974 MNRAS **168**, 491
Helou, G., Soifer, B.T., Rowan-Robinson, M. 1985 ApJ **298**, L7
Kaaret, P., Prestwich, A.H., Zezas, A., Murray, S.S., Kim, D.-W., Kilgard, R.E. Schlegel, E.M., Ward, M.J. 2001 MNRAS **321**, L29
Kennicutt, R.C. 1983 ApJ **272**, 54
Kim, S., Dopita, M.A., Staveley-Smith, L., Bessell, M.S. 1999 AJ **118**, 2797

Klein, U., Wielebinski, R., Morsi, H.W. 1988 A&A **190**, 41
Kronberg, P.P., Wilkinson, P.N. 1975 ApJ **200**, 430
Kronberg, P.P., Biermann, P., Schwab, F.R. 1985 ApJ **291**, 693
Kronberg, P.P., Sramek, R.A. 1992 The 10-year monitoring of the compact radio sources in the nuclear regions of M82. In: *MPI fuer Extraterrestrische Physik, X Ray Emission from Active Galactic Nuclei and the Cosmic X Ray Background* p247
Lynds, C.R., Sandage, A.R. 1963 AJ **68**, 284
Matsushita, S., Kawabe, R., Matsumoto, H., Tsuru, T.G., Kohno, K., Morita, K., Okumura, S.K., Vila-Vilaro, B. 2000 ApJ **545**, L107
Mattila, S., Meikle, W.P.S. 2001 MNRAS **324**, 325
McDonald, A.R., Muxlow, T.W.B., Pedlar, A., Garrett, M.A., Wills, K.A., Garrington, S.T., Diamond, P.J., Wilkinson, P.N. 2001 MNRAS **322**, 100
McDonald, A.R., Muxlow, T.W.B., Wills, K.A., Pedlar, A., Beswick, R.J. 2002 MNRAS **334**, 912
Meaburn, J. 1991 Studies of the Large Magellanic Cloud Using Optical Interstellar Emission Lines. In: *The Magellanic Clouds: Proceedings of the 148th Symposium of the International Astronomical Union*, ed by R. Haynes and D. Milne (Kluwer Academic Publishers, Dordrecht) p421
Melo, V.P., Muñoz-Tuñón, C., Maíz-Apellániz, J., Tenorio-Tagle, G. 2005 ApJ **619**, 270
Muxlow, T.W.B., Pedlar, A., Wilkinson, P.N., Axon, D.J., Sanders, E.M., de Bruyn, A.G. 1994 MNRAS **266**, 455
Muxlow, T.W.B., Richards, A.M.S., Garrington, S.T., Wilkinson, P.N., Anderson, B., Richards, E.A., Axon, D.J., Fomalont, E.B., Kellermann, K.I., Partridge, R.B., Windhorst, R.A. 2005 MNRAS **358**, 1159
Nakai, N., Hayashi, M., Handa, T., Sofue, Y., Hasegawa, T., Sasaki, M. 1987 PASJ **39**, 685
Noguchi, M. 1988 A&A **203**, 259
Notni, P., Bronkalla, W. 1983 M82 - Tilt and Warp of its Principal Plane. In: *IAU Symp. No. 100, Internal Kinematics and Dynamics of Galaxies*, ed by Athanassoula, E. (Dordrecht:Reidel) p67
O'Connell, R.W., Mangano, J.J. 1978 ApJ **221**, 62
O'Connell, R.W., Gallagher, J.S., Hunter, D.A. 1995 ApJ **446**, L1
Ohyama, Y., Taniguchi, Y., Iye, M., Yoshida, M., Sekiguchi, K., Takata, T., Saito, Y. Kawabata, K.S., Kashikawa, N., Aoki, K., Sasaki, T., Kosugi, G., Okita, K., Shimizu, Y., Inata, M., Ebizuka, N., Ozawa, T., Yadoumaru, Y., Taguchi, H., Asai, R. 2002 PASJ **54**, 891
Pedlar, A., Muxlow, T.W.B., Garrett, M.A., Diamond, P., Wills, K.A., Wilkinson, P.N., Alef, W. 1999 MNRAS **307**, 761
Pedlar, A. 2001 IAU Symposium **205**, 366
Rieke, G.H., Lebofsky, M.J., Thompson, R.I., Low, F.J., Tokunaga, A.T. 1980 ApJ **238**, 24
Rodriguez-Rico, C.A., Viallefond, F., Zhao, J.-H. Goss, W.M., Anantharamaiah, K.R. 2004 ApJ **616**, 783
Shen, J., Lo, K.Y. 1995 ApJ **445**, L99
Smith, P.A., Brand, P.W.J.L., Puxley, P.J., Mountain, C.M. 1990 MNRAS **243**, 97
Smith, H.E., Lonsdale, C.J., Lonsdale, C.J., Diamond, P.J. 1998 ApJ **493**, 17
Stevens, I.R., Read, A.M., Bravo-Guerrero, J. 2003 MNRAS **343**, 47

Strickland, D.K., Ponman, T.J., Stevens, I.R. 1997 A&A **320**, 378
Telesco, C.M., Joy, M., Dietz, K., Decher, R. 1991 ApJ **369**, 135
Telesco, C.M., Gezari, D.Y. 1992 ApJ **395**, 461
Thronson, H.A., Walker, C.K., Walker, C.E., Maloney, P. 1989 A&A **214**, 29
Unger, S.W., Pedlar, A., Axon, D.J., Wilkinson, P.N. 1984 MNRAS **211**, 783
Walter, F., Weiss, A., Scoville, N. 2002 ApJ **580**, L21
Watson, M.G., Stanger, V., Griffiths, R.E. 1984 ApJ **286**, 144
Weedman, D.W., Feldman, F.R., Balzano, V.A., Ramsey, L.W., Sramek, R.A., Wuu, C.C. 1981 ApJ **248**, 105
Weiss, A., Walter, F., Neininger, N., Klein, U. 1999 A&A **345**, L23
Weliachew, L., Fomalont, E.B., Greisen, E.W. 1984 A&A **137**, 335
Wilkinson, P.N. 1971 MNRAS **154**, P1
Wills, K.A., Pedlar, A., Muxlow, T.W.B., Wilkinson, P.N. 1997 MNRAS **291**, 517
Wills, K.A., Pedlar, A., Muxlow, T.W.B. 1998 MNRAS **298**, 347
Wills, K.A., Redman, M.P., Muxlow, T.W.B., Pedlar, A. 1999 MNRAS **309**, 395
Wills, K.A., Das, M., Pedlar, A., Muxlow, T.W.B., Robinson, T.G. 2000 MNRAS **316**, 33
Wills, K.A., Pedlar, A., Muxlow, T.W.B. 2002 MNRAS **331**, 313
Yun, M.S., Ho, P.T.P., Lo, K.Y. 1993 ApJ **411**, L17

Active Galactic Nuclei

S.L. Lumsden[1]

School of Physics and Astronomy, University of Leeds sll@ast.leeds.ac.uk

1 Introduction

Active galactic nuclei (AGNs) have been of interest to astronomers for many decades, since the earliest examples such as NGC 1068 identified by Carl Seyfert in the 1940s through the strong emission line spectra they showed (Seyfert 1943). The discovery of radio galaxies and of quasars in the subsequent two decades led to a considerable growth in this interest, to the point that in 2005 alone more than 1000 refereed papers were published on some aspect of their study.

The AGN 'zoo' has expanded greatly in recent years. In addition to the obvious examples such as quasars, radio galaxies and Seyferts, we now know there is evidence for low luminosity AGNs in nearby galaxies that on a global scale look relatively 'normal' (Ho et al. 1997). In addition there are both AGNs with broad permitted lines (type Is, such as Seyfert 1s and quasars) and those with narrow permitted lines (type IIs, such as Seyfert 2s). Obviously, this poses several potential problems to our understanding of AGNs. Why do they have such a large range of luminosity? Why do some have strong line emission but weak radio emission and others strong radio emission but weak line emission? How do AGNs evolve, and what role do they play in the general evolution of the galaxy population? The only aspect where real progress was made in a global AGN unification scheme until recently was in the demonstration that the lack of broad lines in Seyfert 2s was largely due to obscuration of the central broad line region (e.g. Antonucci 1993). The aim of this work is to show how our understanding has moved on in the last decade or so.

There is little doubt nowadays that luminous AGNs are powered through the accretion of material onto supermassive black holes. There is evidence that this is true even at relatively low luminosities, at least in the cases where there is evidence for a compact nuclear X-ray source. This link breaks down only in those LINER galaxies without a strongly nucleated core, where there is strong evidence that shocks and flows related to a burst of massive star formation rather than an AGN give rise to the low ionisation emission lines (e.g. Dudik et al. 2005). Therefore, to gain a full understanding of AGNs we need to study how the black hole evolves, how it is fuelled and how the emission from the AGN itself affects its surroundings.

Furthermore, the permitted line emission from an AGN arises from two distinct regions: the broad line region lies close to the core, which must have high enough density to suppress any forbidden line emission by collisional de-excitation and the narrow line region which has more 'typical' gas densities and both permitted and forbidden lines trace similar structures. Our understanding of AGNs also needs to encompass a full understanding of how these two regions behave as well.

It is no surprise that John Dyson should have worked on AGNs among so many other topics during his career given the importance of shock physics and gas kinematics as a whole to their study. He has contributed in particular to our understanding of how the high and low ionisation lines may arise in the broad line region, how broad absorption quasars may arise and in terms of explaining the overall narrow line emission from AGNs as a result of shock emission. I will touch on these topics below, but the overall aim of this piece is to place such work in the wider context, and I will, therefore, give a review of the current state of AGN research as well.

2 The Central Engine

As noted above, there is overwhelming evidence from a variety of sources that supermassive black holes are the key drivers of AGN luminosity. One obvious example is in the variability of the X-ray emission. In many AGNs this sets limits on the size that are incompatible with any other source than the region near an accretion disk around a supermassive ($10^5 - 10^{10}$ $M_\odot$) black hole. The most extreme examples, such as those AGNs called narrow line Seyfert 1s (which have broad Balmer lines with a velocity width < 2000km s^{-1} – e.g. Osterbrock & Pogge 1985), can show variations of over an order of magnitude on timescales of only a few days (Boller 2001). Reverberation line mapping (e.g. Peterson 1993) is another particularly valuable means of studying the gas near the central engine that will be discussed in more detail in Sec. 3. It also shows results consistent with the emitting gas being only a few light days away from a central supermassive core.

It is relatively trivial to outline the basic system luminosity given that the energy must be released through accretion. The Eddington luminosity sets a limit for spherical accretion from the balance of gravitational attraction and radiation pressure outwards, given by the following if the gas is all ionised hydrogen.

$$L_{Edd} = \frac{4\pi G M_{\rm BH} m_{\rm p} c}{\sigma_T} \simeq 1.3 \times 10^{38} \frac{M_{\rm BH}}{M_\odot} {\rm erg/s} \qquad (1)$$

The actual emitted luminosity of course depends on the accretion rate and how much of the mass can be converted into energy. We have no *a priori* knowledge of the accretion efficiency (which ranges from 6% in a Schwarzschild black hole up to 42% in an accretion disk co-rotating with a Kerr black hole, at least if the energy is released and freely radiated outside the event horizon). It is common therefore in AGN studies to parameterize the accretion rate in terms of the Eddington luminosity, and a nominal 10% accretion efficiency.

$$\dot{M}_{\rm Edd} = 10 L_{\rm Edd}/c^2 \qquad (2)$$

with the resulting dimensionless accretion rate

$$\dot{m} = \dot{M}_{\rm acc}/\dot{M}_{\rm Edd} \qquad (3)$$

Although simplistic this theory demonstrates that two of the key parameters in determining the appearance of an AGN must be the black hole mass and the accretion rate.

Of course, this simple theory does not take account of the actual details of how the accreting matter is turned into radiation. The basic AGN model is based on the assumption of the presence of an optically thick, geometrically thin thermal accretion disk (e.g. Shakura & Sunyaev 1973), which can explain many of the features observed, such as the 'big blue bump' seen in many Seyfert 1s and quasars (see, e.g., the examples and discussion in Shang et al. 2005), since most of the thermal emission emerges in the ultraviolet and optical. A thermal accretion disk does not explain the observed high energy X-ray emission seen in most AGNs, but this is popularly believed to be due to upscattering of the ultraviolet photons from hot electrons in a hot disk corona. This disk+corona model, however, does not seem to apply in all cases.

The main advance in recent years in our understanding of the properties of accretion onto supermassive black holes has actually come from the study of Galactic black hole X-ray binaries. Attempts have been made to explain the different states that these binaries pass through. These range from completely quiescent to high accretion, high luminosity outbursts (see, e.g., Esin et al. 1997). The basis for understanding the behaviour of these systems lies in models of advection dominated accretion flows (ADAFs: see, e.g., Narayan & Yi (1994) for the basic theory and for a brief, simple, introduction to the background to these in an AGN context see, e.g., Maccarone et al. (2003); Falcke et al. (2004); Jester (2005)). The standard thin accretion disk still exists in these models. The main difference to the simpler model is that the matter can be advected into the event horizon without radiating away energy as thermal radiation. Modifications to this theory in which energy can be transported outwards but as outflows rather than radiation (e.g. Blandford & Begelman 1999) also exist – these models help to overcome one flaw of the simplest ADAF models which is that the gas becomes too energetic to remain bound if it is not radiating.

Two ADAF solutions have been studied in particular. In one case the accretion rate is high, and in the other case it is low. The latter state has been studied in detail for AGNs (see the recent review by Narayan 2005). The only necessary condition is that the gas is optically thin, and is unable to radiate efficiently. As a result, the gas itself becomes much hotter than in the standard thermal accretion disk model. There are other possible versions of the low state, radiatively inefficient model (e.g. only the inner disk becomes optically thin, rather than the whole core, or the corona around the disk becomes thin, etc.), but all share the common characteristic that much of the accreted material is advected directly into the black hole without emitting significant radiation. In the low state the accretion disk itself also has a larger central gap than in the normal thermal state (see, e.g., the discussion in Narayan & Yi 1994). The transition between the various states is not absolutely clear, but sources radiating much below 1% of Eddington are probably in the low state and most Seyfert galaxies and quasars are probably in a normal, thermal disk, high state. A schematic of the overall picture is shown in Fig. 1.

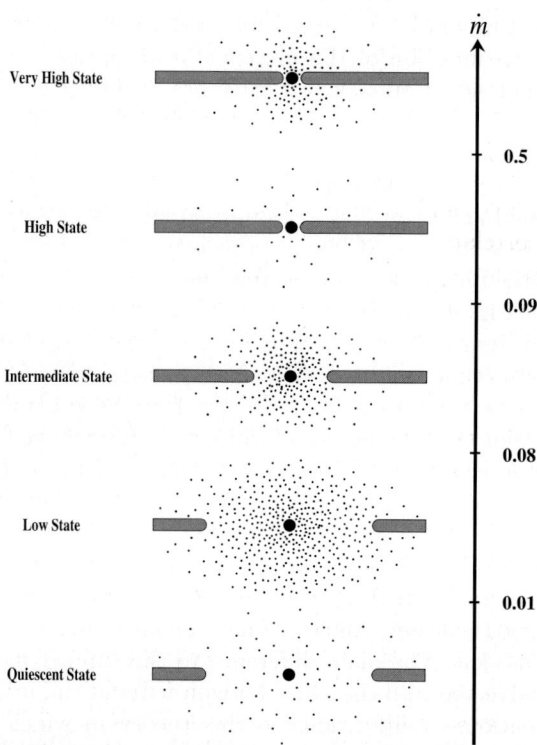

Fig. 1. Schematic from Esin et al. (1997) showing the region near the black hole in the model in which this structure varies as a function of accretion rate. Note how the inner disk vanishes at low accretion rates and a geometrically thick accretion flow replaces it. The normal Shakura & Sunyaev (1973) model corresponds to the high state in this picture.

The high state ADAF infers a highly optically thick gas by contrast, in which the radiation is actually trapped in the gas as it is dragged within the event horizon. It is not yet clear if this mode has a clear parallel in AGNs, since the systems seen with the highest accretion rates, such as the narrow line Seyfert 1s, still appear to be in the thermal/disk dominated state. (e.g. Maccarone et al. 2003). It is known that the thermal accretion disk model should become unstable at very high accretion rates, but it is less clear what exactly we would expect to see in such a class of objects. Such a phase may only apply to the very earliest stages of the evolution of an AGN (Sec. 5).

One of the most detailed studies of how the low accretion model might work has been carried out for the case of radio detected AGNs (Falcke et al. 2004). They suggested that low luminosity AGNs amongst this class (i.e. BL Lacs, FR I radio galaxies, etc.) should be inefficient accreaters. Their preferred model includes the presence of a radio jet along with the ADAF in the low luminosity AGNs (perhaps in keeping with the suggestions of Blandford & Begelman 1999). They showed that all of the presumed inefficient accreters appear to share a clear correlation between their X-ray and radio properties, consistent with a simple scaling with luminosity. By contrast, the efficient accreters such as narrow line Seyfert 1s appear to have suppressed radio emission (Greene et al. 2006).

Finally, we now also have significant observational evidence for the properties of the black holes and their accretion rates in many AGNs. There is clear evidence for a correlation between the black hole mass and the mass of the bulge of the host galaxy (e.g. Kormendy & Richstone 1995; Magorrian et al. 1998; Gebhardt et al. 2000; Ferrarese & Merritt 2000). As a result we can estimate the black hole mass in an AGN from the measure of the stellar velocity dispersion (Tremaine et al. 2002). Reverberation line measurements, discussed in more detail in Sec. 3, show a strong correlation between the broad line width and the estimated black hole mass (Kaspi et al. 2000), allowing us to estimate black hole masses in more distant and/or luminous objects where the stellar continuum is either too faint, or too supressed relative to the continuum luminosity from the AGN itself. It is therefore possible to derive estimates of the black hole masses for significant samples of AGNs. Since it is also relatively straightforward to derive estimates of the AGN luminosities, we are now in the position where we can study the accretion rates of such samples as well.

The most extensive work of this kind (McLure & Dunlop 2004) is based on 14181 quasars drawn from those found in the Sloan Digital Sky Survey (Schneider et al. 2003) which is discussed in more detail in Sec. 5. Fig. 2 shows results from this work. Here it is worth noting that they estimated an upper bound for the accretion rate that are near the Eddington accretion rate, and found that black hole masses also appear to have an upper limit, $M_{\rm BH} \sim 10^{10}\,M_\odot$. The lower limits for these quantities appear to be zero for the accretion rate (since we observe quiescent supermassive cores in nearby galaxies - e.g. Kormendy & Richstone 1995), and less than $10^6\,M_\odot$ for the black hole mass (e.g. Greene & Ho 2006). The mass may extend well below

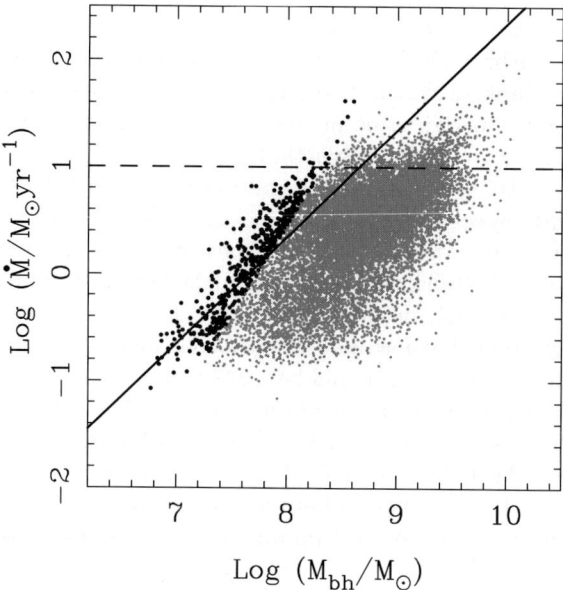

Fig. 2. Accretion rate and black hole mass as derived by McLure & Dunlop (2004) for the sample of quasars from the Sloan Digital Sky Survey. The black dots are the equivalents of the narrow line Seyfert 1s, as determined by the full width at half maximum of the Hβ line being < 2000km/s. The masses are derived using the relationship from Kaspi et al. (2000) and the accretion rates from the ratio of bolometric and Eddington luminosities for these objects.

this value. The Sloan survey data have also been used to find central black holes with lower mass, but only a few dozen have been identified even from that dataset. What is clear is that AGNs with very low mass central black holes are relatively rare.

3 The Broad Line Region

Reverberation line studies provide the best means of determining the size of the broad line region. A recent analysis of the available data yielded results consistent with the size of the broad line region ranging from under a light day in the least luminous nearby Seyfert 1s to a significant fraction of a parsec in highly luminous quasars (Kaspi et al. 2005). The gas present in the broad line region is typified by electron temperatures $T_e \sim 20,000$ K and electron densities of order $n_e \sim 10^{11}$ cm^{-3} (Peterson 1993) in the inner regions. As a result, forbidden and most semi-forbidden transitions are quenched in the inner regions due to collisional de-excitation. Lower density gas must also be present given that CIII] λ1909Å is observed in many AGNs, since this line has a critical density of 3×10^9 cm^{-3}.

Exactly how the broad line emission itself is formed is still a relatively open question. The available evidence from reverberation line mapping observations suggests that the gas within this region orbits in an approximately Keplerian fashion, with higher ionisation species nearer to the centre (e.g. Peterson & Wandel 2000). Models that deviate markedly from this are ruled out at least for the admittedly small sample of AGNs for which reverberation results exist. However a simpler study of the line profiles in the Bright Quasar Survey (Schmidt & Green 1983) shows that not all high ionisation lines have profiles consistent with this model (Baskin & Laor 2005), and for those AGNs with very broad lines the CIV line profile is often narrower than the Hβ one. This has often led to the conclusion that the broad line region has two distinct components (e.g. Sulentic et al. 2000). The low ionisation lines clearly follow the same behaviour as the Balmer lines. The high ionisation lines show similar behaviour to the CIV line.

One thing that is clear is that in the bulk of the AGN population the Hβ line largely traces virialised gas (e.g. McLure & Dunlop 2004). The lack of double peaked line profiles in most AGNs suggests that this low ionisation gas does not lie solely in a disk (e.g. the narrow lines are not just due to viewing the disk pole-on, since then the fraction of single-peaked line emitters would be far less). The clear exceptions to this are a subset of radio loud AGNs and LINERs where a disk model is a good fit to the observed Balmer line profiles (Eracleous & Halpern 2003). This is explicable in terms of the low accretion rate mode outlined in Sec. 2 for the LINERs since these are not low luminosity weakly accreting AGNs, but many of the radio galaxies studied by Eracleous & Halpern (2003) were high luminosity FR II systems. A model roughly akin to the ADAF one where the inner region of the disk radiates inefficiently, with only the outer disk acting as a normal thermal emitting disk, has been proposed for these 'double-peaked' radio loud AGNs (Eracleous & Halpern 2003). Unfortunately the actual accretion rates for these radio galaxies have as yet not been inferred, and it could place these systems at odds with the normal ADAF model. It has been suggested that the best model unifying these different sources is one in which at higher accretion rates the AGN luminosity drives a disk-wind (see, e.g., the simulations of Proga et al. 2000), but that would sit at odds with the idea that the more luminous FR II radio galaxies should be high accretion objects. Further study of such double peaked line emitters is clearly an important aspect of understanding the AGN population as a whole.

Disk-wind models also provide a natural explanation for the broad and narrow absorption line features seen in some AGNs (Proga & Kallman 2004). Broad absorption line quasars were originally considered to be a possible separate class of AGNs differing from normal quasars. However it now seems clear that these objects share very similar properties (e.g. Weymann et al. 1991) with the one exception that the broad absorption line systems as a whole show higher extinction to the broad line region (e.g. Sprayberry & Foltz 1992; Reichard et al. 2003). Hard X-ray observations of these sources also reveal a significant obscuration (e.g. Gallagher et al. 2002). The overall picture is of an AGNs in which a strong outflowing (and ionised) wind must be present. The fact that only a minority of AGNs show such features indicates that the covering fraction of the high column density obscuring material must be low, and we are probably seeing them close to edge on to the equatorial wind. Polarization studies are a useful probe of the geometry of this material. Lamy & Hutsemékers (2004) have shown for example that the observed data agree well with a disk-wind model in which the outflow is largely equatorial but the scattering primarily occurs in the polar regions.

The structure and material forming the broad line region also provide constraints on models for how it arises. Baldwin et al. (2003) have shown that in the most luminous quasars the mass in the broad line region may be more than 10^4 $M_\odot$. However, in such systems the region itself will extend across a significant volume. In such a case the central stellar cluster itself may become encompassed within the broad line region. Several models have been suggested as to how the outflowing material could originate in clumpy stellar processes. One possibility is that the outer photospheres of supergiant stars are further 'bloated' by the ambient radiation field from the core (e.g. Alexander & Netzer 1997, and references therein). Others include stellar collisions with the accretion disk, freeing material (e.g. Zurek et al. 1994) or tidal disruption of stars in the gravitational field of the black hole (e.g. Roos 1992). John Dyson, in particular, has been influential in studying the possibility that wind interactions with nearby obstacles such as stellar winds or supernova ejecta may be responsible for individual clouds in the broad line region (e.g. Perry & Dyson 1985; Pittard et al. 2003). The main goal of such models is not only to explain the presence of broad line "clouds" but also to find mechanisms to ensure that they remain confined and not destroyed. Of course, this assumes the broad line region is made from discrete *long lasting* clouds (as was widely believed twenty years ago). These models have in their favour that they may occur in the whole volume around the core, smearing out the Keplerian rotation curve from gas near the disk itself.

The main alternative that arises naturally in disk-wind type models is that much of the material is simply smoothly ejected from the surface of the accretion disk itself (Murray et al. 1995). This removes the requirement for confinement since the broad line region is continuously regenerated in this model with fresh material. Laor et al. (2006) have shown convincing evidence for the case of NGC 4395, the least luminous Seyfert 1 with the smallest broad line region known, that the observed data agree far better with a continuous model than a clumped one. Such a model is at first sight hard to reconcile with both the observations that the gas appears to act mainly under gravity and does not typically have double peaked line profiles. However, gas leaving the disk will have a velocity typical of the virial speed, since most gas will be freed in the region of the disk where the escape velocity matches the virial motion (since it cannot escape at lower velocities and there is no reason for it still to be bound to the disk at higher velocities). This does not explain how the gas can fill in a larger volume however, and in order to avoid double peaked line profiles in such a scenario gas must be released at a wide variety of radii. Nicastro has outlined a simple theoretical model showing how the observed line widths correlate with the inverse of the accretion rate in such a scenario, naturally explaining how narrow line Seyfert 1s come about for example (Nicastro 2000). It is, of course, possible that both types of models, disk-wind and stellar, may play a role in more luminous AGNs in which the central stellar cluster must surely be partially entrained within the broad line region. One test of the disk wind models is that the broad line region should have a flattened geometry, whereas there is no such natural requirement on the confined clouds models.

Any model must also be able to reproduce the observed correlations between the properties of line and continuum emission from the broad line region, particularly the first principal components found in the spectra of quasars by Boroson & Green (1992). More recent studies do show a general agreement between the data and the concept that the fundamental underlying parameters are the accretion rate and black hole mass (e.g. Boroson 2002). Such studies should also eventually help to settle the issue as to whether there is a single component to the broad line region or not, or, indeed, whether the number of components varies with these underlying physical parameters.

Finally, it is worth noting that the broad line region may actually be completely absent at very low accretion rates. If the Nicastro (2000) model relating line width to accretion rate is correct then at very low accretion rates the line widths should exceed $0.1c$. No AGNs are observed with such broad lines. Laor (2003) has speculated that the broad line region may not exist if such velocities are obtained, either due to interactions of the disk with surrounding material, or simple shear and tidal forces within the disk itself. Another possibility in terms of the Nicastro model is that the region in which the broad line region detaches from the disk surface shrinks within the last stable orbit at low accretion rate (Nicastro et al. 2003).

4 The Narrow Line Region

The narrow line region is now relatively well understood in AGNs. The gas densities observed are typical of clouds in the interstellar medium. The gas kinematics is consistent in many case with purely galaxian rotation (Whittle 1985). This could suggest that we are seeing gas that is passively moving and being ionised by the central engine. This picture breaks down at higher AGN luminosities, however, where clear jet-like motions can be detected in the narrow line region, and more obviously in objects such as FR II radio galaxies where the influence of the jets extends well beyond the galaxies themselves.

The main other diagnostic that the narrow line region provides is as a probe of the spectral energy distribution of the central emitted radiation field. This is particularly useful in those cases where the AGN light itself is difficult to detect directly against the host galaxy, or in wavelength regimes where observations are either difficult or impractical.

In particular, the standard thin accretion disk model implies that there should be a thermal 'bump' in the spectral energy distribution (see Sec. 2). By contrast, models in which the accretion proceeds inefficiently lack this feature (e.g. Ball et al. 2001). This provides a direct testable prediction that at least, in part, seems to be met by the observational data. Low luminosity AGNs in general tend to have spectra that appear very 'LINER'-like, lacking the high ionisation lines often seen in Seyferts, quasars, etc. Simple photoionisation models show that if the spectral energy distribution is more like a power-law, as proposed in Ball et al. (2001), then the natural result is a LINER spectrum (e.g. Ferland & Netzer 1983). Similarly, photoionisation models including the 'big blue bump' naturally explain the emission line spectra of Seyferts and quasars, as well as the strong blue non-stellar continuum. Alternative models to explain these phenomena in the context of fast shocks have been proposed (e.g. Dopita & Sutherland 1995), but it is not clear that the conditions they require are generally met in most AGNs.

5 Orientation Dependent Unified Models

As noted in Sec. 1, the main success in unified models of AGNs has been due to orientation dependent models unifying type I and II AGNs, and in particular in the study of Seyfert 1 and 2 galaxies. Antonucci (1993) presents an early overview of this model, in which an optically thick, compact molecular torus acts to block our line of sight to the broad line region for a wide range of viewing angles. The main tool for studying this model originally was spectropolarimetry. Broad permitted lines can sometimes be seen scattered into our line of sight even when the direct view is completely obscured (cf. Fig. 3).

More recent studies using the same methods have led to contradictory conclusions however. Lumsden et al. (2001) examined a complete sample of infrared selected Seyfert 2s and concluded that the objects in which obscured broad line regions were not present were those with lowest luminosity and highest extinction. From this they concluded the orientation dependent unified model was substantially true for most if not all luminous Seyferts. By comparison, Tran (2003) studied a heterogeneous mix of optically and infrared selected Seyferts and claimed that since most galaxies without evidence for broad lines in scattered light appeared similar to star forming galaxies that there was a separate class of 'pure' Seyfert 2s. However, there is a clear luminosity bias in detecting obscured broad lines, with the most luminous galaxies almost uniformly showing such features (Lumsden & Alexander 2001). Less luminous AGNs selected from an infrared sample are likely to have signifi-

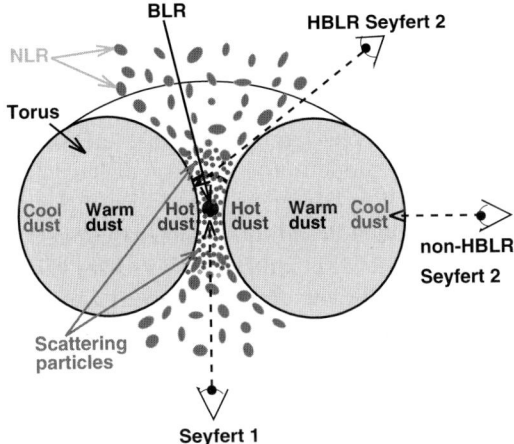

Fig. 3. Schematic of the orientation dependent unified model. The scale height of the broad line region is much smaller than that of the torus, and, hence, over a wide range of angles appears obscured. The narrow line region can extend out to a much greater distance than the broad line region, and, hence, is observed even when the central core is obscured. 'Hidden' broad line regions are those whose presence can be confirmed through techniques such as spectropolarimetry.

cant star formation in any event (since this method effectively picks starburst gaalxies as well). Therefore, the fact that less luminous AGNs without broad lines might look like star forming galaxies is largely a function of the selection. If this is true, then the failure to detect broad lines in the less luminous AGNs in the samples may simply be a matter of insufficient signal-to-noise in the observed data (Alexander 2001).

Other means of examining obscured broad line regions in the type II AGN population are also used. The two most commonly employed involve the use of near infrared spectroscopy to study the inner cores in obscured systems, and in studying weak optical broad features. Near infrared spectroscopy often reveals a broad line region where an optical signature is not clearly present. Veilleux et al. (1997) studied a wide sample of AGNs and found many cases where this was true. Unfortuantely, their work did not directly constrain whether the broad line visibility in the near infrared was related to obscuration or not. More recent work by Lutz et al. (2002) appears to confirm this is the case however. It is possible that this technique is actually more powerful in low luminosity, moderate obscuration AGNs than spectropolarimetry in revealing hidden broad line regions. Ho et al. (1997) used high signal-to-noise optical spectroscopy to study a sample of nearby galaxies to search for weak broad line regions. They showed that it was possible with care to detect such objects in previously unsuspected low luminosity AGNs and greatly boosted the number of such type 1.5–1.9 AGNs known (where 1.5 represents a galaxy in which both Hα and Hβ are weakly visible, but the narrow line region emission dominates all the lines in the integrated spectra, and 1.8–1.9 represent those galaxies in which only weak broad Hα wings are visible).

Overall then there are a variety of techniques that can be applied in studying orientation dependent unification. Unfortunately the exact situations in which each method may work best are rarely considered in studies that concern whether all AGN reveal hidden broad line regions or not. For example it is possible that weak type 1.9 AGN are actually those with significant scattering present. The original hidden broad line region discovered in NGC 1068 by Antonucci & Miller (1985) is of this kind. In other cases the 1.9 classification may be due simply to relatively low obscuration. As a result it is difficult to state with any degree of confidence that the issue of whether the orientation dependent unified model has finally been settled. What does seem clear now, however, is that the simplest form of unification for Seyfert galaxies, namely that in which orientation alone is important, does not hold absolutely. It is clear that the luminosity of the AGN core has an equally, if not more, important role in determining the appearance of an AGN (see the discussion in Lumsden & Alexander 2001).

The other area in which orientation dependent unification has been successful in matching the observation is for radio galaies. For radio quiet AGNs for example, the equivalent of the Seyfert 2s are the FR I radio galaxies. Pole-on radio galaxies with similar luminosity appear as BL Lacs (e.g. Urry & Padovani 1995), with the radio and jet emission from the core relativistically beamed in our direction. A similar scheme is proposed for the luminous radio galaxies, with edge-on systems appearing as FR II radio galaxies and pole-on ones as either blazars or radio loud quasars. Here again doubts have been raised as to the universality of such a scenario in which orientation alone is important (e.g. see the discussion in Marchã et al. 2005).

The main apparent difference between the radio galaxies as a whole and the Seyfert population lies in the host galaxy type. Most low redshift radio galaxies lie in elliptical hosts. The Seyferts, on the other hand, lie mostly in spiral galaxies. Of course, since the classifications are rather subjective there are Seyferts that have elliptical hosts and spiral hosts with formally radio loud central engines (usually taken to be those with high radio to optical continuum ratios). Indeed, there is evidence that the relationship may be more based on the luminosity of the AGN than of its specific type, since both radio quiet and radio loud quasars at the highest luminosities are found to lie in spheroidally dominated systems, with the radio loud quasars in the systems with the most massive black holes (Dunlop et al. 2003). Therefore, the main hope of a greater unified model must be to encompass the fundamental drivers behind these apparent phenomena, namely black hole mass and accretion rate, in addition to the relatively simple effects of orientation.

Finally, it is worth noting here that there are classes of active galaxies which seem to change their appearance from type I to type II with time. This is now a well established phenomena from two different perspectives. First and most obviously are those galaxies in which the broad permitted lines are not only variable but appear or disappear with time. An example of this kind in which the variability appears to be due to a change in the obscuration is NGC 7582 (Aretxaga et al. 1999). Ongoing near infrared observations of the same galaxy show that broad Paβ emission is ever present, so only the extinction to the broad line region can be variable here. The second perspective comes from hard (2–10 keV) X-ray observations of nearby active galaxies. One of the best studied examples is NGC 2992 (Gilli et al. 2000) where the hard X-ray emission has changed by over an order of magnitude in 20 years, and the broad line region which was weakly visible in the optical and clearly visible in the infrared became much less prominent in a manner suggesting it was the intrinsic luminosity of the active galaxy that changed rather than the extinction. The key characteristic in the hard X-ray band is an apparent switch from direct detection of the nucleus to a faint reflection dominated spectrum (Guainazzi et al. 2005). Such variability adds another complicating factor to any attempt to 'unify' different types of AGNs, but may well be important to their overall late-time evolution as the accretion rate drops, or at any stage when there is high but patchy obscuration present.

6 Evolution

There are several strands of evidence pointing to an evolution of the median luminosity of the AGN population with redshift. The most direct of these are the studies of the quasar luminosity function from the 2dF quasar survey (2QZ: Croom et al. 2004). The 2QZ results show that for their optically selected sample of quasars there is a clear trend for the luminosity function to evolve with redshift (Fig. 4), in the sense that the average quasar now is *less* luminous than at $z \sim 1-2$. Heckman et al. (2004) carried out a more general analysis of all kinds of AGNs found in the Sloan Digital Sky Survey, including those from the general galaxy survey, and showed that there is evidence for AGN 'down-sizing' for $z < 1$, so that the most active AGNs today have smaller black hole masses than in the past.

The evidence suggests that there is a relatively hard upper limit to the range of black hole masses both in nearby luminous galaxies and in the more general AGN population of a few times 10^9 M$_\odot$ (McLure & Dunlop 2004). The same study shows that most of the mass of the black holes for the quasars analysed must actually be in place by $z \sim 2$. The results clearly show that it is the smallest black holes (narrowest lines) that accrete most rapidly, and that these objects appear to be limited by the Eddington rate (approximately 100 M$_\odot$ per year in this case). An AGN accreting at this rate for much more than 100 million years, however, would violate the apparent observed upper limit on the masses of the central black holes. As a whole then AGNs accrete matter much more slowly than the Eddington rate for a substantial fraction of their overall lifetimes. This suggests that a different physical limitation applies to the longer term evolution of AGNs, after an initial Eddington limited

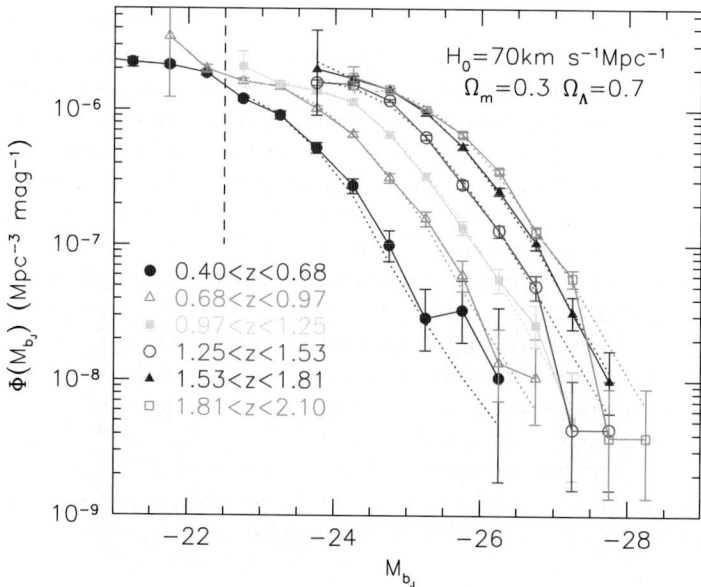

Fig. 4. Evolution of the quasar luminosity function with redshift. Quasars clearly get fainter at lower redshift. From Croom et al. (2004).

growth phase, which could be linked to the physics of the gas supply in the general galaxy formation process, or the accretion disk physics around such supermassive black holes. At present all we can say is that there does appear to be such a limit.

It is worth briefly touching on other areas in which the evolution of AGNs are currently very topical. There is considerable interest for example in the role of AGNs in explaining the behaviour of at least a fraction of the high redshift galaxies that are only detected as heavily obscured sub-mm sources (e.g. Dunlop 2001). AGNs are also invoked as the explanation for the hard X-ray background (e.g. Treister & Urry 2005), and the majority of galaxies associated with hard X-ray sources studied in detail do indeed appear to have clear AGN characteristics (e.g. Silverman et al. 2005). The large galaxy redshift surveys have also allowed us to detect the previously elusive high luminosity counterparts to the Seyfert 2s (e.g. Zakamska et al. 2003). These objects appear to contain hidden broad line regions (Zakamska et al. 2005), though their overall properties may not be what were previously expected (for example, their infrared properties are much more like those of type I AGNs than type IIs - e.g. Sturm et al. 2006).

Of course, we do not have to study objects at $z \sim 1-2$ to constrain models of the accretion rate as a function of black hole mass, or to study evolution. The inferred accretion rates in the nearby AGN population span a similar range to that seen in the quasars studied by McLure & Dunlop (2004).

Indeed, they may provide a more telling challenge to such models, since many of the nearby far infrared bright AGNs clearly have substantial gas supplies still available as evidenced by their extensive nuclear star formation and yet tend to have modest luminosities (and presumably accretion rates). The exact link between the evolution of the galaxy as a whole and its active core is still unclear, even if the observational data imply that the link is definitely there.

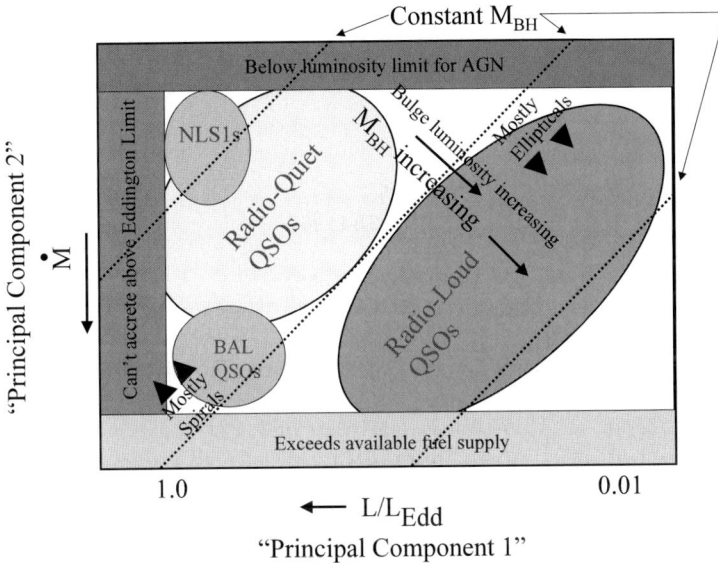

Fig. 5. A new unified model of AGN? From Boroson (2002). Different classes of AGNs appear in different parts of parameter space. The parameters on the axes are the accretion rate and the ratio of the luminosity to the Eddington luminosity, which is the driver behind the first principal component in the analysis of Boroson & Green (1992). Lines of constant black hole mass are indicated.

7 Summary

Much has changed in our understanding of AGNs over the last twenty years. The introduction of the paradigm in which behaviour changes radically with accretion rate, the observation that active galaxies are not only variable but can also appear to change class as they vary, as well as the clear evidence that AGNs do evolve as a class with time all present challenges to our understanding. Many aspects of their study are still not fully mapped out. In particular, the relationship between accretion rate and behaviour only fully seems to work in the transition from low luminosity AGNs to high luminosity ones. For those AGNs with highly variable luminosities it should be possible to follow the transition from one state to another as we see in Galactic black hole X-ray binaries, but no one has produced a convincing example as yet. Despite this, there is now a crude picture of how AGNs behave as a function of accretion rate, black hole mass, evolution and obscuration. The simplest of these, ignoring the issues of evolution and obscuration, is the one outlined by Boroson (2002) and as shown in Fig. 5. Clearly the future lies in both proving the correctness of these outlines of AGN behaviour and in including dynamic evolutionary events into them.

References

Alexander, D.M. 2001 MNRAS **320**, L15
Alexander, T., Netzer, H. 1997 MNRAS **284**, 967
Antonucci, R. 1993 ARA&A **31**, 473
Antonucci, R.R.J., Miller, J.S. 1985 ApJ **297**, 621
Aretxaga, I., Joguet, B., Kunth, D., Melnick, J., Terlevich, R.J. 1999 ApJL **519**, L123
Baldwin, J.A., Ferland, G.J., Korista, K.T., Hamann, F., Dietrich, M., 2003, ApJ **582**, 590
Ball, G.H., Narayan, R., Quataert, E. 2001 ApJ **552**, 221
Baskin, A., Laor, A. 2005 MNRAS **356**, 1029
Blandford, R.D., Begelman, M.C. 1999 MNRAS **303**, L1
Boller, T. 2001 AIP Conf. Proc. 599: X-ray Astronomy: Stellar Endpoints, AGN, and the Diffuse X-ray Background **599**, 25
Boroson, T.A. 2002 ApJ **565**, 78
Boroson, T.A., Green, R.F. 1992 ApJS **80**, 109
Croom, S.M., Smith, R.J., Boyle, B.J., Shanks, T., Miller, L., Outram P.J., Loaring, N.S. 2004 MNRAS **349**, 1397
Dopita, M.A., Sutherland, R.S. 1995 ApJ **455**, 468
Dudik, R.P., Satyapal, S., Gliozzi, M., Sambruna, R.M. 2005 ApJ **620**, 113
Dunlop, J.S. 2001 New Astronomy Review **45**, 609
Dunlop, J.S., McLure, R.J., Kukula, M.J., Baum, S.A., O'Dea, C.P., Hughes, D.H. 2003 MNRAS **340**, 1095
Eracleous, M., Halpern, J.P. 2003 ApJ **599**, 886
Esin, A.A., McClintock, J.E., Narayan, R. 1997 ApJ **489**, 865
Falcke, H., Körding, E., Markoff, S. 2004 A&A **414**, 895
Ferland, G.J., Netzer, H. 1983 ApJ **264**, 105
Ferrarese, L., Merritt, D. 2000 ApJL **539**, L9
Gallagher, S.C., Brandt, W.N., Chartas, G., Garmire, G.P. 2002 ApJ, 567, 37
Gebhardt, K., Bender, R., Bower, G., Dressler, A., Faber, S.M., Filippenko, A.V., Green, R., Grillmair, C., Ho, L.C., Kormendy, J., Lauer, T.R., Magorrian, J., Pinkney, J., Richstone, D., Tremaine, S., 2000, ApJL **539**, L13
Gilli, R., Maiolino, R., Marconi, A., Risaliti, G., Dadina, M., Weaver K.A., Colbert, E.J.M. 2000 A&A **355**, 485
Greene, J.E., Ho, L.C. 2006 ApJL **641**, L21

Greene, J.E., Ho, L.C., Ulvestad, J.S. 2006 ApJ **636**, 56
Guainazzi, M., Fabian, A.C., Iwasawa, K., Matt, G., Fiore, F. 2005 MNRAS **356**, 295
Heckman, T.M., Kauffmann, G., Brinchmann, J., Charlot, S., Tremonti, C., White, S.D.M. 2004 ApJ **613**, 109
Ho, L.C., Filippenko, A.V., Sargent, W.L.W., Peng, C.Y. 1997 ApJS, 112, 391
Jester, S. 2005 ApJ **625**, 667
Kaspi, S., Maoz, D., Netzer, H., Peterson, B.M., Vestergaard, M., Jannuzi, B.T. 2005 ApJ **629**, 61
Kaspi, S., Smith, P.S., Netzer, H., Maoz, D., Jannuzi, B.T., Giveon, U. 2000 ApJ **533**, 631
Kormendy, J., Richstone, D. 1995 ARA&A **33**, 581
Lamy, H., Hutsemékers, D. 2004 A&A **427**, 107
Laor, A. 2003 ApJ **590**, 86
Laor, A., Barth, A.J., Ho, L.C., Filippenko, A.V. 2006 ApJ **636**, 83
Lumsden, S.L., Alexander, D.M. 2001 MNRAS **328**, L32
Lumsden, S.L., Heisler, C.A., Bailey, J.A., Hough, J.H., Young, S., 2001, MNRAS **327**, 459
Lutz, D., Maiolino, R., Moorwood, A.F.M., Netzer, H., Wagner, S.J., Sturm, E., Genzel, R. 2002 A&A **396**, 439
Maccarone, T.J., Gallo, E., Fender, R. 2003 MNRAS **345**, L19
Magorrian, J., Tremaine, S., Richstone, D., Bender, R., Bower, G., Dressler, A., Faber, S.M., Gebhardt, K., Green, R., Grillmair, C., Kormendy, J., Lauer, T. 1998 AJ **115**, 2285
Marchã, M.J.M., Browne, I.W.A., Jethava, N., Antón, S. 2005 MNRAS **361**, 469
McLure, R.J., Dunlop, J.S. 2004 MNRAS **352**, 1390
Murray, N., Chiang, J., Grossman, S.A., Voit, G.M. 1995 ApJ **451**, 498
Narayan, R. 2005 Ap&SS **300**, 177
Narayan, R., Yi, I. 1994 ApJL **428**, L13
Nicastro, F. 2000 ApJL **530**, L65
Nicastro, F., Martocchia, A., Matt, G. 2003 ApJL **589**, L13
Osterbrock, D.E., Pogge, R.W. 1985 ApJ **297**, 166
Perry, J.J., Dyson, J.E. 1985 MNRAS **213**, 665
Peterson, B.M. 1993 PASP **105**, 247
Peterson, B.M., Wandel, A. 2000 ApJL **540**, L13
Pittard, J.M., Dyson, J.E., Falle, S.A.E.G., Hartquist, T.W. 2003 A&A **408**, 79
Proga, D., Kallman, T.R. 2004 ApJ **616**, 688
Proga, D., Stone, J.M., Kallman, T.R. 2000 ApJ **543**, 686
Reichard, T.A., Richards, G.T., Hall, P.B., Schneider, D.P., Vanden Berk, D.E., Fan, X., York, D.G., Knapp, G.R., Brinkmann, J. 2003 AJ **126**, 2594
Roos, N. 1992 ApJ **385**, 108
Schmidt, M., Green, R.F. 1983 ApJ **269**, 352
Schneider, D.P., Fan, X., Hall, P.B., Jester, S., Richards, G.T., Stoughton, C., Strauss, M.A., SubbaRao, M., Vanden Berk, D.E., Anderson, S.F., Brandt, W.N., Gunn, J.E., Gray, J., Trump, J.R., Voges, W., Yanny, B., Bahcall, N.A., Blanton, M.R., Boroski, W.N., Brinkmann, J., Brunner, R., Burles, S., Castander, F.J., Doi, M., Eisenstein, D., Frieman, J.A., Fukugita, M., Heckman, T.M., Hennessy, G.S., Ivezić, Ž., Kent, S., Knapp, G.R., Lamb, D.Q., Lee, B.C., Loveday, J., Lupton, R.H., Margon, B., Meiksin, A., Munn, J.A., Newberg, H.J.,

Nichol, R.C., Niederste-Ostholt, M., Pier, J.R., Richmond, M.W., Rockosi, C.M., Saxe, D.H., Schlegel, D.J., Szalay, A.S., Thakar, A.R., Uomoto, A., York, D.G. 2003 AJ **126**, 2579

Seyfert, C.K. 1943 ApJ **97**, 28

Shakura, N.I., Sunyaev, R.A. 1973 A&A **24**, 337

Shang, Z., Brotherton, M.S., Green, R.F., Kriss, G.A., Scott, J., Quijano, J.K., Blaes, O., Hubeny, I., Hutchings, J., Kaiser, M.E., Koratkar, A., Oegerle, W., Zheng, W. 2005 ApJ **619**, 41

Silverman, J.D., Green, P.J., Barkhouse, W.A., Kim, D.-W., Aldcroft, T.L., Cameron, R.A., Wilkes, B.J., Mossman, A., Ghosh, H., Tananbaum, H., Smith, M.G., Smith, R.C., Smith, P.S., Foltz, C., Wik, D., Jannuzi, B.T. 2005 ApJ **618**, 123

Sprayberry, D., Foltz, C.B. 1992 ApJ **390**, 39

Sturm, E., Hasinger, G., Lehmann, I., Mainieri, V., Genzel, R., Lehnert, M.D., Lutz, D., Tacconi, L.J. 2006 ApJ, in press

Sulentic, J.W., Marziani, P., Dultzin-Hacyan, D. 2000 ARA&A **38**, 521

Tran, H.D. 2003 ApJ **583**, 632

Treister, E., Urry, C.M. 2005 ApJ **630**, 115

Tremaine, S., Gebhardt, K., Bender, R., Bower, G., Dressler, A., Faber, S.M., Filippenko, A.V., Green, R., Grillmair, C., Ho, L.C., Kormendy, J., Lauer, T.R., Magorrian, J., Pinkney, J., Richstone, D. 2002 ApJ, 574, 740

Urry, C.M., Padovani, P. 1995 PASP **107**, 803

Veilleux, S., Goodrich, R.W., Hill, G.J. 1997 ApJ **477**, 631

Weymann, R.J., Morris, S.L., Foltz, C.B., Hewett, P.C. 1991 ApJ, 373, 23

Whittle, M. 1985 MNRAS **213**, 33

Zakamska, N.L., Schmidt, G.D., Smith, P.S., Strauss, M.A., Krolik, J.H., Hall, P.B., Richards, G.T., Schneider, D.P., Brinkmann, J., Szokoly, G.P. 2005 AJ **129**, 1212

Zakamska, N.L., Strauss, M.A., Krolik, J.H., Collinge, M.J., Hall, P.B., Hao, L., Heckman, T.M., Ivezić, Ž., Richards, G.T., Schlegel, D.J., Schneider, D.P., Strateva, I., Vanden Berk, D.E., Anderson, S.F., Brinkmann, J. 2003 AJ **126**, 2125

Zurek, W.H., Siemiginowska, A., Colgate, S.A. 1994 ApJ **434**, 46

Astrophysics and Space Science Proceedings

Diffuse Matter from Star Forming Regions to Active Galaxies, edited by T.W. Hartquist, J.M. Pittard, S.A.E.G. Falle
Hardbound ISBN 978-1-4020-5424-2, December 2006

For further information about this book series we refer you to the following web site:
www.springer.com